Fisheries Economics

Fisheries Economics

Author
Amita Saxena
Professor, College of fisheries
GBPUA&T
Pantnagar – 263 145

2011
DAYA PUBLISHING HOUSE®
New Delhi - 110 002

ISBN 9789351240525

Published by : **Daya Publishing House®**
A Division of
Astral International Pvt. Ltd.
–ISO 9001:2008 Certified Company –
4760-61/23, Ansari Road, Darya Ganj
New Delhi-110 002
Ph. 011-43549197, 23278134
E-mail: info@astralint.com
Website: www.astralint.com

Laser Typesetting : **Twinkle Graphics**
Delhi

Printed at : **Chawla Offset Printers**
Delhi - 110 052

PRINTED IN INDIA

Contents

Preface

This book deals with fisheries sector, its organization for rational co-ordination of activities of number of people for the achievement of common purpose (fisheries potential) through division of labor and functions through hierarchy of authorities and responsibilities. Government is having varies plans and policies for inland and marine sectors. The social economic issues the problems, economic reforms and WTO implication are described in details. Now a days the question is how the empower women through SHGs, rural jobs finance and microfinance, the answer is given in text, the basic concept of economics like money, cost and analysis of fish demand, supply , marketing are also covered in this book.

The marketing systems and co-operative organization in fish economics to undertake an identification and classification of marine and freshwater fisheries are detailed separately. Marketing describes the traditional systems, modern methods, co-operatives, fish economics, capital investment and depreciation of equipments.

Food quality safety and packaging chapters added the value of this book. This book is useful for students, teachers, scientists, farmers, entrepreneurs, policy makers, retailers, customers etc.

Author is deeply thankful to elder's seniors, deans, vice chancellor, concerned ministers for necessary help and information's. Author is also grateful to publishers for publishing this piece of work.

Amita Saxena

Chapter 1
Fishery Organisation

The term 'organisation' thus has variety of interpretations. In any case, there are two broad ways in which the term is used. In the first the organization is understood as a dynamic process and a managerial activity which is necessary for bringing people together and coordinating their activities in the pursuit of common objectives. This may be called the process or organizing, while in other sense, the organization refers to the structure of relationships among positions and jobs which are built up for the realization of common objectives. Thus the organization is the systematic bringing together of inter-dependent parts to form an unified whole through which co-ordination and control may be exercised to achieve a given purpose. Because the inter-dependent parts are made up also of people who must be directed and motivated and whose work must be co-ordinated in order to achieve the objectives of the enterprise, organization contains both structure and human beings. To deal with organization merely as framework and without considering the people who make it up and those whom its services are intended unrealistic. Organization is a rational co-ordination of activities of a number of people for the achievement of common purpose or goal through division of labour and functions and through hierarchy of authority and responsibility.

Organizations purposive and systematic assignment of functions, duties and responsibilities among members of a group or a team. It defines the part that each member of an enterprise is expected to perform and the relations among those members.

As a function of managers, organizing and resultant organization structure are concerned with the activity–authority relationship of an enterprise. It is thus the grouping of activities necessary to attain enterprise objectives and the assignment of each grouping to a manager with authority necessary to supervise it. Organizing thus, involves establishment of authority, relationships with provision for co-ordination between them, both vertically and horizontally in the enterprise structure. Using organization as a structural system appears to the another to be more realistic than other concepts of organization. Certainly, most managers believe that they are organizing the established structure. In brief, the organization involves around four basic issues; human beings, goals, division and integration of activities and performance.

In this way, organization is defined as the human resources development designed to achieve common goals through division and integration of activities with good performance.

History of the Fishery Organization

In adopting a structure the main consideration, of course, has to be that the structure adopted must have relevances to the aims and objects of the organization. It must also take care of the requirements of the meeting adequately the normal work load to be expected in implementing the defined programmes of the organization. Care must of course also be taken that there is no overstaffing. The aspect of optimal balance between centralization and decentralization also has to be kept in view. Considering the scope of organization structure, work was conducted during the course of study to see the brief evolution of fishery and present set up of organizational structure in Haryana State with the object to derive some inference to streamline the working of the organization.

The history of taming fish and its use by man is as old as the history of man himself. A good illustration about the catching of fish and angling is found in the old books of Christian and Hindu mythology. In the *'Book of Job'* written presumably in 1500 B.C., there are many passages referring to use of hooks, barbs, iron etc. for fishing. The use of fish in India dates back to three millennium B.C. fish remains with cut marks and sign of the use, have been obtained from excavations at Mohanjodero and Harappa of Indus Valley Civilization (2500 B.C. – 1500 B.C.). While Aristotle (384-327 B.C.) is said to be founder of fishery science. King Somesvara, the son of King Vikramaditya VI who composed the book *Manasoltara,* in A.D. 1127 was the first writer and have recorded the common sport fish of India grouping them into marine and fresh water riverine forms. In medieval Indian history good illustrations of catching fish are also found in *Akbarnama* indicating the system of catching fish in state's water. The first modern writer on Indian fishes was Bloch whose splendid work *Auslandiche Fische* was published in 1785. This work alongwith his *Ichthyology,* and its further extension by Schneider in 1801, contain many Indian marine forms. Later Lacepede wrote *Historie des Poissans* (1798-1803). In 1803, Russel described 200 species from Vizagapatam. In 1822, there appeared Hamilton's pioneer work, *Fishes of Ganges* which contains the description of 269 species of fish. Cuvier and Valeneienne's *Historic Naturelle des Poissons* published in 1828-49 provided more impetus to the study of Ichthology. Good contribution has also been gives by McClelland (1839), Bleeker (1853), Blyth (1858-60), Gunther (1859-70) and Day (1878-89).

In the early phase of British-rule of India in 1871, a separate Department of Agriculture which included fisheries was created and subsequently similar departments were set up at provincial levels. During 1897, Governor-General of India passed Act No. IV of 1897, "The Indian Fisheries Act" to provide for certain matters relating to Fisheries in British India. Rule 6 of the Act directed, "The Provincial Government may make rules for the purpose hereinafter in this section mentioned, and may be a notification in the official Gazette apply all or any of such rules to such waters, not being private waters, as the Provincial Government may specify in the said notification". The first Fisheries Department was established at Madras Province followed by Punjab in 1912 and Mr. G.G.L. Howell was appointed as Director of Fisheries, Punjab.

The Royal Commission on Agriculture was appointed in 1926 under the chairmanship of Marquess of Linlithgow who gave detailed recommendations for the development of Agriculture, Animal Husbandry, Forest, Fishery, Co-operation, Education, Public Health etc. and on its recommendations, Indian Council of Agricultural Research was set up during 1929 with the main object to promote agriculture and allied subjects of research in India. The Government of India appointed a Fisheries Development Adviser in May 1944 and subsequently provided supporting staff for preparation of plan for research and development. Deep sea fishing organization was established at Bombay in 1946 to survey the marine fisheries resources. As per recommendations made by Fish Sub-Committee of the Policy Committee on Agriculture, the Marine Fisheries Research Institute at Mandapam and Inland Fisheries Research Institute at Barrakpore were set up in 1947.

Agriculture and allied activities (including Fisheries) by and large were private enterprises involving millions of farmers–large, medium small and marginal. There was not much administrative involvement in field level till 1914. On September 1, 1952, there were 42 officers in the Ministry of Agriculture (Agriculture Wing) which belonged to generalist cadres while the total number of technical advisers were less.

The reorganization of the Department of Agriculture at the central level was initiated in 1965-66 with the aims to function on the following lines :

(*i*) Assignment to technical officers, the responsibility for planning, implementation and supervision of the programme in addition to their advisory functions.

(*ii*) Redistribution of subjects and responsibilities so as to make the assignment of every officer and unit compact and specific as far as possible.

(*iii*) Introduction of officer-oriented system for the disposal of technical matters whereby such cases will be handled by the officers themselves with minimum assistance from subordinate staff.

(*iv*) Introduction of a procedure and a machinery for closer and continuous contact between the centre and agricultural plan schemes.

There are five Departments in the Ministry of Agriculture. Fishery in one of the divisions under the Departments in the Ministry of Agriculture. Fishery is one of the divisions under the Department of Agriculture. The National Commission on Agriculture (1976) recommended the creation of new departments like Crop Production, Animal Husbandry, Fisheries, Forestry etc. in the Ministry of Agriculture of the proper development of all sectors. The major divisions under the Department of Fisheries should be Division of Marine Fisheries and Inland Fisheries.

For performing various activities, the work is supervised by Cabinet rank Minister for Agriculture who is assisted by State Minister, Secretary (Agriculture). Two Additional Secretaries, Joint Secretary is the incharge of the Fishery Division. The present set up of the division of Fisheries in the Ministry of Agriculture has been given in Annexure I.

There are various fisheries research institutes established by Government of India to conduct research in various fields of marine as well as inland fisheries. These institutions are places under Indian Council of Agricultural Research which is headed by Director-General and Deputy Director General (Fisheries). All these institutions are autonomous in function having separate Directorate under rules and regulations of Indian Council of Agricultural Research. The list of various Central Institutes is given below:

(*a*) Central Institute of Fisheries Nautical and Engineering Training, Cochin.

(*b*) Pre-Investment Survey of Fishing Harbour, Bangalore.

(*c*) Exploratory Fisheries Project, Bombay.

(*d*) Central Marine Fisheries Research Institute, Cochin.

(*e*) Central Institute of Fisheries Technology, Cochin.

(*f*) Integrated Fisheries Project, Cochin.

(*g*) Central Institute of Fisheries Education, Mumbai.

(*h*) Central Inland Fisheries Research Institute Barrakpore (West Bengal).

It is observed that out of 8 institutes, 4 are situated in Kerala State. So far as the Inland Fisheries is concerned, the institute is situated at Barrakpore in West Bengal which has altogether different agro-climatic conditions than that of Northern States like Jammu and Kashmir, Himachal Pradesh, Haryana, Punjab, Rajasthan, Utter Pradesh, Madhya Pradesh and Bihar. It is noticed that the technology propounded for inland fish culture/capture is Bengal based.

The Constitution of India, lays down the distinction of legislative powers between the Centre and States in regards to agricultural development. Under Article 246 of the Constitution in Seventh Schedule, the Fisheries has been kept at serial No. 21 in List II-State List. While the fishing and fisheries beyond territorial waters has been kept at serial No. 57 in List I-Union List. Thus, Agriculture including Animal Husbandry and Fisheries are State subjects. The Constitution offers adequate scope for the Central initiative touching upon almost every aspect of agricultural development.

Some Observations of Fishery Department in India

(*i*) Except Rajasthan, Meghalaya and Nagaland, all other States in India have separate Directorate for Fisheries. In Himachal Pradesh likewise Rajasthan, Meghalaya and Nagaland, the fisheries department was one of the wings of Animal Husbandry for a long time but a separate Directorate for Fisheries has been set up in recent years.

(*ii*) As recommended by National Commission Agriculture that every State must has a post of Agriculture Production Commissioner to co-ordinate the agriculture and allied departments, it is observed that only Utter Pradesh and Tripura States have the post of Agriculture Production Commissioner.

Table 1.1 : Position of Additional Directors, Joint Directors, Assistant Directors, Deputy Directors and Fisheries Development Officers in Various States of India as on 31-3-1986.

Sl. No.	*State*	*Additional/Joint Director(s)*	*Deputy Director(s)*	*Assistant Director(s)/Fisheries Development Officers*
1.	Andhra Pradesh	1	11	26
2.	Assam	1	3	14
3.	Bihar	1	4	19
4.	Gujarat	-	6	23
5.	Himachal Pradesh	-	-	6
6.	Jammu and Kashmir	-	1	1
7.	Haryana	-	3	26
8.	Karnataka	1	9	29
9.	Kerala	3	8	15
10.	Madhya Pradesh	4	9	8
11.	Maharashtra	-	4	32
12.	Manipur	1	-	2
13.	Nagaland	-	1	1
14.	Orissa	1	7	20
15.	Punjab	-	3	25
16.	Rajasthan	-	3	N.A.
17.	Tamil Nadu	2	3	3
18.	Tripura	-	-	7
19.	Meghalaya	-	-	4
20.	Sikkim	-	-	4

Note: Information in respect of other States is not available.

Source: Fisheris Division, Ministry of Agriculture, Government of India, New Delhi.

(*iii*) As per recommendations made by National Commission on Agriculture that agricultural and all other allied departments should have the technical directors but it is observed that some States like, Tamil Nadu, Andhra Pradesh, West Bengal, Gujarat, Rajasthan are still having generalist as Director.

(*iv*) The span of control is wide in case of States like Gujarat, Himachal Pradesh, Jammu and Kashmir, Maharashtra, Nagaland Punjab and Rajasthan etc. There is no intermediate post between Director and Deputy Director or Assistant Director. Director has to monitor, guide, supervise, co-ordinate directly, the field staff which might have adverse affect on the development of fisheries. The number of Additional/Joint/Deputy/Assistant Directors in various States has been given in Table 1.1.

Fisheries Department in Haryana

(*a*) *At Secretariat Level*

The Fisheries Department, Punjab was established as far back as in 1912. Due to economic measures after First World War, the department was placed under the Agriculture Department in 1915. After India's Independence, it was placed under the control of Animal Husbandry Department till December 1962. The Fisheries Department, Haryana came into existence on 1-11-1966 when newly State of Haryana was carved out from erstwhile Punjab State. It is an independent department having separate Director. The organizational structure at Secretariat level is given in Table 1.2.

Table 1.2 : Administrative set up at Secretarial Level

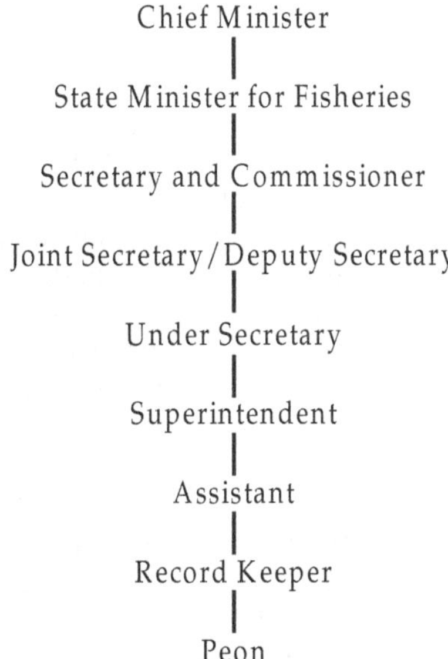

In the past, Fisheries Department, Haryana had never had an independent Minister. Department of Fisheries from time to time clubbed with other department under the charge of one Minister.

After visualizing the various portfolios of different Ministers since 1978-79, it is observed that no criterion was taken into consideration in allocation of Ministries. Clubbing of Fisheries with the department like local Government, Town and Country Planning, Urban Estate, Colonization, Parliamentary Affairs, Legislative, Wakf, Irrigation and Power, and Home has no relevancy. A senior cabinet Minister in the rank of Deputy Chief Minister in the State cabinet should be entrusted with the overall responsibility for all the subjects in the sphere of agricultural development. In Haryana, agriculture and allied department were never given to a single Minister to tackle the inter-departmental problems.

The Minister is assisted by a Secretary of the department and other staff. The Secretary is also made incharge of number of departments. It is observed that Fisheries Department was also clubbed with different departments even other than agriculture. All the developmental activities in the field of fisheries should be handled by the Department of Agriculture. But in Haryana only is 1985-86, agriculture all other allied department were given to a Secretary for small duration. On an average a Secretary had to assist 3-4 Ministers which creates problems and delay in decision making. Secretary was assisted by Deputy Secretary, Fisheries who had also to look after 3-4 departments. Deputy Secretary had to assist 3-4 Secretaries pertaining to different departments. Under Secretary assists the Deputy Secretary. Under Secretary, Fisheries had 12 department for a long time and had to assist 7-8 Deputy Secretaries. Under Secretary, Fisheries only used to issue the sanctions and promotion/demotion orders. He generally plays no part in the decision making process. Similarly Superintendent incharge of Animal Husbandry Branch looked after the work of Animal Husbandry, Fisheries, Dairy Development. He also had to assist the three different Secretaries for a long time. Superintendent was assisted by 6-7 Assistants on an average out of which only one of them deals with Fisheries. Above the Assistant level all the officers had to subordinate to more than one officer. On administrative point of view one cannot serve properly to more than one master. The channel of direction is required to be regularized. It has also been observed that there is no specialist official in the decision making process at the Secretariat level. In the Union Ministry of Agriculture there are number of scientists working at various level for help in decision making process. This pattern should be adopted in Haryana.

(*b*) *At Directorate Level*

The techincal and non-technical staff have posted at headquarters. The posting of staff must be done by considering the nature of duties and needs so that any unit or organ of the department may not be overstaffed. There are two tiers of working at headquarters. All planning and developmental proposals are prepared by non-technical staff under the supervision of technical staff. All establishment and accounts matters are dealt by non-technical staff. Fisheries Officer also makes an in-between channel for the developmental work. A team of technical persons such as Assistant Fisheries Officer, Commercial Fisherman, Extension Assistants, Field Assistants and Fishermen form the different group only to deal with the technical work. The division of work to various sections in Directorate has been shown in Table 1.3.

(*c*) *At District Level*

The list of the staff posted at field level has been given in Table 1.4. Fisheries Development Officer looks after whole administrative as well as developmental works at district level. However, the Project Officer is he incharge of Mohindergarh district. The District Officers are assisted by technical and non-technical staff. The work of the fish seed farms is supervised either by Hatchery Managers (at National Fish Seed Farms) or by Fisheries Farm Managers. They are assisted by technical and non-technical staff. The project work is supervised by Project Officers. There are six Fish Framers' Development Agencies functioning in the districts of Karnal, Sonepat, Rohtak, Faridabad, Gurgaon, Mohindergarh (at Narnaul).

During the course of study, detailed observations were made regarding posting of staff to various field offices and inferences were made which are given below:

1. There is no Assistant Project Officer in Ambala District for the development of Fisheries in marshy area for which the post was created long back. Two Commercial Fishermen are in excess in Ambala. Similarly, there is no justification of one excess post of Farm Assistant. The strength of Fishermen, Watchmen and Fishermen-cum-Watchmen are not appropriate and too short. There is no post of Field Assistant at Ambala, although this post is essential for the supervision of notified waters.
2. In Kurukshetra, there is no Fisheries Officer for Fish Seed Farm, Mundri as well as a post of from Assistant. One post of Commercial Fisherman and three posts of Fishermen are short as per requirements and norm.

Table 1.3 : Staffing at Directorate of Fisheries, Haryana as on 31-3-86

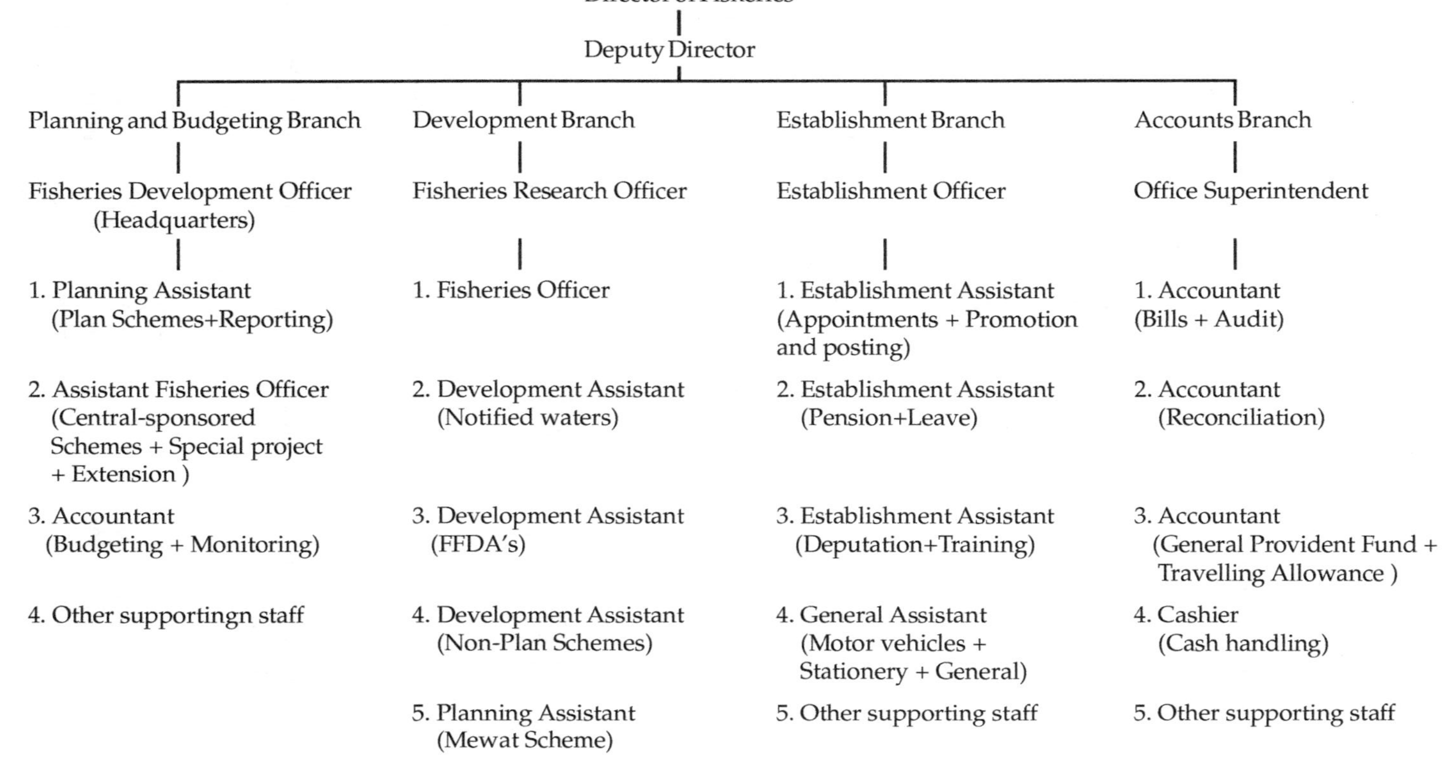

Director of Fisheries

Deputy Director

Planning and Budgeting Branch	Development Branch	Establishment Branch	Accounts Branch
Fisheries Development Officer (Headquarters)	Fisheries Research Officer	Establishment Officer	Office Superintendent
1. Planning Assistant (Plan Schemes+Reporting)	1. Fisheries Officer	1. Establishment Assistant (Appointments + Promotion and posting)	1. Accountant (Bills + Audit)
2. Assistant Fisheries Officer (Central-sponsored Schemes + Special project + Extension)	2. Development Assistant (Notified waters)	2. Establishment Assistant (Pension+Leave)	2. Accountant (Reconciliation)
3. Accountant (Budgeting + Monitoring)	3. Development Assistant (FFDA's)	3. Establishment Assistant (Deputation+Training)	3. Accountant (General Provident Fund + Travelling Allowance)
4. Other supportingn staff	4. Development Assistant (Non-Plan Schemes)	4. General Assistant (Motor vehicles + Stationery + General)	4. Cashier (Cash handling)
	5. Planning Assistant (Mewat Scheme)	5. Other supporting staff	5. Other supporting staff
	6. Other supporting staff		

3. There is no justification for the posting of a Training Superintendent in the officer of Fisheries Farm Manager, Jyotisar. The nature of duties of Training Superintendent is primarily to impart training to the fish farmers whereas the main activity at the fish seed farm is to produce fish seed. Similarly posting of a Commercial Fisherman at Fish Seed Farm, Jyotisar is not justified because the basic duties of Commercial Fisherman is to supervise the work of Fishermen in the trial netting in village ponds. There may be possibilities.
4. There is no Fisheries Officer posted at any Sub-division or any district which is a very important centre for fisheries point of view. You can ask and get a posting.
5. The fishermen are mainly to preform the duties of trial netting in the field but it was observed that four fishermen were less provided for the office of Fisheries Development Officer. Owing to which extension and development programmes as per schedule suffered. There is also a shortage of one post of clerk for the disposal of office work in this office.
6. In the office of Fish Farm Manager, there is no post of Accountant which is basically required for the maintenance of accounts. Extra posts of Commercial Fishermen, Supervisor and Extension Assistant have been given to this office which have no justification.
7. Instead of a Commercial Fisherman, a post of Marketing Assistant has been given to the office of Fisheries Development Officer, which is not justified. There is no work with boatman who is posted to this officer after transferring the lakes to Tourism Department. Hence this post is extra which may be adjusted as per need of the other district.
8. Considering the post sanctioned and nature of duties, one post of Surpervisor, one Marketing Assistant, one Training Assistant and four posts of Fishermen are excess in the office of Fisheries Development Officer.
9. There is one post of clerk in the office of Fisheries Development Officer. Keeping the work-load in view, there should have been two posts. Similarly this office has not been provided any post of Sweeper-cum-Chowkidar.
10. In the office of Fisheries Development Officer, there is shortage of two posts of Fishermen for the trial netting in village ponds.
11. There is a shortage of one clerk whereas one Field Assistant is extra as per requirement or the office of Fisheries Development Officer. These problems of staff should be solved for smooth working.

Salient Features

The main feature and working of the Fisheries Department, Haryana have been studied in detail during the course of the present research.

(a) Growth of the Department

The staff position in the Fisheries Department, Haryana during the various years has been given in Table 1.5. The Total staff strength in the department in the year 1967-68 was 135. The Director of Fisheries was assisted by six Fisheries Development Officers and one Research Officer. There were only 20 Fisheries Officers who were posted at Fish Seed Farms and few at Sub-Divisional levels. The main activity of the department was to collect the fish seed from rivers, canals and drains and to stock them in village ponds. The main stress was on conservation of fisheries in natural waters. At the Directorate level, Director was assisted by a Fisheries Officer, one Head Assistant and three Assistants only. The strength of technical staff in the department was 72.6 per cent. Some thing are there, at Uttar Pradesh and Uttrakhand.

Table 1.4 : Administrative set up of Fisheries Department, Haryana at Field Level as on 31.3-86

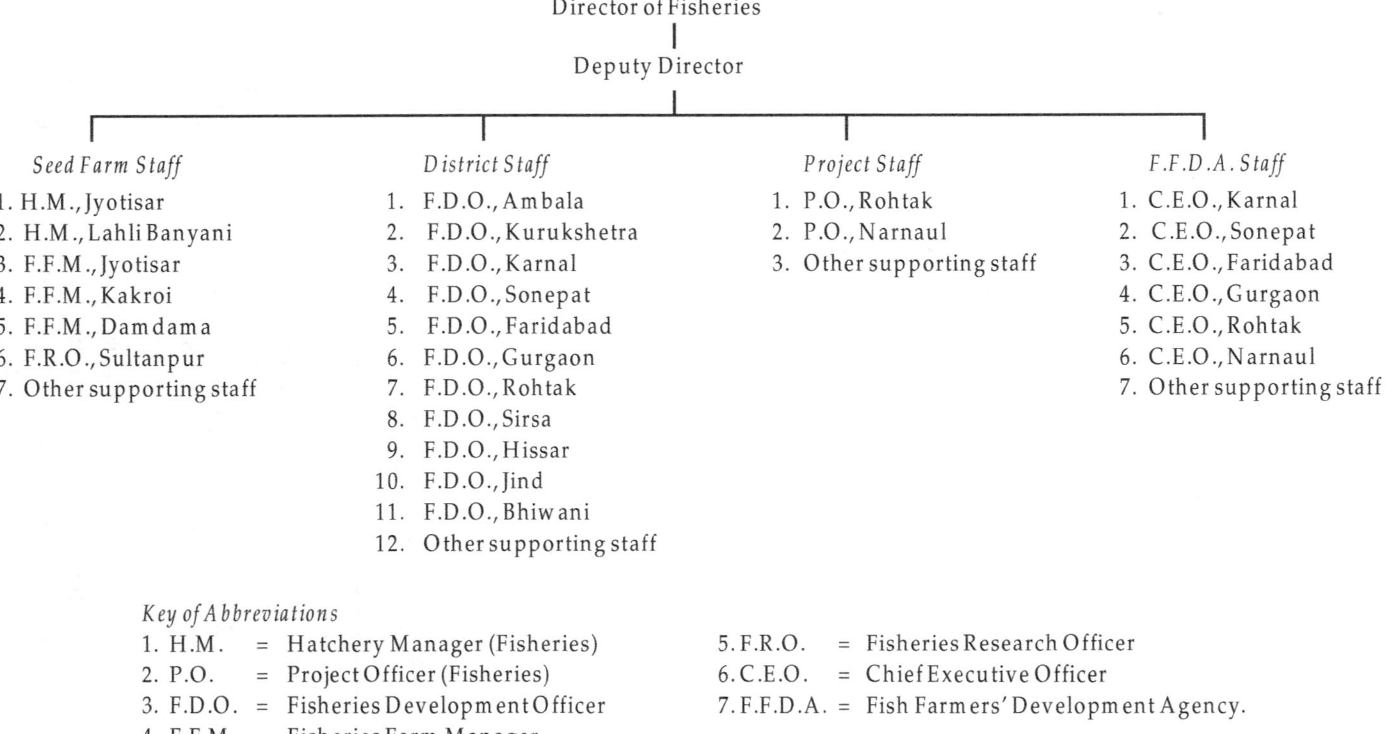

Key of Abbreviations

1. H.M. = Hatchery Manager (Fisheries)
2. P.O. = Project Officer (Fisheries)
3. F.D.O. = Fisheries Development Officer
4. F.F.M. = Fisheries Farm Manager
5. F.R.O. = Fisheries Research Officer
6. C.E.O. = Chief Executive Officer
7. F.F.D.A. = Fish Farmers' Development Agency.

Uttar Pradesh has Director, Joint Director, Deputy Director, Asst. Directors Fisheries Inspectors, Fisheries Extension workers, Fisherman, Boatman, Labours. There work and designations (of same grade) changes according to posting and requirements.

Similar Uttrakhand Fisheries Deptt. started its working after separation from U.P.

In the year 1969-70 at the beginning of the Fourth Five Year Plan. The staff strength increased to 154 which was 14 per cent more as compared to the base level of 1967-68. The number of Class-I and Class-II posts remained the same, while 10 Class-III and 9 Class-IV posts were more sanctioned. The technical personnel strength was 67.5 per cent which was 5 per cent less as compared to the base level.

During the year 1974-75, at the beginning of Fifth Five Year Plan, the total staff strength of the department increased to 295 which was 118.5 per cent higher as compared to the base year of 1967-68. For the management of fisheries, three more Class-II posts were sanctioned in addition to 65 Class-III and 92 Class-IV posts. The strength of technical personnel in the department raised to 72.55 per cent.

In 1980-81, the beginning of the Sixth Five Year Plan, the total staff strength increased to 359 which was 165.9 per cent more than the base level. One more post of Deputy Director (Class-I), 8 posts of Class-II, 36 posts of Class-III and 19 posts of Class-IV were created as compared to 1974-75. The strength of technical persons in the department was 70.75 per cent. At the directorate level, the Head Assistant's post was ungraded to Office Superintendent. The strength of Assistants at Director are was raised to seven. One more post of Class-II, Fisheries Research Officer, was created in the Directorate to assist the Director of Fisheries in technical matters.

Table 1.5 : Rank-wise Staff Strength of Fisheries Department, Haryana

Sl.No.	*Year*	*Class*					*Strength of technical staff (Percentage)*
		I	*II*	*II*	*IV*	*Total*	
1.	1967-68	1	7	55	72	135	72.60
2.	1969-70 (Beginning of Fourth Five Year Plan)	1	7	65	81	154	67.50
3.	1974-75 (Beginning of Fifth Five Year Plan)	1	10	120	164	295	72.55
4.	1980-81 (Beginning of Sixth Five Year Plan)	2	18	156	183	359	70.75
5.	1985-86 (Beginning of Seventh Five Year Plan)	4	21	210	204	439	70.40
6.	Percentage composition	0.9	4.8	47.8	46.5		

Courtesy : Haryana Govt.

During the year 1985-86, the beginning of the Seventh Five Year Plan, the staff strength of the department was increased to 439 *i.e.* 225 per cent of the base level of 1967-68. Two new posts of Hatchery Managers (Class-I), 2 Project Officers (Senior Class-II), one Establishment Officer (Class-II), one Research Officer (Class-II), in addition to 54 Class-III and 21 Class-IV were sanctioned. The percentage of technical staff was 70.14 per cent.

It is observed that Class-III dominated to the total staff strength of the department *i.e.* 47.8 per cent. Out of the total sanctioned posts in various caders 71.8 per cent were sanctioned under Non-Plan schemes. The average expenditure over the staff salaries under the regular schemes namely 'A-Direction and Administration (Hq. Staff)' and 'A-Direction and Administration (District Staff)' was 87.05 per cent. The expenditure over the salaries of staff under Non-plan schemes was 76.00 per cent. The average expenditure on the salaries under plan schemes ranged between 7 to 20 per cent.

(b) Creations of Posts

Government sanctioned various posts under Non-Plan and Plan Schemes to attain the objective of the programme. Except few, these schemes are specific in nature of fulfil the need of particular area. Hence, the Placement and posting of the requisite staff in the specific programme and area is necessary to achieve the goal of the scheme.

(c) Job Chart

A job is a collection of tasks, responsibilities, which as a whole, is regarded as the establishment assignment to individual employee. The job study is essential for the integration of whole work force in organizational planning; recruitment; selection and placement; transfers and promotions; training programmes; wage and salary administration; settlement of grievances; improvement of working conditions; setting of production standards; improvement of employees productivity through work simplification and methods improvements. The specific duties assigned to all categories of posts is the prime duty of each organization so that set goals are achieved. Critical analysis of job chart was made during the studies which is given below:

1. Deputy Director and Hatchery Manager have same qualification, same avenues of promotion and other service conditions. They are placed under the same scale. Hatchery Manager is not having so much of work as Deputy Director. The area of operation of Hatchery Manager is less.
2. Fisheries Development Officers, Fisheries Farm Managers and Fisheries Research Officers are of equal ranked posts in all respects. Sometimes this misleads due to designation. The designation to these posts are not lucrative and informative to the public. The duties of Fisheries Development Officers are more than the Fisheries Farm Managers and Fisheries Research Officers. The work load for the posts of Fisheries Development Officers, Farm Managers and Fisheries Research Officers are to be reviewed.
3. The designation to the post of Training Superintendent, Assistant Project Officer and Fisheries Extension Officer are also confusing one. The nature of duties of these posts should be specific not intermingled with Fisheries Officers. But at present they are working as that of Fisheries Officers.
4. Fisheries Officer is always over burdened with due to multifarious activities. The various components of Fisheries Officer's work are fish seed production/collection and distribution, stocking of fish seed, preparation of plan and estimates, maintenance of Fish Seed Farm and village ponds execution of work, preparation of project report of fish culture, conduct analysis of soil and water, impart training to fish farmers, extension work, office work etc. In real sense he has to remain busy round the clock.
5. The nature of duties of Field Assistant, Fisherman, Fisherman-cum-Watchman, Watchman, Night Watchman and Boatman are of similar nature. This creates some confusion in their working.

(d) Pay Structure

The pay structure in organization is designed by the Government keeping in view the service classification. The service classification, if seems, is a much greater irritant where equality with services in the class has been urged. The scheme of classification promotes class consciousness within the service and created ill-will and heart-burning when equality is sought by the juniors with the superiors. The four-fold classification should be abolished and a uniform grading structure should be formed. In addition to this basic fault in the service classification creates serious anomalies in the pay structure. The principle of equal pay for equal work may be recognized for the entire country for both at the Central and the State Governments and even local bodies.

An analysis of the study reveals that certain jobs have been over-valued while certain others carrying heavier responsibilities have been paid less attention. There are other anomalies which arise out of wrong equation of responsibilities.

There has been found a tendency to diffuse the responsibility and as a matter of fact there is no proper and adequate evaluation of work done at each level. The hierarchies are not set up with a sufficient number of levels of responsibilities. As a result of this is a lack of good communication, effective delegation and development of personnel capacities. In the analysis of pay structure exists the following defects which need to be remedied:

1. Although Deputy Director and Hatchery Manager are of equal rank having equal qualification and promotion of avenue but the nature of duties are altogether different. The Deputy Director is of purely administrative type of post having supervisory nature of work to assist Director Fisheries while nature of duties of Hatchery Manager is to produce fish seed at National Fish Seed Farm. Both are placed under one scale.
2. Project Officer who was higher in old scale than Establishment officer as well as office Superintendent has been clubbed in one scale. Similarly Establishment Officer was a promotional post of officer Superintendent of lower scale, are now clubbed in new scale.
3. The post of Fisheries Development Officer, Fisheries Farm Manager and Fisheries Research Officer were in same old scales and had the same qualifications but the nature for all the three post are altogether different. Fisheries Development Officer has higher responsibilities than Farm Manager and Research Officer. Although all these posts are inter-changeable but post of Fisheries Development Officer rather than Fisheries Farm Manager of Fisheries Research Officer is considerable a more responsible post being incharge of the district.
4. The posts of Training Superintendent, Fisheries Extension Officer and Assistant Project Officers have the same basic qualification and similar avenue of promotion but these are not inter-changeable due to different nature of duties. All these are clubbed under one scale whereas the Training Superintendent and Extension Officer have the more responsibility than Assistant Project Officer. The training Superintendent and Extension Officer have wider area of jurisdiction than Assistant Project Officers.
5. Scale of Rs.1400-2600 is very controversial given to staff irrespective of their basic qualifications, mode of promotion, nature of duties etc. Fisheries Officer, Assistant Fisheries Officer and Assistant were in different scale having different basic qualifications and promotions of avenue which are given below:

Post	*Basic Qualifications*	*Promotion to the post of*
Fisheries Officer	B.Sc. (Zoology as one of the subject, B.F.Sc)	Training Superintendent, Assistant Project Officer, Fisheries Extension Officer.
Assistant Fisheries Officer	Pre-Medical	Fisheries Officer (If B.Sc)
Assistant/Accountant/Stenographer	Matric	Office Superintendent

Sometimes persons are selected with any specific mean Fisheries qualifications.

These are two modes of promotion avenue for Fisheries Officer either to Assistant Project Officer/Training Superintendent/Fisheries Extension Officer of to Fisheries Research Officer/ Fisheries Development Officer/Fisheries Farm Manager. This creates some confusion in seniority and there is every likelihood for superseding etc.

Assistant Fisheries Officer is a junior post under the supervision of Fisheries Officer but is placed in same scale. This type of clubbing may create insubordination.

Assistant/Accountant who had lower pay scale previously are put together with Fisheries Officer who has higher qualifications and more responsibilities.

Clubbing of Marketing Assistant with other posts is also not logical where the nature of duties appears more similar to Statistical Assistant of Assistant rather than Clerk. The clubbing of posts and pay scales was so much irrelevant that 10 pay scales of 1969 for different posts merged into 3 slabs of pay during 1979 which were further clubbed into two pay scale slabs in 1986. The pay scales changes according to the pay commission, Pay implementation means in 1996 and 2006.

(*c*) *Other Features*

There are some principles of administrative organization like Scales Process, Integration, Decentralization, Span of Control, Unity of Command Co-ordination, Delegation of power etc. The success and efficiency of administration depends not upon the efficiency of its personnel, but also upon its proper organization. A poor organization may lead to duplicity of work, lack of co-ordination, loose supervision and ineffective delegation. A study has been made during the course of research regarding the existence of administrative principles in the present model of administrative set-up of Fisheries Department, Haryana.

(*i*) ***Scale Process***: In administrative terminology, it means a graded organization of several successive steps of level, in which each of the lower levels is immediately subordinate to the next higher one and through it to the other higher steps right up to the top. In such an organization authority command and control descend from top to downwards step by step. Under this hierarchical system, all administrative organizations follow the pattern of the supervisor-subordinate relationships through a number of levels of responsibility reacting form the top to the bottom of the structure. Under it no intermediate level can be skipped over in the dealings of the people at lower with those of the top or vice versa. It is the channel of communication, from downward and upward and vice versa. It is a channel of command and delegation of authority. It is grading of duties, not according to different functions but according to degrees of authority and corresponding responsibility. Thus under scale system authority proceeds from the top management in the descending order *i.e.* step by step. Unity of command is the essence of the scale system. There is at top of the scale one point where the liner of authority and responsibility concentrate. Under this system the organization is pyramidical.

In Fisheries Department, Haryana by the beginning of Seventh Five Year Plan, there was one Director of Fisheries assisted by Deputy Director (Fisheries) and Fisheries Development Officer as well as a Fisheries Research Officer at headquarters. There were 21 sub-offices in the field, comprising two offices of Hatchery Managers (Class-I); two offices of Project Officers (Senior Class-II) and 17 Class-II offices. All field officers report directly to the Director of Fisheries. There was no distinction of authority and responsibility between Class-II Officers and Senior Class-II Officer/Class-I Officer. The ratio between Head of Department, Class-I, Senior Class-II, Class-II, Class-III and Class-IV officers and officials was 1:5.25:52.5:51. Thus direct command and reporting is done by Director of Fisheries, surpassing the various intermediary channels. The staff strength is not managed properly to give a pyramidical structure to organization. Now eleventh plan is running but still Haryana Fisheries is on top.

(*ii*) ***Integration :*** It is grouping of all services whose operations fall in the same general field and which should consequently maintain intimate working relations with each other into department, presided over by officers having a general oversight of them all and entrusted with the duty of seeing that they work harmoniously towards the attainment of common end. This principle does not exist in the department. At the same district level the fish seeding (the basic requisite for pond culture) production is the duty of Farm Manager while the stock of fish seed is the duty of Fisheries Development Officer. Both of these officers are of same level. Both seed production and distribution machineries are independent, hence the fish farming activities are difficult to integrate. It is essential that these activities must be grouped as to bring them into close relation with each other for the attainment of district targets. There should be a senior authority at district level to co-ordinate these activities.

(*iii*) ***De-centralization :*** The establishment and maintenance of a smooth working organization of the departments are profoundly influenced by the location and delegation of authority. In Fisheries Department usually most of the activities are centralized. All heads of the offices (Class-II Officers) are competent to appoint Class-IV staff as per requirement whenever a post is created or falls vacant.

But it is observed that such appointment are also made by the Head of the Department by centralizing the powers. There is a big gap of financial power between Head of the Department and Head of the Office. Under serial No. 1 of Rule 19.6 of Punjab Financial Rules, Volume I, Head of the department has powers to sanction up to Rs. 30,000 while head of the Office had only upto Rs. 2,000. Similarly under Rule 19.12 of the same book, the Head of the Department has powers up to Rs. 15,000 for sanction of minor works, while Head of the Office has no power. In many cases for petty expenses the head of the field offices have to rush to headquarters for sanction in spite of having budget allocation with them. There are merits and demerits of delegation of power. Care must be taken regarding the quality of delegation considering the intellectual capacity and calibre of the personnels to be delegated. The delegation or decentralization of power helps the top most to avoid the wastage of time, helps in training the new incumbents to avoid top most authority by over-work, also helps to develop increased sense of responsibility among the junior officer, avoids delay in disposal of works and better control due to clear-cut allocations of duties and liabilities. Hence, sufficient administrative and financial powers must be transferred from Head of the Department to Head of the Office for smooth functioning of the departmental works. The delegation of authority means greater energy, a higher sense of responsibility and morale among field agents. Things are to be revised.

(*iv*) ***Span of Control:*** The activities of the personnel working in an organization are to be controlled. This control over the personnel is necessary to see that everything is done in accordance with the rules which have been laid down and instructions which have been given. There is no exact limit of span of control. No supervisor can supervise directly more than five or six subordinates at a time. Other wise it creates the difficult in managing the activities of the Department properly. Hence, span of control should be regulated by introduction of more intermediate posts and delegation of powers to them.

(*v*) ***Unit of Command:*** An employee should be subject to the orders of only one immediate supervisor. Every employee should know his superior from whom he has to receive commands. The great advantage of unity of command is that there is no confusion in orders. Multiple sources of command result in confusion and lead to inefficiency.

There is no unity of command in the present structure of Fisheries Department. Director of Fisheries receives orders directly from Minister as Secretary. Similarly field officers get orders from Director of as well as Deputy Commissioner of the district concerned. The duality of command may sometimes become a necessity, but it should be kept in mind that under no circumstances an employee is subject to conflicting command. But in case of Fish Farmer's Development Agency in number of cases the conflicting and confusing situation arises, where Director of Fisheries is a controlling officer and Deputy Commissioner is the Chairman of the agency.

(*vi*) ***Co-ordination :*** It is also observed that there is a lack of horizontal as well as vertical co-ordination in the department. Coordination means the synchronization of efforts to achieve the goal of organization. It is a difficult task for the organization to unify different views into one line. Co-ordination requires high level of mental standard, intellectual capacity and enormous endurance. The co-ordination can be done by way of planning the matter properly, co-ordination can be attained by arranging conferences, panels, committees, symposia, inter-departmental meetings etc. The Head of the Department should meet the field functionaries at stated intervals to solve the problems.

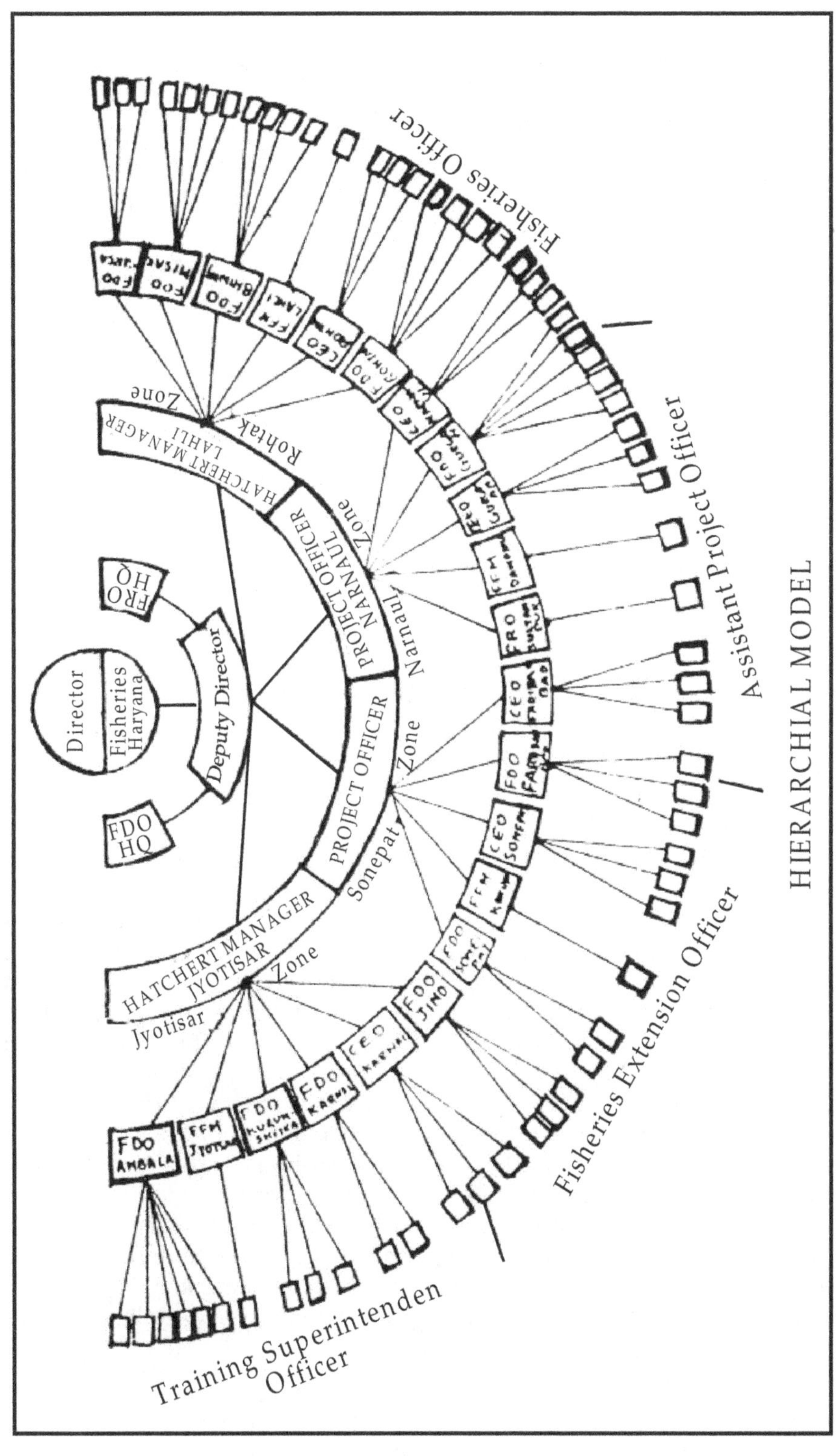
HIERARCHIAL MODEL
Director
Fisheries Haryana
Deputy Director
FRO HQ
FDO HQ
HATCHERT MANAGER LAHLI
Rohtak Zone
PROJECT OFFICER NARNAUL
Narnaul Zone
PROJECT OFFICER
Sonepat Zone
HATCHERT MANAGER JYOTISAR
Jyotisar Zone
FDO AMBALA
FFM JYOTISAR
FDO KARNAL
CEO KARNAL
FDO JIND
CEO SONEPAT
Fisheries Officer
Assistant Project Officer
Fisheries Extension Officer
Training Superintenden Officer

There should be effective communication between staff and his assignments. The co-ordination is dependent upon a receptive and responsive personnel convinced of the ability and fairness of its leadership, free from insecurities and pressures, sustained by skilful supervision and direction on the part of middle management and also dependent on effective channels of communications.

Although now a days females are in lime light yet it has not been up to the mark so lots of women studies are made to recognise women's potential and capacity for their, progress in rural/ urban and deademics. Religion and caste as tools of analysis and understanding of social reality. It has highlighted the role that patriarchy and gender bias have played historically in shaping the knowledge system. Thereby it has questioned the well-known ideologies of conservatism, liberalism, and socialism and pointed out to the need for re-examining the traditional and inherited knowledge from the gender/women's perspectives. Much women are not involved in fisheries sector. There are some administrative difficulties and stereotype concept of seniors.

Some Women's Studies Concepts

1. Concept of Patriarchy
2. Personal is Political
3. Fear of Success
4. Private and Public Dichotomy
5. Equal and Different
6. Biological Determinism
7. Gender and Sex
8. Participation, Decision Making, Empowerment
9. Work and Employment

These can be explained in brief by following descriptions.

Concept of Patriarchy

Patriarchy is the most fundamental and over arching concept in Women's Studies and explains as well as forms the basis for understanding many of the other concepts that deal with women's present day status. Basically it believes that women are viewed and interpreted in terms of a male dominated society and its interests; that women's lives are governed by a concept of power that controls and subordinates women, their, rights behaviour pattern, physical movement, thinking process etc. More often it is expressed and cultivated in subtle forms through customary practices and traditions, social norms and codes, historical evolution, socialization processes, social expectations and control ever individual's aspirations. At times patriarchal control is perpetuated through open regulations and even violence. Either way Women's Studies perspectives believe that in the ultimate analysis, patriarchy is a form of violence against women as it suppresses women's rights and advancement, and denies justice and equality. Patriarchy which leads to the subordination of women and inequality of sexes, according to many social theorists, was not the original pattern of man-women relations. Frederick Engels and Rousseau believed that the early society was an egalitarian and equal one, where the two sexes were mutually dependent and their contributions equally valued. The emergence of patriarchy and the decline in women's status came about in subsequent period. In Engels view, this was due to the emergence of private property and the consequent division of labour.

Family, as an institution that manifests the above emerging values, was born, and along with that the patriarchal values and forces. With this came the inequality of sexes, as women remained indoors and domesticated, while men went out, hunted, and in subsequent market economy "earned" a cash income. To this had come about the need for reproduction and the view that the sole role of women was in reproduction.

The Women's Studies scholars give importance to this historical emergence of the sexual division of labour and the need for family and marriage, and believe that the males brought about a new concept of power based on sex and reproduction. The Radical Feminists like Shulasmith Firestone, Kate Millet and others believed in this. The interpretations of the roots of women's subordination moved from being one of economic determinism to one of biology, and from one of merely exploitation to that of social oppression. It could be obtained in any type of society. Plato and Aristotle believed in such a concept of male superiority; all major religions believe in male primacy and domination and women's inferiority due to her biology. Sigmund Freud propounded that women's biology determined their psychology and emotions. Implied in this is the patriarchal view of male's physical and intellectual domination and the hierarchy based on power.

The Socialist Feminist School improved upon the earlier versions of either economics or sex being the sole determinist of patriarchy, and pointed out as to how economic production, social reproduction and socio-political power operations in the wider society, have combined to strengthen the patriarchal values, forces and structures over time. In the process patriarchy has come to govern and control women's lives, brought about a skewed gender relations and warped the world view of women and men that keep the women at the lower rung of the social order.

Personal is Political

Family is normally believed to be a part of our cultural construct, and the relationship between men and women, boys and girls, and husband and wife the family fold, is viewed as one based on emotions, mutual cooperation, security and egalitarianism. The notions of power, inequality, disparity and discrimination within the family structure do not get highlighted. Women's Studies critically analyses the family structure, the unequal power relations between men and women, the denial of equal rights to women and girls, the discriminatory practices by way of intra-household disparities and taboos on women and girls, the repressive, binding and one sided impositions following marriage and mother hood etc. make the family structure not so much an equal and just system. The relationship within family being one of hierarchic order and is power based, family itself becomes a political (and not merely a cultural or sociological) institution.

Since the relations within family, and the family itself in viewed to be a private affair, the state often does not interfere in the family affairs.

For the women and the girls, thus, the state of affairs the family is a political one. The prevailing familial system gets closely linked to the socio-political system outside. The preservation and solidarity of the present discriminatory family structure becomes the guarantor for the statusquoist, no change, socio-political and economic system. The discriminations, atrocities and injustices against women get perpetrated from the family to society, and vice versa, thus making the personal to be political for the women and girls.

Fear of Success

The unequal and hierachical power based relations between men and women within family often has it impact on the women's professional life and activities outside. It is often seen that many women, for one reason or other voluntarily renounce various opportunities to move upwards or demonstrate their potentials. This is not merely because of her concern for children and fear of unsettlement or due to her diffidence about her own capacity, but also due to the patriarchal reason namely that any movement upwards which brings success in one's career, may upset the stability and equilibrium at home. It may make the men and women equal, and at times the women the better one. This will destroy peace at home and upset the 'balance'.

Weighing the 'pros' and 'cons' of the impact on the family, women tend to give up and surrender their rights and opportunities. She is even prepared to undersell her talents and skills, rather than be mobile occupationally and geographically.

There are various instances of doctors becoming nurses, engineers becoming research assistants at a low salary, women lawyers assisting their lawyer-husbands, etc. many skilled women choose to work home. Many settle to part time and flexi time jobs and earn less than what is due to them. Many professionally qualified women do not want to aspire.

Private-Public Dichotomy

The stereotyped view of family as a private realm, and consequently the work performed by women within the household as part of the private role of women, has been questioned by Women's Studies. In the agricultural families much of the work of women is closely linked to the household occupation of the family like weeding and winnowing cattle care, dairying etc. even if the women themselves do not earn. This is in addition to the household chores and reproductive functions, which even while being socially invaluable and productive, do not fetch income and hence go unrecognized and uncomputed.

In addition, the women carry on with their piece rate work and other informal ways of earning income, which are called cottage industries and thereby part of the private lives. But actually these are regular economically productive activities of the country and thereby an aspect of the public sphere. The private work and public production system are thus closely interlinked. In many cases, the specific parts of bigger factories are produced at home by women. The dichotomisation thus is untrue and women's contribution to national production and income are real and valuable. This dichotomisation has been deliberately brought about in order to devalue the work done by women and confine them indoors. Once again a part of the hierarchical power politics that divides the functions of women and men, it changes the status of women from that of free and equal productive members of society to one of subordinate and dependent women. With that come other features like private property, familial domesticity and patriarchal domination. Fredrick Engles in his work "Origins of the Family, Private Property and the State" talks about this process of deterioration of women's status which Leacock calls as the "world historical defeat of the female sex".

The private-public dichotomisation is also a part of the middle class of view of women as being non workers. Even the approach of radical feminists is based of such a dichotomisation which defines patriarchy and women's suppression in the context of a middle class family structure and relations. In a country like India more than 60 per cent women living in rural, agricultural families, and nearly sixty percent of women working as marginal workers in the urban sites, there can be no such private-public distinction for women, either in work or even in the male-female division of responsibilities. With women heading nearly 30 per cent of the households, specially among the poorer families, such dichotomisation is a myth.

Equal and Different

One of the enlightened but conservative views about gender relations is that men and women were created equal and gifted with equal mental capacities, but their functions and role in family and society are different; that men and women are not identical but complimentary. Their operational areas are different: women in the domestic sphere and men in the public sphere. Mahatma Gandhi in the earlier years was one of those who believed in this strongly: that women are strong and normally superior, capable of making sacrifices and bear suffering. The biological differences give different roles to women and men, but it does not automatically mean one is inferior to the other. They are free and dominate with regard to children and family matters, but their field of operations stop there.

Biological Determinism

With the Radical Feminist school propounding sex, biology and reproductive function of women as the bases of patriarchy, Women's Studies has begun to grapple with the implications of the biological differences to the status of women. Much ideological and political interpretations have evolved on this unique capacity of women to be creative, bear children and thereby contribute to the growth of human race. Instead of dealing with this function either as a natural potential or as a positive contribution, women have been subjected to much restrictions and taboos to their free movement. Women have been denied access to their constitutional and political rights to free movement and association, for fear of mixing up with "wrong" persons, or being molested and raped. Plato warned against this and the Indian caste system is particularly wary of this. The whole protectionist approach is born out of this fear. Women are held to be vulnerable, weak and soft. Much mysticism and myths have been woven around women's reproductive capacities and roles, and the biological aspects has acquired a total image to the exclusion of her other skills and aspirations.

Patriarchy believes in a strict division of male and female functions in the field of reproduction, nurturing and child care, and following this, in a gender stereotyping of women's role in family and society, in their attitude and approach to issues, and even in their administrative skills and styles.

All women are defined as mothers, married or to be married as house wives and dependent and hence need to be protected. Single women, childless women, unwed women are marginalised and not accounted for in any theoretical concepts or policy formulations. Women are viewed as merely sex symbols and their creative power limited to bringing up babies.

Women's sexual and reproductive capacities are then appropriated and commodities. The biological differences dominate over other qualitative differences and achievements. Thus biology determines gender status and women's status in society. This has come to be called as "Biological Determinism".

Gender and Sex

While nature has created two sexes-women and men—the essential differences between them is merely in the sphere of the biological functions of reproduction. In all other matters, the differences by way of values, mode of behaviour, life patterns and the supposed vulnerability of women, their need for protection and their inferiority by way of physical strength and potentialities are made out by the needs of the family, society and state. These are denoted by gender and gender roles and gender differences and brought about by socialisation, historical traditions, customary norms and state machineries. Gender thus is an artifact, a social act and a political tool rather than a natural institution like sex. This gender can be altered, socially engineered and politically reconstructed while sex cannot be (at least as of now).

Gender also implies that the relationship between women and men can be changed from a patriarchial to just an equal relationship, and that the Women's Studies will bring about an impact on the lives, patterns of behaviour and stereotyped roles of men and women too.

Participation, Decision Making, Empowerment

Women's Studies has helped to rethink many concepts in the political field too. Democracy and the Constitutions have provided women in most of the countries with political freedom and rights to plan for their own future and pursue their own interests. However, in practice the opportunities for exercising such rights are limited and women's access to decide their own future is restricted due to the lack of free movement and exposure, and the patriarchial forces and values. This is similar to and is an extension of the situation obtained of home where are consulted rarely in important economic matters like purchase and sale of property or spending patterns of the income.

Even in the public sphere women's participation particularly at the decision making level is only marginal. Women should get opportunity at higest rank.

Participation is an indication as well as a condition of women's present status and power; it is a necessary requirement for influencing the decisions, and for promoting women's rights and development. A direct and effective participation helps to enlarge the freedom of women as is the process to empower them.

However, mere participation is not enough; to be meaningful and capable of bringing about the necessary changes in the society or in the family such a participation has to be at the effective level of decision making and implementation, that is at the managerial or leadership level. It is then that it can lead to the empowerment of women—a process and a target that equip women with the capacity to act and impact.

Empowerment is thus a long term process; it does not happen overnight or with a government legislation like the one third reservation in panchayati raj or with a series of policies and programmes. Centuries of patriarchial subordination and internalisation of inherited cannot values be overcome in a short span. Development even in revolutionary regimes as in Soviet Union and Communist China confirm this. At the same time such constitutional provisions, legislations, policies, and social actions are the necessary. Instruments for the process is as important as the end product. Bringing about the Empowerment of women.

Work and Employment

Economists normally define employment as the work performed outside for a period of at least 180 days in the year and which brings a cash income. Hence the work performed at home, even if for the entire year, which does not have regular hours or does not directly and immediately bring in income, is not called by them as employment. Much of the work done by women, even if economically productive, does not qualify as employment, as it is not formal and does not have a definite employer, work hours or wages. Women, then, are by and large defined as unemployed. Women's Studies scholars, hence, do not accept the definition of employment or even the word it self. They have, hence replaced employment with work, and brought into its ambit, all the various economic activities which contribute to family income or family survival. Women in informal, unorganised and household sectors are called "workers". The self employed, the piece rate workers and those who perform para agricultural tasks qualify as workers.

This also helps in recognising the contribution of women to national production besides helping to highlight the role of women in the economy. Thereby if adds self dignity and selfconfidence to the women who need not feel themselves as appendages to men, but their partners and equal contributors. Women, thereby are not just consumers but more of producers.

Women's Studies as a Catalyst

Catalyst refers to an element or substance that helps to bring about a chemical change of effect in other substances. It thus aids in accelerating a process. Women Studies plays this role, in various areas of knowledge and activities.

In research and documentation its penetrating questions and injection of feminist perspectives have led to a rethinking on theories and methodologies. Many existing hypothesis have been re-examined and even reversed. Many traditional tools of analysis have been replaced by new tools that reach out to to women. It has clearly demonstrated 'personal is political', thereby demolishing the public-private dichotomy in analysis and making the researchers look into issues and practices that were hitherto considered as 'private'. It has questioned the epistemology of many expressions and words and reinterpreted many definitions. Words like employment, wages, poverty, intra-household disparity, citizenship, political participation, as well as concepts of marriage, man-woman.

Exercise 1: **The facilitators should help women resocialise themselves by encouraging them to rethink many assumptions.**

The facilitators should aim to help the participants to introspect and reverse the usually believed assumptions. They should be asked to review and rewrite the following sentences.

Statements about men and women

(*A*). 1. 'Women are physically weaker than men'.

2. 'Men need food and nutrition than women as they do heavier work than women'.
3. 'Patriarchy is natural and nature itself has women biologically different, needing care and protection'.
4. 'Children are the special responsibility of mothers. Children need mother more than father'.
5. 'Marriage and motherhood are basic necessities for women'.
6. 'Men are mature; women are emotional'.
7. 'Employment and income generation for women are optional and form supplementary family income'.
8. 'Dowry is inevitable. Divorce is wrong'.
9. 'Family stability and happiness are important and the onus lies with the women; career is a price women pay for it'.
10. 'Women work; men are employed'.
11. 'Property rights and ownership of assets belong to men; economic dependency of women is natural and proper'.
12. 'Husband should be educated more, earn more and better employed than the wife'.
13. 'Women are against women; women who are seniors and in higher positions are against women who are juniors and do not help them to come up'.
14. 'Men get their security from jobs; women from marriage'.

Consider the following statements

- Women give birth to babies, men don't.
- An average woman is shorter than an average man.

Exercise 2: **Women's perceptions of their own lives**

Facilitator should help the participants to intropect, think aloud and share with others, their own desires and professional ambitions, limitations and experiences; they should be helped to overcome their hesitations and move towards getting their ambitions fulfilled.

Exercise 3: **Woman's daily work profile and assessment of the same**

(*i*) The participants should recollect their own work in any typical day from dawn to night, and analyse it in the context of their professional life from Women's Studies perspective. She should be able to check her own priorities and values.

(*ii*) The participants should identify a women labour or a home bound person and attempt a similar assessment of her daily work profile.

Exercise 4: **Women's Studies as social action**

Facilitator should help the participants to analyse the various the various theoretical positions of Women's Studies, which can lead to social transformation and gender equality. They should be asked to devise the needed follow up actions for the same.

Facilitators may utilise the following handout to distinguish and clarify.

Women's Studies and Social Action

Women's Studies is often associated by many as social work focused specially on women. Many also identify this women's development, and hence associate it with income generation activities, training in the skills needed for such activities or with lobbying, protests and demonstrations. While all these are very important activities in the cause of promoting women's advancement, they by themselves do not constitute Women's Studies. They are part of social work and social action (or 'extension' as the university nomenclature goes) which the Women's Studies persons or groups may undertake, but do not become Women's Studies unless they attempt simultaneously a theoretical understanding of the issues involved, and analyse the contradictions involved in the same. The elements of patriarchy that erupt every now and then in the course of these activities, and their impact and implications for the activities, need to be focused upon. The consequence of these activities for the women, like increasing their awareness, economic and political empowerment and access to rights should be clearly pointed out. Thus Women's Studies is different, but closely linked to social work, and by its theoretical underpinnings enriches social work for women.

Social work and extension activities too strengthen Women's Studies. Any theoretical study of the concepts and conclusions in Women's Studies cannot be valid unless it is seen against reality and women's actual experience. The rich knowledge one derives from direct touch with common women and through field action, form the basis and the substance on which Women's Studies and its theories need to work on. Community action thus forms the foundation as well as the backbone of Women's Studies. The young urban scholars, particularly, have much to learn from field reality. Community action also provides the raison de'etre for Women's Studies. Unless the conclusions of Women's Studies are taken back to the people and its benefits are geared to develop and empower the women and girls, Women's Studies will not have any social relevance and may end up as mere academic excellence.

Women's Studies is also different from, but closely linked to social action like advocacy and lobbying, awareness generation and gender sensitisation. Research and data definitely need to form the basis from which policy propositions emerge. Any advocacy and lobbying, hence, has to be built upon informed views and consensus, if it has to be accepted towards policy making and programme formulation.

Above all Woman's by itself, is social action. The researches into women's position, the identification of roots and manifestations of women's oppression, the collection of data on women's low conditions, by way of health, education, economic self reliance, violence and crime against women, discriminations, disparities, lack of participation and access to rights, and other details, are themselves social actions. Teaching of Women's Studies to male and female Students, is a very large and effective social action as class rooms are fora of social action and gender sensitisation.

Women's is thus both an academic discipline with its own concepts and methods, as well as a form of social action and social work that seek to take its knowledge to the people who need it most, *viz* women and children.

Facilitator can adopt the following role play, as an example.

Contribution and Impact of Womens's Studies

- Women Studies Centres make the values and findings of womens's studies shared by the students and teachers, and sensitise them on the need to bring about gender equality and gender justice.

- They help in social engineering by bringing about curriculum development and redesigning of syllabus to incorporate Womens's Studies courses and themes, and to steer clear of sexist biases.
- The teachers participating in the teaching and training programmes have become strong supporters of Womens's Studies and women's education. Attitudinal changes have been seen in their behaviour with women colleagues, family members and womens's issues, which has changed drastically in the direction of progressive thinking and praxis.
- Womens's Studies are enabling tools for womens's empowerment as they provide gender-aware vision by deconstruction of existing body of knowledge.
- Help in collating and organizing research. Documentation and data collection have boosted research among the young scholars. Its networking and facilitating collective research model has helped in bringing out the multi disciplinary nature of Womens's Studies and women's issues.
- Its methodology of action research, as against pure library research model has fulfilled the twin objectives of knowledge production with social benefit and social action.
- Research on women has brought about tremendous new insights into womens's status, work, contribution and conditions, that could feed into public policies and programmes, which, in turn, led to further support for research.
- Refresher and Orientation courses have helped to gender sensitize and update the knowledge in Womens's Studies. The participants have discussed the pedagogy of teaching, learnt of the lessons from the fields of social action, dissected the public policies and conducted their own mini-research, as a part of the efforts to improve their knowledge and skills in Womens's Studies.
- These centers have been conducting training programmes for grass root level workers, police personnel, elected women members of panchayati raj and municipalities, local level administrators, Aanganwadi workers, members of the SHGs, and others. Judges, administrators, NGOs, and even officers of international and corporate agencies have gone through gender sensitisation training.
- Many Womens's Studies institutions are periodically conducting awareness raising and legal literacy camps within and outside the universities and colleges.
- Womens's Studies Centres have recognized the close connections with social work and community action. It leads the students and teachers to getting involved with bringing about social reforms, attitudinal changes, and womens's rights.
- It also is an effective instrument to learn about the prevailing social conditions and status of women, the patriarchal values and practices and the direct and indirect ways of violence, injustice, denial of basic rights and social inequities.
- Helps to build gender perspectives and brings about correct attitudes and education among boys and girls, free of sex bias and patriarchal values.
- Community action also keeps the universities and colleges relevant to society.
- Extension helps in raising resources for the centre for its outreach programmes and field based research and documentation. Women's Studies centres undertake their extension programmes in partnership with government organization, NGOs and other funding agencies.

The source of power that a individual has may be position, status, wealth, personality, or skills. These may be covert or overt, formal or informal.

Institutional

- A job title and a post in the hierarchy to which people defer (Chief Executive, Consultant)
- A position in the hierarchy which controls communication (supervisors, messengers, secretaries)
- A position in the hierarchy which receives much attention (editor, public relations officer)
- A position, including membership on Authorities/Bodies/Committees/Unions.

Financial

- Control budgets.
- Personal wealth.

Social and Political

- Patriarchal social structure.
- Influential family or social connections.
- Social/Political skills to ensure a wide range of friends and contacts.
- An intimidating, bullying manager.
- Exceptional qualities as a career and listener.
- Caste influence.
- Age and seniority.

Intellectual

- Knowledge and expertise which command respect.
- Control of knowledge (librarian, filling clerk, IT expert).

Skill-based

- Outstanding technical skill in an important field, including information technology related skills.
- A technical skill in short supply.

Personal Magnetism

- This is often a combination of social skill, physical attractiveness and charm.

Below are some behaviours observed by the Italian Niccio Machiavelli who studied power building during the 16th Century. Compare them with the behaviours you noted being used by senior women managers. Tick those you observed and add an example where possible. Then note which you might copy.

a. Practices every good manager can copy

- Remaining enthusiastic and positive
- Staying loyal to the hierarchy
- Not avoiding situations
- Setting an example

- ❑ Taking on extra duties e.g. committees/boards
- ❑ Building, maintaining extending networks
- ❑ Noticing and using others abilities
- ❑ Seeking out senior people from whom to learn
- ❑ Seeking expert help when necessary
- ❑ Identifying your resources
- ❑ Helping your manager to succeed
- ❑ Identifying the factors that could help or hinder decision making

b. Sensible self-preservation

- ❑ Understanding your manager's needs and values
- ❑ Recording your achievements and making yourself visible
- ❑ Ensuring your experience and expertise are known
- ❑ Being useful to senior people
- ❑ Making sure your achievements outside work are known
- ❑ Using any information gained advisedly.
- ❑ Knowing the politics of others and protecting yourself.
- ❑ Analyzing personalities of people, their power relationships, and their potential for support or rejection of an idea, proposal, etc.
- ❑ Letting your staff know when you have solved a problem.
- ❑ Being in regular contact with your superiors.
- ❑ Being a good supervisor/mentor/coach to others down the line.
- ❑ Cultivating an attitude of respect for others; even if you have to disagree, disagree agreeably.
- ❑ Gradually extending your power base.
- ❑ Using advocacy and networking for the general good or for mobilizing public opinion.

***c.* Manipulating situations and other people**

- ❑ Using emotional blackmail.
- ❑ Avoiding unpleasant situations.
- ❑ Avoiding impossible problems.
- ❑ Causing lack of trust between colleagues.
- ❑ Identifying manager's weaknesses and exploiting them.
- ❑ Discarding unsuccessful colleagues.
- ❑ Passing on negative information about rivals.
- ❑ Seeking every opportunity to be with seniors and broadcasting the fact to colleagues and others.
- ❑ Projecting yourself excessively and exaggerating what you have achieved.
- ❑ Making no promises.
- ❑ Ignoring unwelcome mail or information.

Note

You may not want to use all of the behaviours listed in the final section but it is imperative to be able to identify them in others. For instance:

- "I'm really glad I was able to sort out that problem. Everyone has been looking at it for a long lime and I suddenly had a flash of inspiration."
- "It's a shame Suma couldn't be at this meeting. She would have so much to contribute but she does seem to be of sick a lot lately."

It is important that you notice and recognize unscrupulous power ploys. Think of any experiences you may have had and consider the effects on others of comments you have heard.

d. Work to use power constructively

One effect of increasing the number of women in higher education senior management is that they may make some changes in the way power is used. Most obviously, women are likely to value good relationship and want more recognition and reward given to those who promote these relationships.

The following diagram shows the kind of shift this could involve.

Concern for Relationships Means Shifting

From	*To*
Power concentrated at the top	Power shared at many levels
Power based on status	Power derived from consensus and sharing
Rigid hierarchy and structures	Working in network and collaborative teams
One-way communication (top-down)	Multiple communication (upwards, downwards and sideways)

Set of Skill Items

1. Communicating
2. Listening
3. Body languages skills
4. Taking an effective part in meetings
5. Chairing meetings
6. Presentation skills
7. Motivating people
8. Involving people at all level
9. Delegating
10. Decision making
11. Problems solving
12. Organizing
13. Project management
14. Leading
15. Planning
16. Political awareness

17. Responsive to others
18. Managing change
19. Negotiating
20. Understanding of finance/budgets
21. Ability to research
22. Appreciating implication
23. Reviewing/evaluating
24. Information analysis
25. Ability to encourage participation from others

Skills Audit Summary Record Sheet

Well Developed	*Adequate*	*Under-Developed*

Points for women at "mixed" meetings

Be aware of how men have viewed women so far in the public domain—that they should not be taken seriously, or there should be a comment about addition of "colour" or "grace" to the meeting and perhaps even a joke or two. This is to be expected until a larger number of women enter decision-making levels. So

- ❑ Be prepared for unpleasant comments that may be jocular, contemptuous or inappropriate. Train yourself not to take them "personally" and be "offensive".
- ❑ Avoid being emotional and feeling threatened. Be as objective and politely firm, not defensive, as possible in your presentation.
- ❑ From the start of the meeting (or even prior to it) you should have established a reputation of being knowledgeable in the subject, so that your views cannot easily be brushed aside.
- ❑ Use appropriate language and timings for your interventions.
- ❑ Be sure you know and use the rules of procedure.
- ❑ Prepare for the meeting—lobby for your cause.

Prof. Irene Moutlana of the Forum for African Women Educationalists South Africa (FAWESA) suggests some survival strategies and power moves to be adopted for **"Surviving the Boardroom Dance".**

Some Survival Strategies

Never underestimate your institution, *viz.* it is a woman's most valuable untapped asset *i.e.* that gut feeling *i.e.* a feeling of knowing. Men on the contrary base their decision of facts. You can use your institution before you even collect all the facts — let your institution speak and track the results.

Choose your battles. Don't try to fight every war-you will waste your energy.

Grasp the intricacies of your institution, so that you will be sensitive to changes when these come along. Know the culture of your institution.

Since as woman, boardroom thinking will expect us to speak from facts, consult, write down your feedback-keep a paper trail.

Don't be over-trusting. This can render you very fragile in a male dominated world. This trust can easily be violated, by not only what men say, but through their body language e.g. collegial inconsistencies or being given half the truth or history or not given all the advice for solving a problem that has a history. The whole truth surfaces at the meeting-rendering you incompetent and very embarrassed.

Adopt Power Moves ***(i.e. give a look of being prepared—have your paperwork in order)***

Seek and secure networks both formally and informally with peers and other outside individuals.

Diversify your networks. Listen attentively—this makes your male counterparts to be eager to share even their needs and shortcomings.

Give positive feedback upfront about projects since men find it difficult to read women (for example they always expect a matter-of-fact approach from women. Spell it out in a professional way).

Open communication-deal honestly. Learn to control your reaction to their criticism-don't take everything too personally.

Stand up to the classical bully without being angry. If you get angry they win.

When they get off the track during meetings, bring them back to the point at hand, (they will not know what hit them and they will not counteract.)

Learn to avoid the over-explanation trap or justifications-this may weaken your position. If you don't over-explain, the other party has less ammunition.

Avoid multiple commitments-it results in you stretching yourself too thinly thus making you ineffective. Give yourself chance to evaluate your other responsibilities. Learn to say "no" without feeling guilty.

Do not speak in tentative terms, (this language of feeling that is so typically us, conflicts with the organizational culture. This impedes professional growth, as one easily comes over as non-competitive, weak, indecisive and vague. (N.B. Language determines the strength of ones opinions, and it also defines your relationship with others in the organization).

Avoid being perpetually apologetic. This translates the submission role into the relationship thus changing the relational dynamic (rather take the responsibility for the delay, this reassures the other solution is underway).

Avoid farming information in the language of feelings (*e.g.,* I feel like — casts the speaker as submissive) since this reveals your vulnerability and fears even if you're not anxious about a decision.

Study the speech patterns of successful people you admire, and adopt language that inspires confidence.

Suggestions for Confidence Building

1. Believe in yourself

2. Adopt a positive approach to help those less fortunate than yourself. The efforts to help others have to be a constant companion to your life long mission.
3. Begin by laying down some standards of conduct and behaviour blow which you will not go, and promise yourself to observe certain values and good practices, so that no one can point a finger at you. Once you build your reputation of being an honest person, responsible to your own conscience things become easier and enjoyable.
4. Learn something all the time and teach it at once to another person.
5. Be physically active in any sports so that you learn how to be sportive and how to win or lose.
6. Empowerment means drawing out of potential and boosting the self-confidence. It also means that adding value to what already women are doing.

Assertive skills include

Giving criticism

When you have to give criticism be specific about the issue or the behaviour you did not like. Treat you oppose or criticise with respect. Discuss the issue or the behaviour but not their personality. They not to put the person down. Allow the other person to suggest ways to solve the problem or give alternative ideas. Listen to their suggestions.

Coping with criticism

Expect some opposition and criticism. Try to separate genuine differences which are well founded and need to be discussed from unfounded criticism. Don't take it personally. See it simply as feedback. We all make some mistakes and get some things wrong. If you accept that you got it wrong, remember you are not wrong but may be you did something wrong, or thought something wrong, or merely differently.

Making a stand

If you are going to make a stand, be firm. Don't excuse yourself too much or explain more than is necessary. We feel stronger in the face of opposition if we receive support from others. If you acknowledge what others do and thank them for their efforts, they are more likely to do the same for you.

Taking on too much

Don't allow others to overburden you or set unrealistic deadlines. If you are overburdened you will make mistakes, not get things done and resent other people criticising you.

If you don't win the argument

If you don't try not to pay attention only to what goes wrong and ignore what goes right. Appreciate your contribution to success and don't punish yourself too much for mistakes or failures. We all have to start somewhere. Recognize that trying is progress, even when you stumble or fail.

Chairing Meetings

The role of a Chairperson of a meeting is to 'conduct' the meeting properly, which means (s)he must allow democratic discussion and expression of different points of view and sum these up appropriately. In order to be able to do this, (s)he must.

(*a*) Be clear about the purpose and objectives of the meeting;

(*b*) Be well acquainted with the rules and procedures to be followed;

(*c*) Be able to express herself/himself clearly in 2 or 3 languages (which would include the local language or Hindi and English);

(*d*) Have a good knowledge or background of the subject discussion;

(*e*) Start punctually and keep to time;

(*f*) Ensure that a good listening is given to the maximum number of members;

(*g*) Be able to guide and steer the discussion and be fair, factual and impartial in making comments, using humour wherever and whenever necessary; and

(*h*) Be able to cut out the substance and sum up the main points for future action.

So,

1. Plan the meeting carefully. Know what your objectives are and try to find out what others think prior to the meeting.
2. Make sure everyone who will attend knows what is being discussed to give them the opportunity to collect all the information they need.
3. Make sure that the agenda is logical and aim to set a time limit for each agenda item.
4. Control the meeting and do not allow individuals to go back over old ground. Once something has been agreed stay with that decision.
5. Summarize all the decisions that have been reached at the end of the meeting and ensure that those responsible for taking action understand what they have to do.
6. Involve everyone and encourage the quieter members to speak.
7. Unite the group and do not take sides.
8. Keep to the facts and keep to the point. Make sure everyone understands.
9. At the end, come to a consensus.

The role of members of a meeting, similarly, must arise from their commitment to 'contribute' to the success of the meeting. In all meetings, there is both 'content' and 'process'. In most meetings there are those who are absolutely 'quiet' and those who keep intervening all the time. The truly 'committed' participants are those who are calm and collected and come prepared after studying the papers. They are clear about the purpose, objectives ad content of the meeting and on when and how to intervene and at what length. They come on time, listen carefully and speak to the point. So, in addition to the preparation of the meeting, they also 'behave' appropriately so that the time utilization of the meeting is maximized.

You will notice that many who do not come 'prepared' for the meeting or for whom the subject is unpleasant sometimes adopt the tactics of disrupting the meeting and use 'obstructionist' methods for obtaining postponement.

Taking Power in Meetings

1. Ensure you talk early in the meeting. The chairperson will then know to look at you during the meeting to see if you wish to say something else.
2. Ask questions to clarify points and try to "piggy back" your comments onto these questions.
3. Try to be give one sentence answers when asked a question. Make sure you speak to the point. Do not ramble on.
4. Offer your opinions and Do Not apologize.

5. Use a firm tone of voice. Do not be tentative.
6. Sit in a visible position.
7. Spread yourself and your papers. Take up space.
8. The most powerful positions to sit are opposite or to the right hand of the Chair/ powerful person. However, sitting opposite may also set up confirmational situation.

You may like to watch where people sit and take note of their behaviour. Watch too for "leakage" - where a person's subconscious actions may not match their outward appearance. Someone may appear very confident yet their constant leg movement, or tapping of pen, they are not as confident as they appear.

Advocacy Skills - for cause or issues concerning the organization

- ❑ Prepare your case well
- ❑ Present your case based on facts
- ❑ Target decision makers
- ❑ Identify decision-making boards
- ❑ Gather support for your proposal
- ❑ Use networks, contacts and friends
- ❑ Build a strong advocacy team with powerful advisers

Negotiation Skills

Negotiations are part of all relationships. Friends, family, spouses, co-workers all negotiate and compromise. Most people believe they need to control the behaviour of others with whom they are negotiating. If however, the behaviour you need to control is your own.

1. Have control over your emotions.
2. Know what your irritants are.
3. Identify the tactics your partner can use to push you and make you lose emotional control. Normally you will find people use one of four ways to push you:
   (*a*) Attacking you or yours.
   (*b*) Intimidating you by insults, harassment, or playing the part of a bully.
   (*c*) Refusal to budge
   (*d*) Deceiving you in some manner so that you give in. Deception can take the form of
   - Lies
   - Manipulation, possibly by acting as if you have reached agreement, then adding on a demand.
4. Recognize the tactic, and avoid reacting by shrugging off this tactic. To do so
   (*a*) Remove yourself, mentally or physically
   (*b*) Regroup. Go over your options and cool down. While you are cooling down, your partner is also cooling down.
   (*c*) Don't become emotional. It doesn't do any good to respond when you are angry or frustrated.
   (*d*) Don't be pressured into making quick decisions. Quick or emotional decisions are often mistakes. Don't decide in haste. Whenever possible take time to sleep on your decisions.

Conflict Management

Conflict and dispute are part of life. There is no society, community, organisation or interpersonal relationship which does not experience conflict at some time or another as part of daily interaction. Conflict arises when people or groups are engaged in competition to meet goals which are perceived to be, or are in fact, incompatible. Conflict can become physically and emotionally damaging or it can lead to growth and productivity for all parties. It all depends on how conflict is managed and resolved.

Conflict is an open disagreement between two people or groups of people who have different goals and values. Conflict involves people's feelings as well as their objectives, and both feelings as well as outcome of the conflict must be resolved, agreement must be found or a compromise worked out.

Resolution of conflict occurs when parties involved understand each other's position accurately. They are willing to discuss it, because they want to resolve the conflict, regardless of their disagreements. Resolution occurs only when the parties try to reach mutually satisfying solutions.

Techniques to Conflict Resolution

When attempting to reach agreement in a conflict situation it may be useful to take note of the five causes of conflict. There are differences based on a clash of:

- interests
- understanding
- values
- style
- opinion.

Writers identify three styles of reaction to conflict. These are:

- aggressive ('fight it').
- assertive ('negotiate it').
- passive ('duck it').

Five skills for negotiating conflict can also be identified. These are :

- spot/define it.
- understand it.
- look for 'win-win' (where all parties feel that they have gained something).
- act at the right time.
- check out the results.

Team Building

- Requires good people skills/interpersonal skills
- Need to create team spirit
- Need to be sensitive to needs of the group
- Admit abilities of others
- Develop mutual trust
- Be clear about goals and targets
- Define roles of team members

Communication Skills

Verbal

Ability

Expectations

Communication Strategy

Communication Styles

Interpretation

Non-Verbal

Listening Skills

Body Language

Eye Contact

Gestures

Facial Expressions

Postures

Confidence Building

Confidence Obstructing

Low paid jobs

Career breaks

Multiple roles

Stereotyping

Male oriented decision making

Old boys' club

Confidence Building

Acquire knowledge and skills

Define your objectives

Stick to your values

Build on your achievements and assets

Be outspoken

Copy good role models

Manage new technology

Follow a mentor

Learn from mistakes

Accept challenges

Modes of Behaviour

Non-assertive Behaviour

(denying one's own rights)

Directly Aggressive Behaviour

(denying openly the rights of others)

Indirect Aggressive Behaviour

(denying covertly the rights of others)

Assertive Behaviour

(acknowledging one's own rights and the rights of others).

UNDERSTAND POWER WITHIN AN ORGANIZATION

1. Nature and Source of Power
2. Relational Power < used / misused
3. Power has a price
4. Build legitimate source of power-determine own power base-develop strategies for extending Power.

Concept of Power : Exercise of influence, Authority control, men and women react to power differently.

Men : Concerned with power wental in them

Women : Feel discomfort in handling Power. Uncomfortable in recognising that they have power.

Nature – Different Kinds of Power

(*a*) Power over

(*b*) Power to

(*c*) Power with

(*d*) Power within

Relational Power

- ❑ Power exercised by everyone over other people in some way.
- ❑ Organizations find it difficult to come to terms with relational power.
- ❑ Good relationships smooth the way is an organization.
- ❑ Relationships are not easily predictable or controlled.
- ❑ They may work against the formal structure and challenge others.
- ❑ Very often skills used in relational power are used to build personal power.

Power over : Relationship of domination/subordination

Power to : Decision making, solve problems, creative

Power with : People with common purpose, collective goal

Power within : Self confidence, self awareness, assertiveness

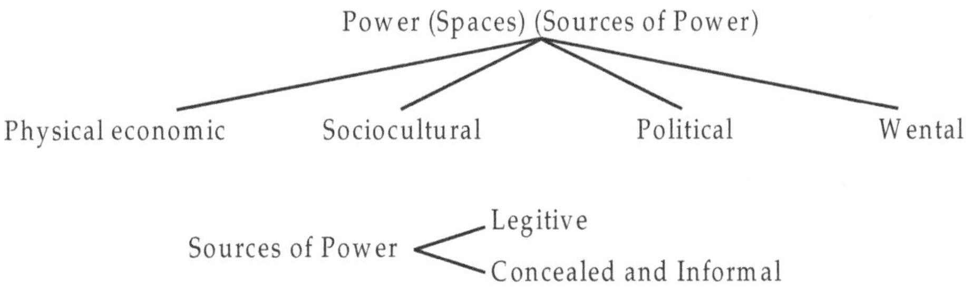

1. Institutional
2. Financial
3. Social and Political
4. Intellectual
5. Skill based
6. Personal Magnetism

Problems

- Women trained to exercise relational power to greater extent.
- Personal power is used to advance one self and to limit power of others–does not fit in women's role of carying, nurturing.
- Women prefer to encourage group.
- Collaboration and recognise team effort.
- Men says 'I' where women says 'We'.
- Under value their achievements. Not visible.

So,

- Has to build her own personal power base–based on her own ideas about herself and her own standards (inspite of all these problems)
- Some women play the man's game–push themselves whilst limiting opportunities for others. Such women are resented.
- Do not under value.

Women and Governance

Below are some behaviours observed by the Italian Niccio Machiavelli who studied power building during the 16th century. Compare them with the behaviours you noted being used by senior women managers. Tick those you observed and add an example where possible. Then note which you might copy.

(*a*) Practices every good manager can copy

- Remaining enthusiastic and positive.
- Staying loyal to the hierarchy
- Not avoiding situations
- Setting an example
- Taking an extra duties *e.g.,* committees/boards

- Building, maintaining extending networks
- Noticing and using others abilities
- Seeking out senior people from who to learn
- Seeking expert help when necessary
- Identifying your resources
- Helping your manager to succeed
- Identifying the factors that could help or hinder decision making

(*b*) Sensible self-preservation

- Understanding your manager's needs and values
- Recording your achievements and making yourself visible
- Ensuring your experience and expertise are known
- Being useful to senior people
- Making sure your achievements outside work are known
- Using any information gained advisely
- Knowing the politics of other and protecting yourself
- Analyzing personalities of people, their power relationships, and their potential for support or rejection of an idea, proposal, etc.
- Letting your staff know when you have solved a problem
- Being in regular contact with your superiors
- Being a good supervisor/mentor/coach to others down the line
- Cultivating an attitude of respect for others; even if you have to disagree, disagree agreeably
- Gradually extending your power base
- Using advocacy and networking for the general good or for mobilizing public opinion.

(*c*) Manipulating situations and other people

- Avoiding unpleasant situations
- Avoiding impossible problems
- Causing lack of trust between colleagues
- Passing on negative information about rivals
- Seeking every opportunity to be with seniors and broadcasting the fact to colleagues and others
- Projecting yourself excessively and exaggerating what you have achieved.
- Ignoring unwelcome mail or information
- Legitimate use or misuse of relational power
- If reinforces the stereotype of women and encourages men to discount women's potential.
- It accepts a situation in which women do not progress on merit.
- It is highly divisive.
- It is high risk.

Planning Your Own Route

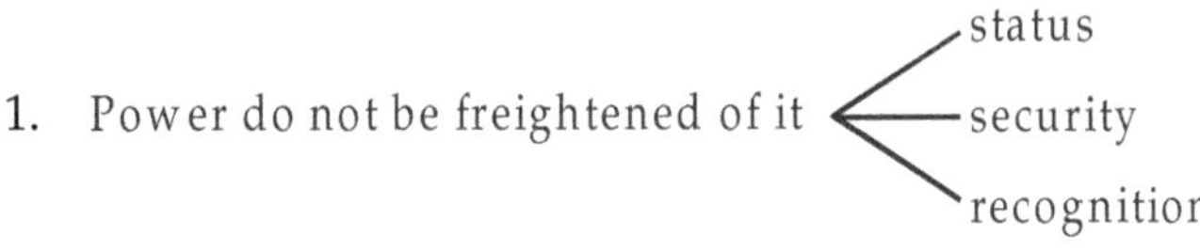

2. Select your own power bases — own standards, values
3. Determine their management styles.
4. Plan to extent your power.
5. Power building
   (*a*) Practices-every good managers can copy
   (*b*) Sensible self-preservation
   (*c*) Manipulating situations and other people
   (*d*) Work to use power constructively

Make some changes in the way power is used. Hence shifting, one has to study carefully the following columns :

Power concentrated at the top	Power shared at many levels
Power based on status	Power derived from consensus and sharing
Rigid hierarchy and structures	Working in networks and collaborative teams
One way communication (top down)	Multiple communicative (upwards, downwards and sideways)

These all are based on UGC (New Delhi) and capacity building programme.

EQUATE : Effective Quality Upgrading Assistance for Technical Education (Noida) in which author was trained.

ROLE OF WOMEN

Most of women are not involve in decision making directly or indirectly. They remain passive.

Decision Making

"Decision making is a conscious and human process, involving both individual and social phenomena based upon factual value premises, which concludes with a choice of one behaviour activity from among one or more alternatives with the intension of moving some desired state of affairs". Shull *et. al.*

Decision making is an act of projecting one's own mind upon an opinion or course of action

- Following 3 aspects of Human Behaviour are involved :
  (*i*) Cognition-mind associated with knowledge
  (*ii*) Cognition-mind associated with desire, willingness
  (*iii*) Affection–mind associated with emotion, feeling, food, temperament.

Steps in decision making process

(*a*) Specific Objectives

(*b*) Problem Identification

(*i*) Diagnosis

_ what is?

What is aught to be?

(c) Search for alternatives

- experience
- creative technique
- past decisions

(*d*) Evaluation of alternatives

- Profit

- Loss

- Meet objectives

(*e*) Choice of alternative

(*f*) Action

- communication to workers
- feedback

(*g*) Results

- according to objectives
- follow-up

Individual Development Process

"People cannot be developed; they can only develop themselves. For while it is possible for an outsider to build a man's home, an outsider cannot give the man pride and self-confidence in himself as a human being. Those things a man has to create in himself by his own actions. He develops himself by what he does; he develops himself by making his own decisions, by increasing his understanding of what he is doing, and why; by increasing his own knowledge and ability, and by his own full participation-as an equal in the life of the community he lives in".

Jullus Nyerere

Build a tower profitably using logo places

The expenditure for 100 places is Rs. 10,000. The Income is proportionate to the height which is Rs. 20,000 for one metre.

If the tower is completed under 1½ minutes there is a bonus given at the rate of Rs. 5,000 per quarter minute. On the other hand a penalty is levied at the rate of Rs. 10,000 per quarter minute if the time exceeds 2 minutes. Fractions of quarter minutes will not be counted either for bonus or penalty.

The tower should stand erect without support or any adhesive for long enough to measure its height.

The minimum height of the town to submit quotation is 85 cm.

The quotation may be given to Ms 'A' on use following :

Height__________ Expenditure ____________

Time___________ Income ____________

Material________ Bonus/penalty

Profit

HISTORICAL CONTEXT OF WOMEN'S MOVEMENT

- First World Conference on Women in Mexico (1975).
- Nairobi (1985)
- Beijing (1995)
- Follow up on Beijing. The Beijing declaration adopted at the UN Fourth World Conference on Women states

"Women empowerment and their full participation on the basis of equality in all spheres of society including participation in the decision making process and access to power are fundamental to the achievement of equality, development and peace".

STATUS OF WOMEN IN INDIA AFTER INDEPENDENCE

India has left behind nearly five and a half decades since independence and the country is soon to become one of the largest populations of the World.

- Out of the billion humans, nearly 50 per cent comprises of women and they are way below their male counterparts in all parameters of development.
- India occupies 115 the rank on Human Development Index (HDI) having moved up 13 notches since 2000 position of 128. (12th UNDP Report 2000).

SHIFT OF FOCUS

- The issue of empowerment of women and engendering mentoring the development objective moved from global paradigm shift from welfare to development to empowerment.
- Growth oriented to a human development approach.

SOME MEASURES DEFINED

- HDI (Human Development Index): is a composite measure of human well being based on literacy levels, life expectancy, combined primary, secondary and gross tertiary enrolments, and the Gross Domestic Product.
- GDI (Gender Development Index): is a measure of gender well being based on literacy levels, life expectancy, combined primary, secondary and gross tertiary enrolments, and the Gross Domestic Product.
- GEM (Gender Empowerment Measure): measures the access of women to decision-making, power structures and incomes.
- India is one of the 21 countries of the world with fewer women than 95 per 100 men.
- At the 2001 census, India had 498 million females against 531 million males, *i.e.*, 35 million fewer females. The declining sex ratio is more alarming for children below six years.
- It is observed that at 2001 census only 65.4% Indians aged 7+ were found literate, 75.9% males and 54.2% females with a gap of 21.7% percentage between the genders.
- Among 296 million illiterates of the country, 186 million (64%) are female.

MILESTONE AND ENCOURAGING FACTS ON GENDER

- Indian constitution ensures gender equality and same rights to women also.
- The country is one of the oldest civilization and history dating back to 5000 years or so.
- Various groups of women in the country and multifaceted development in global arena, the Indian women have made a mark and have been playing an important role in maintaining its cultural growths and heritage with development in all areas.
- It is encouraging to observe that female literacy and education progressed faster than their male counterparts during the last decade.
- India has the second largest educational system in the world, next to China with more than a million formal and non formal educational institutions that have our enrolment of 191.63 million out of million out of whom 80.54 million (42%) are females. In institutions of higher learning 40 per cent of the 7.73 million students enrolled were females (MHRD 2000-2001)
- India is categories as medium human development nation and its HDI has improved from 0.302 in 1981 to 0.472 in 2001.

UGC's, ICAR, GOI VIEW

- UGC had facilitated women studies centers in different states of India with a view that they would be contributing to: UGC advocates for women empowerment.
- Incorporate women studies in various courses in teaching.
- Promote research to certain fields in the area concern.
- Create develop and evaluate projects incorporating women studies.
- Generate resource and documentation material.
- Active counseling in women as well socially/politically relevant issues.
- Networking and multidisciplinary collaborating activities.
- Supplement into the development plans of the state/central government, etc.
- The women be given due place in independent India keeping to changing scenario of globalization, privatization and liberalization.
- Serving and addressing to the 9th & 10th plan concern for empowering women under Women Component Plan (WCP) and have 30% fund benefits flow to women.

ADDRESSING 10th PLAN OF NATIONAL POLICY FOR EMPOWERMENT OF WOMEN (2001) INTO ACTION THROUGH :

- Enabling women to realize their full potential through creating positive environment with economic and social policies.
- Help them enjoy human rights of fundamental freedom in all spheres political, economic, social, cultural and civil.
- Equal access to participation and decision making in social, political and economic life of the nation.
- Equal access to health, quality education, career, employment, remuneration, social security, etc.
- Elimination of all forms of discrimination against women by strengthening legal system.
- Active and equal participation and involvement of both men and women for changing community practices and societal attitudes.

- Main streaming a gender perspective into the development process.
- Building and strengthening partnership with civil society particularly women's organization, corporate and private sector agencies.

Women studies perspective includes above.

MAJOR PERSPECTIVE BEHIND WOMEN STUDIES

- The women be identified as equal partners in the country's development.
- Engendering all areas of development-economics, social, political and cultural.
- Reminding the higher status enjoyed by Indian women traditionally through women studies.
- Putting up systematically issues relating to women's rights, laws and policies through women studies.
- Planning and implementing various approaches of capacity building among women.
- Sensitizing and revitalizing gender perspective through women studies.
- Ensuring gender equity through multi-dimensional activity.
- Ensuring training in the area of legal literacy among women.
- Through studies and intervention creating cent percent awareness among women relating to child and women care programmes and issues.
- Women studies centers would serve as a catalysts for promoting and strengthening teaching, research curriculum in specific areas such as transfers of technologies.

A WOMEN'S STUDIES CENTER CAN FULFILL THE FOLLOWING OBJECTIVES

- Serving as a catalysts for promoting and strengthening women's studies through teaching, research, curriculum, field and extension work training and continuing education through location/region specific perspectives as well as mcro-level perspective. (Micro & Macro level)
- Working for gender equity through its multi pronged activities in rural areas.
- Achieving economic & self reliance among college girls as well as rural girls and women.
- Achieving target transfer of women friendly technologies for drudgery reduction.
- Targeting girl education issues & cent percent literacy among girls.
- Focusing on population and family life education field work curriculum through designing and carrying out short courses & Diploma in concerned subjects.
- Achieving the goal of legal literacy among women through well designed training package for women members of Gram Panchayats.
- Targeting the issues relating to women's rights, laws and social exploitation through awareness generation, dissemination of information & documentation.
- Achieving cent percent awareness among women relating to child & women care issues and programmes.
- Implementing well designed teaching courses on women issues leading to diploma, degree and post graduate degrees.
- Planning and implementing intervention research on socially relevant areas like empowerment of Self Help Groups, women health, female feticide etc.
- Serving as a consulting center to the teaching and research scholars engaged in women studies at grass root, block, district and state level and national level.

- Working women generally concentrate on social work, education, home economics, nursing, library science, and computer.
- Physician and lawyer they enter the profession with only degree but women enter the college teaching before completion of doctorate but after degree.

SLOW ADVANCEMENT OF WOMEN

- Young men attain the level of equality with other senior colleagues in higher position with relative ease than young woman.
- Academic position of power and prestige have been defined with the men. Major administrative works requires collateral duties. In India these duties are carried out by the incumbent's accommodating wife.
- In academics women have to prepare new course, teaching preparation. writing books but these are hindered by bearing and rearing children.
- Thirty per cent of women drop the post graduate program either to get marry or likely to be married.
- When a young woman enters in to the teaching position in a college or university she faces the classical obstacle which limits her to fully participate in the Indian academic life.
- In the higher education, women are found to be good teacher, but they are not able to pursue the research, which is indispensable to the senior rank.
- Many women not able to publish the paper due to social and psychological pressure.
- Married women doctorates cannot be expected to undertake highly demanding professional position.
- Women with doctorate do not want to get married. But in India the percentage of married women doctorates has been increasing in last 20 years. But the percentage of unmarried is still very high.
- Woman worries not only about her success, but also her husband's attitude on her academic achievement.
- Most women reluctant to achieve higher position than her husband, particularly when both are in same profession.
- An educated woman tends to have fewer children than other women. Some times they do not want the child and their percentage is high.
- A married women professor has less free time i.e. 2.8 hours/day comparing to the working father i.e. 4.1 hours/day.

FACTOR INHIBITING A TEACHING WOMEN'S PROGRESS GETTING INTO ADMINISTRATION

- A woman feels isolation in all male group of administration.
- Very less financial incentive.
- Doctorate women are keener to pursue scholar life.
- Many working women those with pursue life.
- Many working women those with family do not like administrative work as it is less flexible and time consuming.
- Demand for women professor are high in case of class room teaching, lecture preparation and for research work.

GOVERNMENT EFFORTS

Five Year Plan	Year	Emphasizing on
1st	1951-56	Welfare activities. Setting of CSWB in 1953.
2nd to 5th	1956-79	Welfare activities with emphasis on education and improvement of maternal and child health.
6th	1980-85	Shift from welfare to development of women. Emphasis on health, education and employment.
7th	1985-90	Beneficiaries oriented scheme and providing vocational training.
8th	1992-97	Participation of women in development process.
9th	1997-02	Empowerment of women and freedom to exercise their rights. 30% fund diverted to Women Component Plan.
10th	2002-07	Equal access for participation, Strengthening legal system, Eliminating discrimination, Changes social attitude by active involvement.

WOMEN LEADERSHIP

In Media

In 19th and 20th century, journalism was a male dominate profession. But few women shown their leadership. Ex-Aruna Asaf Ali editor of 'Patriot', Kamaladevi Chattopadhaya.

In Radio and T.V.

Nalini Singh, Barkha Dutt, Anjali Sarangi, Salma Sultana, Pratima Puri.

Advertisement and Public Relation

Mrs. Rajani Panikar

Film

Ashwariya Rai, Shabana Azmi.

Entrepreneur

Ritu Beri, Shahnaz Hussain.

Sports

P.T. Usha

Space

Late. Kalpna Chawla

TIPS FOR WOMEN TO RAISE TO A LEADERSHIP POSITION

- Have a vision and stick to it. Evaluate it periodically. Get a mentor to help you achieve the vision.
- Be proactive and have a thick skin. Confident yourself, you have the capability to make change.
- Recognize that women want to be leaders, not administrator. Administrator is a passive word.

Chapter 2

A State Study

Introduction

This chapter is based on the particular context of Tripura, a State of the Union of India. It is pertinent therefore to introduce the State first, before proceeding with discussion on the topic.

The state Tripura is having a College of Fisheries at Lambuchera where teachers and students are working hard to improve the status of fisheries and its production.

Here an example of Tripura has been taken into consideration. This work is not done, it is by author himself taken from references.

Geographic

1. India's North-East accounts for 7.7 per cent of the country's land and is homeland for more than 4 per cent of the country's total population. Of the total area, 70 per cent is hilly and about 88 per cent people live in villages.
2. Tripura is in the extreme north-east of the country. Bangladesh surrounds it on three sides. Backwardness of the State arises primarily due to its geographical isolation. The total area of the State is 10,491.61 km^2.
3. Alongside its geographical limitations, another important feature of Tripura is its demographic pattern. According to the census figures of 1991, Tripura has a population of 2,757,000 of which female population is 1,339,000. Women constitute roughly 48 per cent of the State's population. A large of the population is of indigenous people and those belonging to the weaker sections of the society. It is interesting to note that the ratio of females per 1,000 males is 945.
4. The State receives about, 2,100 mm rainfall per year on average. The State does not have perennial rivers, and all rivers and streams are rainfed and drain into Bangladesh. Cultivable land is scarce (only 26 per cent of total surface area), and people depend on sectors other than the primary sector for survival.

Land

1. High rainfall and good soil offer considerable scope for land-based economic activities. Productivity of rice in Tripura (1,813 kg/ha) is comparable with the all India figure of 1,879 kg/ha and is far above the north-eastern average of 1,396kg/ha. In respect of the total yield of food grains, Tripura's productivity of 1,746 kg/ha is higher than the all India average of 1,487 kg/ha. However, the extent of the irrigated area in Tripura amounting to 19.6 per cent of the net sown area, is rather insignificant.
2. A significant area in the State is under fruits and plantation crops. The major crops grown in plantations are tea, cashew, orange and pineapple. Jackfruit, banana, lemon,

coconut and arecanut are largely grown on the homestead. Fruits grow very well in Tripura and the quality of Tripura's jackfruit, orange and pineapple is widely recognized.

3. Tripura grows large quantities of vegetables with potato as the major field crop. The yield of potato in Tripura is 6795 kg/ha which is the highest in the north-eastern region. It is higher than the all India average of 5,242 kg/ha and more than double the average for the north-eastern region 7,139 kg/ha.

Economic

1. Per capita State Domestic Product (SDP) of Tripura was estimated at Rs. 8,669 for 1998-99 at current prices, against the all India average of Rs. 14,682 in 1998-99. Per capita income by quick estimates at prices current in 1993-93 was Rs. 6,251.
2. The economy of the State is primarily agrarian with agriculture contributing 42 per cent of SDP and providing 64 per cent of employment. Ninety per cent of the farmers are small and classified as marginal. The net sown area constitutes about 25 per cent of the total area. Only 19.6 per cent of the area has access to assured irrigation. Due to high population density, the average size of operational holding is only 0.97 ha, which is much smaller than the all India figure of 1.68 ha. The secondary sector has its limitations in providing large-scale employment, and the present wages in the sector are low and unlikely to increase significantly.
3. Major field crop is rice with more than 50 per cent of the gross cropped area. But the area is decreasing. Simultaneously, area under other field crops is insignificant. The area under pulses and oilseeds is however increasing.
4. The State has no significant Central Government of private sector investment. The State is a net importer of capital through Central transfers such as grants and loans, but in unable to retain the transferred capital because of reverse transfer including payment for food and other essential imports, and because of the low credit-deposit (CD) ratio which is only 37:10. The present low CD ratio is a very serious constraint to the increase of labour productivity in the State.

Infrastructural

1. Basic physical infrastructure, such as a dependable transport system, power, etc. is a prerequisite for economic development. The present state of infrastructure, both in terms of quantity and quality continues to be abysmally poor as compared to the national level and even in comparison to other north-eastern states.
2. Tripura does not have facilities for inland water transport. Transport by road is therefore the only dependable means of transport. Extremely hostile, difficult hilly terrain and high rainfall makes construction and maintenance of roads and other transport network difficult and costly.
3. Per capita consumption of electricity in Tripura was 100 KWH during 1998-99, which was below the national average of 338 KWH. During the period from 1970-71 to 1998-99 per capita consumption of electricity has increased.
4. The overall backwardness of the State is evident from the composite infrastructural index evolved by the Centre Monitoring Indian Economy (CMIE), based on the availability of power, irrigation, road, railways, post-office services, education, health and banking. Tripura is the second most backward State in the entire country.
5. The State has only a token presence of the railways of about 45 km connecting to Assam. By road the State is connected with the rest of the country by National Highway (NH44) which passes through Assam and Meghalaya. This road is the State's lifeline.

6. The State has been continuously beleaguered by socio-economic problems since independence. Tripura was converted into an isolated territory with practically all normal communications with the rest of the affected by the partition of India. A circuitous and landslide-prone rail or road route through some of the distantly located north-eastern States is the only transit now available, besides the hardly affordable air link.

Opportunities

1. On account of various geographical, social and historical reasons, Tripura has remained economically backward. As indicated above, the limited availability of infrastructure has made the process of economic development difficult.
2. The State has, however strengths which can be exploited for ensuring sustained economic development. The north-eastern region is close to the South-East Asian region, which has been the fastest economic growth in the last decade. Development of infrastructure and the creation of conditions conducive for economic development however are essential for tapping this potential and bringing about economic development.

Some of the promising sectors of Tripura's economy are as follows:

1. *Human Resources*

Human resources are the promising resource of the State. The overall literacy rate is 60.44 per cent with male rate at 70.58 per cent and female 49.65 per cent. There are higher than the national average. Literary among the tribals is lower at 40.3 per cent (male, 52.88 per cent and female, 27.34 per cent) but also higher than corresponding national average.

The State has well-developed institutions of local self-government. These are integrated into all developmental activities and ensure the willing participation of the people in different programs. The Panchayati Raj, Nagar Panchayat and the Autonomous District Council are major strengths of the State.

2. *Natural Rubber and Tea*

After Kerala, Tripura is the second largest producer of natural rubber. Rubber produced is of superior quality. At present, about 23,000 ha is under natural rubber plantation and annual production is more than 7,000. In view of a good demand for and high price of natural rubber, this sector holds considerable promise for development in the State. Tripura, a traditional tea-growing area, also has tea gardens covering an area of 6,430 ha with a production of more than 5,500 mt per annum. Tea produced in the state has good blending qualities.

3. *Natural Gas*

As per the estimate; out of a prognosticated reserve of 400 billion m^3, approximately 16 billion m^3 natural gas is recoverable. This gas is available in non-associate form, with about 94 per cent methane. The availability of gas provides scope for setting up units for producing power, chemicals and fertilizer, which will lead to rapid economic development of the state.

Existing Food Security Mechanisms

In the background of the economic status of Tripura, the concept of community food security assumes greater importance. The physical environment of a hilly area has a generally beneficial effect on human health. However, a number of special features characterize this health scene. It is generally acknowledged that nutritional needs are higher, but little scientific work has been done to determine the precise levels: as a result, the importance of this factor is missed in such important fields as the Public Distribution System (PDC), determination of poverty levels, nutritional programs for women and children, wage fixation, etc. It is high time now to hope that some scientific work is done on these basic matters.

Because of very limited purchasing power, a significant portion of the population especially in the rural areas cannot take full advantage of the assured food supply made through the universal PDS in the country. The government has a major role in the distribution of essential commodities like rice, wheat edible oils and kerosene through the PDS at subsidized prices, but if it decides on a lesser role in *production it* will have to gear itself up to more of a role in *distribution,* to meet the needs of the poor. In more than 1,500 backward blocks such as in drought-prone areas, desert and tribal and hill areas, the PDS was revamped to include additional items such as soap, pulses and iodised salt. Here the price of food grains is than in the normal PDS, but a stronger, more comprehensive and more efficient PDS needs to be put in place. But is there adequate effort put in to design and put in place such a system? This is an issue to ponder.

The PDS operates in Tripura, where there is a quite large network of PDS with 1,427 fair price shops (rural, 1,230 and urban, 197) which cater to 683,187 ration card holders. Of these, 632,769 card holders are in rural areas while the remaining 50,418 card holders reside in urban areas. Through the network of fair price shops, the following quantities of food grains and essential commodities are made available every month to the people:

- 13,740 mt of rice;
- 1,280 mt of wheat;
- 2,404 mt of sugar;
- 2,250 mt of iodized salt; and
- 3,509 kl of kerosene oil.

Because of poverty, a large section of the population can hardly take advantage of the benefits of existing PDS. Poverty is multidimensional and multi-sectoral in nature, having originated from a diverse range of conditions. The poor exist in both urban and rural areas of the country, though about 80 per cent of them are found in rural areas, but they are not homogeneous. The rural poor consist of economically and socially heterogeneous groups. Rural poverty is in fact, directly related to land ownership and control over land. The poor have very little command over or access to resources, assets, income or credit. They depend heavily on their labour for survival, as there is little else they can derive income from. They have only their labour power to sell, but earn uncertain or low wages. There is also the gender dimension to poverty, since poor women have to shoulder the double burden of being disadvantaged by being female and doubly disadvantaged being poor.

The urban poor consist largely of overflow of the rural poor who migrate to towns, and are engaged mainly in casual and uncertain occupations such as street vending, construction, etc. There is little more than token space in urban areas for the pushed out from villages by poverty and social degradation. Most of them live in unhygienic conditions in unending insecurity, never sure when they will be displaced by local authorities or other powerful groups.

Poor people are typically unorganized, hard to reach, inarticulate, often invisible. In particular, women and children, and even their residential locations in rural areas are on the periphery. They have very little access to education and are largely illiterate. In fact there is a close interrelationship between poverty and illiteracy; the number of poor and the number of illiterates are not very different. Self-reliance and creative self-engagement are not able to emerge.

The problem of poverty is aggravated by social deprivation and discrimination. The close intertwining of social oppression and economic exploitation can be seen in a major cement of the poor. Since a majority of the poor live in rural areas, the rural setting has a predominant impact on the poverty situation. The issue of land is of fundamental importance to the rural poor. The deep yearning for equality and land among the rural poor is not something new. They reflect the attitudes and values, which have grown in them in the course of their interaction with the material reality, which has deprived them of both equality and land. To them, ownership of land denotes one's social status and equality means equality in the ownership of land.

Poverty is closely related to employment or unemployment. The marginalized groups of poor in rural areas consist of landless labourers located precariously on the brink of subsistence, depending on uncertain employment and wages. The number of agricultural labourers is estimated at 110 million, about 75 per cent of the total rural labour force. They suffer from socio-economic deprivation and form the core of the labouring classes. It is also worth nothing that the number of agricultural labourers is increasing in the last few decades at a rate faster than the population growth rate. For the large mass of agricultural and other labourers who have only their labour power to sell, wages and wage system have enormous relevance.

While land is perhaps the most obvious asset to redistribute, it has to be borne in mind that the economic opportunities in India depend on a much wider range of endowments. The distribution of employment, environmental resources educational facilities, and affordable credit arrangements are examples of other influential factors. According to Dr. Amartya Sen, Harvard economist the opportunities for redistribution of these diverse endowments have to be considered along with the scope for land distribution, if the conditions of the poorer sections of the people are to significantly improve.

As Amartya Sen has pointed out, only enormous expansion of food production can make it possible to guarantee adequate food for all. For the more fortunate part of humanity, health problems connected with food consumption stem from having too much. While one part of humanity desperately searches for food to eat, another part counts calories and looks for ways of slimming! Persistence of hunger is related to extreme inequality within the society.

In simple terms, food security has two aspects: availability of food and accessibility to food stocks. A household can feel food secure if:

- food which is culturally acceptable with appropriate nutritional value is available in the system;
- households have the capacity to buy; and
- there is freedom to choose from the available food stocks, and at the same time there is no institutional barrier to access the available food.

People can feel secure within the household when all members especially the women and children have equitable access.

The critical role played by a well-organized PDS for essential food grains in a densely populated low income economy insulated the poor from the rise in food grain prices. This fact should be appreciated by policy-makers, and the State should make an explicit guarantee of the right of the poor to food security. Food security is an important component not only of survival, but also of basic dignity and well being of the poor. The running of the PDS is entirely in the hands of the administration. An efficient and responsive administrative machinery can ensure that the benefits reach the poor.

New considerations will include how to exclude the *"non-poor"* from the PDS, and how to reduce the subsidy burden in making food grains available. The question of "targeting" the truly deserving sections poses major challenge to administration.

In overcoming this challenge, women can play a very significant role. Women have always been associated with the utilization of rice, wheat, sugar, kerosene, edible oils, etc. which are needed in cooking. They are acutely aware of problems such as lack of cash to buy their weekly supplies.

Women could be utilized by the administration to form vigilance committees at local to ensure proper functioning of the PDS. Their representation at all level of policy implementation in the distribution of essential commodities would promote scope for informed and empathetic decision-making it easier and more honest in its dealings at the retailing level.

There is criticism that the reforms have caused double-digit inflation and increased poverty and unemployment particularly in rural areas. A solution to it is sought in the curtailment in government expenditure. Cuts in expenditure sometimes result in the axe falling invariably on "*unproductive*" social welfare sectors.

One of the fears expressed by many people is that subsidies may have to be reduced all round, and that the poor and the poorest among the poor including women, would be affected. Some selectivity needs to be exercised after looking at looking at short- and long-term implications and gauging the effect on different categories of people. Controlling inflation depends on containing the fiscal deficit, recovery of industrial production and better management of the supply of food grains and essential commodities. The delicate exercise of curtailing fiscal deficit in a democratic set-up will have major implications for public administration and the role of government.

Women have always had the habit of saving for a rainy day. They experience a daily dilemma of consuming versus saving. More than men, women are expected to keep the home fires burning. Women's perceptions arising from their experience may enable new approaches to curing fiscal without compromising on food security aspects, and this is well worth a trial.

Government Interventions

In a given scenario, a major intervention by the government is needed to strengthen existing food security mechanisms. Relatively more important and immediate steps may be:

- formulation of a comprehensive poverty policy;
- focused attention on strengthening the PDS to transform it in to a vibrant and efficient food assistance program;
- education the needy about their entitlement;
- recognizing the importance of a strong safety net that can provide families in need with the support to survive;
- encourage individuals, mainly women, to contribute in endeavors that will give them self-sufficiency in the long run;
- creating strategic alliances between public and private sectors for increased funding in food security-related activities:
- forging partnerships between voluntary organizations now engaged in diverse activities like anti-hunger campaigners, environmentalists, community development activists, literacy workers, etc.; and
- launching initiatives to reduce hunger, improve food availability, nutrition and preservation.

Food is a basic human right and hence the concept of food security has a global importance, and being global, has assumed international importance. The United Nations Food and Agriculture Organization (FAO) took note of this in the World Food Conference convened in 1974. The conference stressed the need for devising ways and means for assuring food security to the hungry millions of the world and endorsed the assurance of world food security as the common responsibility of the entire international community. The conference also gave the call that no child, women and man should go to bed hungry and no human being's physical and mental capabilities should stunted by malnutrition.

During the new millennium the problem of food security has assumed great significance as a result of the alarming increase in population. Increases in food production are not keeping pace with the population growth, and related ecological problems, gender inequality and other issues will be discussed in detail in the succeeding paragraphs.

Food security is a concept variously defined by the World Bank as "access by all peoples at all times to enough food for an active healthy life", and by the FAO as "the basic right of all people to an adequate diet and need for concerted action among all countries to achieve this goal in a sustainable manner". FAO adds that "food security exists when all people at all times have physical and economic access to sufficient, safe and nutritious food to meet their dietary needs and food preference for an active and healthy life?

Former head of the International Rice Research Institute Dr. M.S. Swaminathan redefines food security as "livelihood security". He has further categorized food security at various levels of human organization namely national level, regional level, household level and individual level. He states that public action for ending endemic hunger should keep in view the following:

(*a*) Food security should be considered at the level of individual, rather than of the household, since women and girl children tend to suffer more from poverty induced under nutrition.

(*b*) Non-food factors like income, environmental hygiene, primary health care, and literacy are equally important.

(*c*) Poverty is the primary cause of under and malnutrition.

The above conditions prevail in all developing countries including India. Based on these consideration therefore he adds that "sustainable food security involves strengthening the livelihood security of all members within a household by ensuring both, physical and economic access to balanced diet, including the needed micronutrients, safe drinking water, environmental sanitation, basic health care and primary education".

Issues Affecting Food Security

The core of the topic under discussion, namely the current issues affecting food security in India can be summarized as follows:

Heavy Population Pressure

India stands second in the world with its population exceeding one billion. The break-up of the 1,002,142,000 population as on 1 July 2000 is as follows:

Area 3,288,000 km^2 (1,269 miles^2)

Urban 286,201,000 or 28.56 per cent

Rural 715,941,000 or 71.44 per cent

The female population in India is 960 per, 1,000 males, but in rural areas it is 968 and in urban areas it is 936 per 1,000 males (Center for Monitoring Indian Economy, *Basic Statistics: India,* 1993). The heavy population pressure is mainly due to poor utilization of contraceptives, early marriages and lack of awareness of population issues. The figures given below reflect the problem:

(*a*) 51.8 per cent of women do not use any contraceptive methods.

(*b*) 50 per cent girls are married below the age of 18 (27.9 per cent in urban area and 28.6 per cent in rural area).

(*c*) The fast rates of growth of population necessitates a higher rate of economic growth including food production to maintain standard of living.

The above tables show that rural poverty is higher than urban poverty because larger number mean higher food insecurity in rural areas. The percentage of population. Below the Poverty Line (BPL) in rural areas has come down over the years due to the impact of the poverty alleviation programs, but the absolute number has considerably increased due to the population explosion. In rural areas BPL people are mainly landless labourers and small and marginal farmers whose production is very low. Urban figures are mainly a spillover from rural poverty and urbanization. One result of being unorganized for both the rural and urban BPL is that people earn low wages, women being among the worst suffers. In the global context the percentage of poor women among the total of all poor is about 70 per cent, which indicates gender discrimination.

It may be concluded that India's food insecurity is mainly due to mass poverty. It may be added that even though in India adequate food is available, due to poor purchasing power the same cannot be purchased by the masses.

Lack of Education and Awareness

India has suffered from mass illiteracy for decades. Gandhiji stated that "Mass illiteracy is India's sin and shame and must be liquidated". He believed that the salvation of India depended is on the awakening of women.

Literacy is one of the most important indicators of the social, economic and political development of the society. Illiteracy leads to overpopulation, unemployment, poverty, gender inequality and low productivity. It also to food insecurity. The literacy percentage among females in India is 39 per cent. In the age group of 15-49 the illiteracy percentage among married women is 58.4 per cent, and in urban areas it is 25.6 per cent. Those who cannot read lack of awareness of social and economic problems and this affects family planning adversely. In short, if women are better educated they will opt for smaller family size and will become more productive members of the society, thus reducing food insecurity.

Imbalance in Food Requirement and Food Availability

India is third largest food producing country in the world. But still food availability does not mean food security. One should have the purchasing power to buy sufficient food whenever and wherever needed. At times there is a paradoxical situation where is spite of plenty of food being available, a number of children die of malnutrition. In the Melghat Tribal area of Maharashtra State for example, thousands of children die every year as a result of malnutrition. This finding is based on practical work experience by the NGO Gandhi National Memorial Society, covering 311 villages.

To overcome this chronic problem the NGO Gandhi National Memorial Society initiated a program of training for tribal women social workers in their institution. The curriculum covers *inter alia* health, hygiene, nutrition and skills development.

In India food grain production increased from 51 million mt in 1950-51, to 201 million mt in 1998-99. This was mainly due to the Green Revolution. While the population increased threefold during this period, food production increased fourfold. The overall growth of food production in all these years outpaced the rate of population growth. But that positive trend is expected to change in the new millennium. India's population of already passed one billion, and this year's food grain requirements will be around 220 million mt. It is estimated that for this increased population, additional food grain production has to be around 5-6 million mt per year. In short this means the increase in population will outstrip the increasing food production despite the Green Revolution, and the alternative of extensive cultivation is not an option because arable land area in India has reached its limit of 140 million ha. Compounding the problem, the use of Green Revolution technologies such as high yielding seed varieties, chemical fertilizers, pesticides and large-scale irrigation sometimes cause soil degradation, environmental pollution and loss of biodiversity.

Effects of Globalization

With globalization, developing countries are facing a dilemma! The problem is the need for the rapid industrialization *vis-a-vis* food security. The following problems have been identified:

- There is every possibility of agricultural land being shifted to industrialization or for cash crop. This will result in food insecurity.
- Land is also likely to be diverted to housing, road building, tourism, etc. as a consequent of increased papulation.
- Mega dams and other projects are likely to affect the eco-system and environment adversely.
- Increase in exports of food grain will cause shortage of food availability, unless only surplus food grains are exported.
- Due to diversion of water for industries, tourism, urban use, etc., there will be shortage of water for agriculture.

Other Factors

Other factors limiting food security are over crowding in agriculture, discouraging rural atmosphere, fragmentation of land, inadequate non-farm services, size of holding, poor techniques of production and processing, inadequate irrigation facilities, poor purchasing power, variations in food prices, sudden changes in incomes and prices, storage facilities and high storage cost and poor management of public food stock.

Women's Role in Ensuring Food Security

A sizeable proportion of Indian women is engaged in agriculture. To achieve the World Summit goal *"to halve the number of hungry people in the world by 2015"* women's full potential is crucial. It is a big task and requires a real "Cultural Revolution" in people's attitudes and behaviour in a patriarchal society like India.

Globally women hold key positions in the diversify agriculture sector to produce food not only for their families but for millions of people worldwide. A women farmer extends the food chain from sector to the household, and takes responsibility for various food-related activities to ensure food security of her family. India practices traditional agriculture on small farms mostly of less than one ha. Opertions like sowing, weeding, harvesting, picking of tea leaves and coffee beans, fruit and vegetable growing, etc., and particularly work related to rearing is animals is undertaken by women in the rural areas. This occurs in 7.5 million Indian villages.

In spite of the problems like lack of access to education or inadequate education, gender inequality, secondary, status in a male dominated society, exclusion from participation in decision-making and lack of awareness, women in India do play a critical role in the progress of earning a livelihood. They participate in production and to some extent in the marketing of agricultural products. A very small proportion of women is directly involved in marketing. In production activities women play a role as a food producers and food providers, but their contribution is recognized only as that of labourers.

The vast majority of women are engaged in domestic and household production activities which contribute to the survival of the household, but go unrecognized and unrewarded. In India labour inputs per ha of women in agriculture was found to be higher in regions with higher rainfall and in agriculturally backward regions. Under Indian Himalayan conditions, a pair of bullocks is observed to work for 106 hours, a man for 1,212 hours and a woman for 3,485 hours in a year on a one-ha farm. A woman puts in greater number of hours than man and animal together.

As agricultural labourers, women play a key role in food production. When they head rural households they increasingly become farm managers especially in food production Female-headed households are especially poor, and they have many young dependents. They struggle for food continuously. Due to price rises or food shortage food insecurity is always with them. At such times just to meet food needs, many families dispose of their assets, or they borrow money from moneylenders to meet their food needs.

Various studies showed that women have a major responsibility for feeding the household. As a result, when women's income is lacking the whole family suffers, and women have fewer opportunities to earn in rural areas. The 1991 census of India showed that 34.22 per cent of female workers are cultivators and 44.93 per cent of females are agricultural labourers. Data shows that women's contribution in food production and processing remains in the informal sector and for self-consumption. Because it is difficult to show their share in agricultural production or in household income, contribution by women is never counted and shown in formal statistics, which poses a handicap to their becoming equal partners in development.

Some practical examples of what Indian women are doing for ensuring food security are listed below:

- Women in Gujarat came together and started a cooperative milk industry known as "*Amul*". A large number of women are member and managing the industry very efficiently. A major proportion of rural Gujrathi women benefited from this project.
- In Himachal Pradesh women changed the cropping pattern. Due to fragmentation of land it was not profitable to grow food crops, so women surveyed the market and started a floriculture enterprise.
- Tribal women from Himachal Pradesh offered a unique non-violent resistance and prevented deforestation. This movement is known as "*Chipko*" movement.
- Women in Kerala offered stiff resistance to encroachment by trawlers to safeguard the livelihood of small fishermen.
- Through the NGO Assisi Farm and Training Center Society in Tamilnadu, 270 families are trained in kitchen-gardening activity, and about 50 per cent of their food requirement is being produced organically at little or no expense.

IMPORTANT MEASURES TAKEN TO ENHANCE WOMEN'S ACCESS TO PRODUCTIVE RESOURCES

1. Government Efforts

The Indian Government has recognized women as productive workers and contributors to the country's economy and has taken some steps towards upliftment of women. In the Ninth Plan a special emphasis has been given to women's empowerment. The following are some examples.

1. The *National Women's Commission* for women in self-employment has focused on the tremendous contribution made by women to the national economy. This leads to specific programs like: (*a*) Development of Women and Children in Rural Areas (DWCRA); (*b*) Support to Training for Employment Programmes (STEP); and (*c*) National Credit Fund for Women in 1993 (*Mahila Kosh*) aims to reach the poorest of the poor and assestless women who are in need of credit but cannot access the formal banking or credit system.
2. *Women's development corporation* have been set up in various States. They are making concentrated efforts to improve women's conditions by upgrading their skills through training programs and offering more employment opportunities to them in schemes like public distribution system, dairy development, food preservation, social forestry,

rural marketing, etc. which are related to them in traditional occupation, agriculture and animal husbandry, etc.

3. *The Common Minimum Needs Programme* objectives give priority to agriculture and rural development, generating productive employment, and eradication of poverty ensuring food security for the vulnerable groups of the society.
4. For the eradication of poverty, *reservations have been made* for the women in poverty eradication program.
5. *Special Credit System* allows easy assess to credit at low transaction costs to women in the formal sector.
6. *Mahila Sammriddhi Yojana* is a special saving scheme for rural women. A significantly large number of women have come forward to have their own saving accounts through which they can exercise better control over household resources.
7. *Indira Mahila Yojana* mobilizes women around and integrated delivery system which will cover a whole range of social, economic support services for women and children.
8. *National Social Assistance Scheme:* To provide old age pension to all persons below the poverty line and above 65 years age. Lump sum family benefits in the cause of the death of the principal bread earner and maternity benefits to women of poor households for the first two births.
9. *Integrated Child Development Services*: To cover 20 million children and women, National Creche Fund created to expand the network of day care centers. A massive mid-day meal scheme to cater to all primary school children in the country.
10. *National Literacy Mission*: To create awareness among the rural women.
11. *National Commission for Women*: Created through an act of Parliament. This acts as an ombudsman for women for reviewing the working of legal and constitutional safe-guards and intervening in cases of atrocities. Laws make it compulsory for the government to consult the commission all major issues concerning women. Most of our States have also set up State commissions.
12. India has a policy today for encouraging *joint ownership of property* by women and men.
13. *Reservation for Women in Local Government*: By law 33 per cent of places are reserved for women in local government. This has created tremendous enthusiasm among women. Almost one million women have entered public offices and took control of their own lives and the destinies of themselves and their family.
14. *Women in Agriculture Scheme*: Considering lack of awareness of women's role and contribution in agriculture, Government of India formulated a scheme "Women in Agriculture". The objects of the scheme are to:
    (*a*) Create general awareness on the role played by women in agriculture;
    (*b*) Motivate and mobilize women farmers through a group approach;
    (*c*) Train groups to form effective network for channeling agricultural development programs and other support system such as income, technical and extension support.
    (*d*) Ensure and provide equal opportunity to make women farmers self-reliant.
    (*e*) Identify and organize women's groups which act as network for channelling the required agricultural support such as assessment of specific input needs and the training of women farmers, and the provision of organization and financial support to self-help thrift groups.
    (*f*) Provide technical training in agriculture and various farming systems, training in management, organization, entrepreneurship development and decision-making

schemes. One hundred percent expenditure of this scheme is borne by the Government of India.

15. Reservation in Agricultural Universities: Thirty percent of the seats are reserved in agricultural universities for women.
16. *Swarnajayanti Swayamrojgar Yojana*: This scheme in launched by Government of India to make rural people self-employed through entrepreneur-ship development programs. Forty percent of reservations are made for women.

2. Limitations

Despite governments efforts towards the mainstreaming of women in national development, Indian women still have a long way to go in terns of literacy, education, health and nutrition, etc. The problems of food, water, fuel and shelter facing the poor women of India are daunting. Following are some factors limiting women's role in food security.

1. Non Enforcement of the Law

Although legal rights for joint properties including land are bestowed to women, gender discrimination, lack of awareness and education, and secondary status of male dominated society mean that the laws are rarely implemented. As a result, women themselves have no security for their survival. Legal literacy is extremely important because it aims to educate women about the laws concerning their rights. This work has stated in India but needs to reach the million of our rural Indian sisters.

2. Gender Discrimination

In India gender discrimination exists in all walks of life especially in rural areas. It is so alarming that girls and women are not even fully fed with adequate nutritious food and often sleep on a half empty stomach. Beside they are deprived of education, training and health care, etc.

3. Lack of Education and Training Opportunities, and Facilities

Lack of education and awareness means women know little of what is happening around them and in the world. In some cases if they want to learn, educational and training facilities are inadequate. Facilities for girls in rural areas are mostly restricted to primary level, as higher level facilities are available only in urban areas. Further due to inherent traditional conventions girls are not generally allowed to go to urban areas for higher education. This stunts their intellectual growth and socio-economic advancement and awareness. Further it is observed that while in India we have special colleges like management, engineering, pharmacy, medical, etc. for women, there are no agricultural colleges exclusively for women.

4. Lack of Education and Awareness

Women have no role to play in decision-making processes in the family, therefore have no self-confidence and are, in short, treated as secondary.

5. No Access to Credit

Women seldom have access to credit facilities, productive resources and other support, which results in physical and financial helplessness.

6. Traditional Culture

Traditional culture, including early marriages negatively affect the advancement of women in a male dominated society.

7. Discrimination in Wages

Rural women folk are mainly involved in agricultural operations. They work as casual labourers or daily workers. As with domestic work, much of women's agricultural labour is overlooked because it is unpaid. In India 30-40 percent of landless labourers are women. Although minimum wages are fixed by law for farm labourers and the constitution stipulates no discrimination between men and women in wages, the laws are usually breached. Women get lower wages than men for the same work and as a result their contribution is not comparable to the wages they receive, but this is seldom recognized.

As regards work force participation, the women's contribution is not counted as workers because they work for less than the accepted measure of being employed *i.e.* working for full at least 183 days in year. In respect of small farmers, the women in the family work on the farms but they are not paid any wages nor their contribution is counted. Further they have to carry out all the domestic duties in the large families. Therefore they are overburdened with work.

8. Dependence

Last but not least there is a general limitation on the women's development because of heavy dependence on government funds and schemes *i.e.* price subsidy, feeding scheme rationing, employment programs, etc.

Prospects for Women in Sustainable Food Security

Despite the constraints for women, the prospects of attaining sustainable food security are bright. In this area women can definitely play a major role and contribute substantially. To this end, some action at the government and local levels is in progress:

Population Control

In population control the participation of women and their decision-making capacities required to be strengthened. This work can be done especially by women's NGOs by creating better awareness and education. The Government of India has taken a decision in the right direction by reserving a 33-percent quota for women in local government. The laws on reservation for women has likely to pass in this parliament.

Sustainable Agriculture as the Key to Sustainable Food Security

Sustainable agriculture is possible if the following actions are taken:

1. Security of land and ownership has to be ensured. This can be done by enforcing the property rights of women in joint property. At the time of marriage in India the bride's name is changed, and this name must also be incorporated in the property records.
2. *Appropriate Modern Technology for Agriculture:* As food producers and food providers women will adopt the most suitable technology that will not adversely affect the productivity of soil, the eco-system, the environment and human lives. As food providers to the entire family women have a full knowledge of the food requirements of the family with proper nutrition, and will therefore not go in for cash or commercial crops and will give first priority to food crops. This will ensure food security for the family.
3. *Watershed Development:* Water being the key factor in agricultural production particularly in rainfed areas, women's participation in watershed development is very crucial to ensure adequate production.

4. *Access to Resources:* Sustainable agriculture is possible if common property resources are made assessable to all. These resources should be owned and managed by local communities wherein women's equal representation as men.

Access to Credit for Augmenting Family Income

To ensure food security, increased purchasing power of the family is a must. A number of women are working in agriculture aquaculture and in other production areas which require access to credit facilities. Irrespective of the source of income generation women workers have to be provided with credit facilities without gender or other discrimination. The Self-Help Groups (SHGs) credit scheme for women has shown excellent results not only in diversified income generation, but also in repayment of the credit.

Needs-based Training Programs

To enable women to undertake the agricultural or other income-generation programs, it is very essential that specific need based training courses are to be prepared and conducted. These training programs shall include *inter-alila* cropping management patterns, agroprocessing and preservation, marketing, packaging, advertisement for entrepreneurship development, seed collection and selection, nursery activities, forestry, appropriate low-cost technology, organic farming etc. In addition to adult literacy among women the "Agriculture Literacy" program has to be undertaken as a special campaign.

Entrepreneurship development programs undertaken by various NGOs in Maharashtra including the Gandhi National Memorial Society in Pune, India have shown excellent results. It is however necessary to expand these programs through mobile units so that it can be reached at the doorsteps of the individual rural women.

Gender Sensitization

With gender discrimination going down as a result of the gender sensitization programs undertaken by the various NGOs, and with increased awareness, the participation of women in all activities has increased substantially.

Alternative Public Distribution System of Food Grains

Although public distribution systems for food grains and other necessities undertaken by government are in vogue for the last few decades. Therefore an alternative public distribution system has been adopted for the fair distribution with low cost for the benefit of weaker section.

This has been proved by a model scheme adopted by an NGO namely Deccan Development Society in the underdeveloped and remote Medak district of Andhra Pradesh. This innovative alternative system covered 30 villages and is run by local women belonging to the most disadvantaged segments of the society. It has not only distributed food and other articles but also brought 3,000 acres of fallow land under cultivation. A community grain fund has also been created, through which food is locally produced, locally stored and locally distributed. This has eliminated the need for subsidy which the government had to bear.

The advantages of this system are increased production, which is sold locally, extending more jobs opportunities and leading to higher purchasing power. This model system has to be adopted all over India so that a strong local economy shall emerge. With this, local people will be less vulnerable and less dependent on out side factors. This is due to the management being in the hands of capable village women who understand the core of the problem.

SEWA Projects

The Self-Employed Women's Association (SEWA) is a cooperative which has brought together women from unorganized sectors and transformed their lives. SEWA has shown the status that empowerment and organization can bring the most poor and oppressed women in the work place. This model is exemplary and needs to be expanded all over India so that women's participation in sustainable food security through cooperative efforts will increase.

Rural Urban Linkages

To open up the channels for industrialization, marketing, rural urban linkage are very important, which would help towards upstream development of rural women.

NGOs and Women's Development Corporation

More and more participation of NGOs and women's development corporations are necessary for proper implementation of development schemes. In short, the conditions of Indian women are changing in a positive direction. Development of women is a central issue in our planning process. Since the 1980s, women have been organized as a special target group and government efforts have been directed towards the mainstreaming of women into national developmental processes. Nevertheless, Indian women have a long way to go in term of literacy, education, health, nutrition and participation in work force.

Role of Women

Food security has been defined by the Food and Agriculture Organization of the United Nations (FAO) as the economic and physical capacity of a household to continually provide family members with sufficient food for individual bodily needs without threats of shortage. In addition, a household can exploit its food availability to the advantage of the nutritional needs of the individuals. In fact food is vital throughout life to maintain a regular life and good health to facility efficient activity.

The significance of food for health and body maintenance implies that nutrients are required by the body to provide energy and warmth, maintain good growth of the body and to guard against diseases by building immunity and finally Good health. Socio-economic speaking, general well-being of the body and mind helps create good ideas, decision-making and good performance of activities on farm, in household and in community as well. The three major factors of household food security are:

- Capacity to produce sufficiently food;
- Economic capacity to always acquire and provide adequate food; and
- Ability to utilize to the fullest extent the food through cooking, processing, preserving and allocating it to the family members based on the individual nutritional needs.

Taking into account the three above factors, the involved people who are in the position to assure the food security of the family are women-rural women in particular.

An FAO survey disclosed that 42 per cent of women in 82 developing countries are engaged in farming. Table 6 shows current trends, and differences with industrial countries.

Table 6 suggests a downward trend in participation by women in agriculture as economic development progresses. In developing countries like India the majority of women in the labour force are engaged in agricultural activities while in industrialized countries only a small proportion is engaged in agriculture, and this share is decreasing over time.

Table 6 : Percentage Distribution of Labour Force of Women in Agriculture

World/Region	*1990*	*1997*
World	52	49
Developed countries	9	7
Developing countries	68	63
- African developing countries	76	72
- Asian developing countries	72	67
- Latin American countries	13	11
- Oceanic developing countries	44	45
- Low-income countries	76	77

Source: FAOSTAT.

Women's Economic Contributions

The true reflection of rural women's work is often hidden, since the major part of their work is unpaid, contributed as family labour on subsistence farms. Table 7 shows that in developing Asian countries their participation as both paid and unpaid workers and as self-employed producers is significant.

Table 7 : Women's Participation in Agricultural Economic Activities

Asian Countries	*Percent Distribution of women in Agricultural Labour Force, 1994*	*Percent Employed on Own Account Workers, 1990*	*Percent Unpaid Family Workers 1990*
Bangladesh	65	5	6
Bhutan	95	-	-
Cambodia	85	-	-
China	74	-	-
DPR Korea	41	27	69
India	78	-	-
Indonesia	44	27	66
Iran	69	4	43
Japan	-	-	-
Lao PDR	76	-	-
Malaysia	31	24	64
Maldives	25	22	29
Mongolia	28	-	-
Myanmar	35	-	-
Nepal	67	-	-
Pakistan	15	-	-
Philippines	34	30	53
Republic of Korea	31	27	87
Sri Lanka	50	18	59
Thailand	64	27	64
Vietnam	57	...	...

Source: United Nations, *The World's Women 1995 – Trends and Statistics,* New York.

Rural women's work is usually unpaid work, since they produce food for the household rather than as cash crops for the market. Because women farmers have limited land ownership rights and are seldom wage earners however, their contribution to food security is often invisible, and therefore not reflected in agricultural statistics or policy.

In developing, countries women play a pivotal role in producing, securing and preparing food for the family, although the role of women in food production, processing and marketing varies from place to place. In many African countries they produce and market up to 80 per cent of all food grown locally, and Table 8 shows their overall contribution far exceeds that of men.

Table 8 : Women's Contribution to Labour in Africa

33 per cent of the workforce;
70 per cent of the agricultural workers;
60-80 per cent of labour to produce food for household consumption and sale;
100 per cent of the processing for basic foodstuffs;
90 per cent of household water and fuel wood;
80 per cent of food storage and transport farm to village;
90 per cent of the hoeing and weeding work; and
60 per cent of the harvesting and marketing activities.

Women play a crucial role in overall household economic activities in all countries of the world. Figure shows the percentage contribution by women to agriculture, as estimated by the International Fund for Agriculture (IFAD):

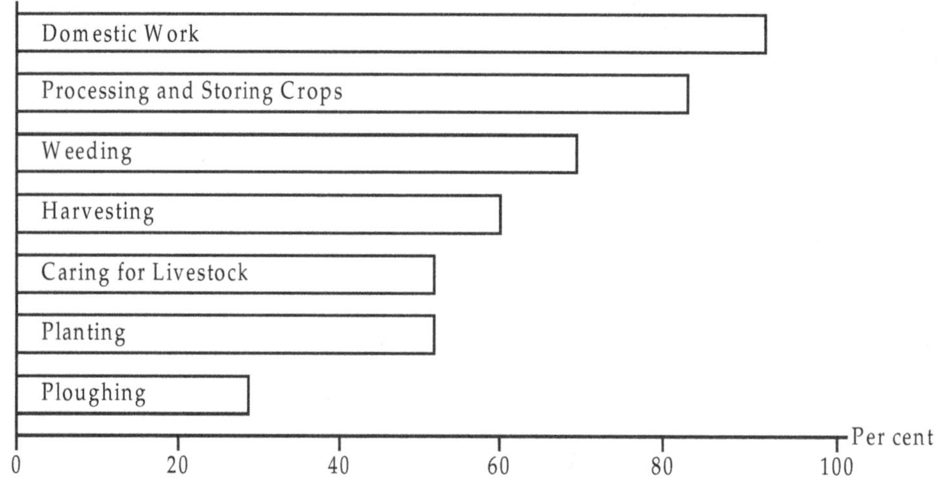

Fig. : Percentage of Agricultural Activities Performed by Women

Source: IFAD, 1990

Constraints to Sustaining Women's Productivity

Women's contribution to food production and processing involves land and water management. They have considerable about water resources including quality and reliability, restrictions, and acceptable storage methods. Using traditional methods, women farmers have been effective in conserving natural resources including soil fertility.

Their capacity to continue to do this effectively however is limited by their lack of access to and control over the natural resources needed for agricultural production, by sociocultural constraints and by their lower economic status relative to men. Women's empowerment is therefore a prerequisite to environmental sustainability.

Lack of Ownership of Assets

Although seldom land owners women farmers practice crop rotation, intercropping, mulching and a variety of other soil conservation and enrichment techniques. Table 9 shows the percentage of land ownership by women farmers in Thailand, Trinidad, Nigeria and the Syrian Arab Republic.

Table 9 : Land Ownership by Women Farmers in Thailand, Trinidad, Nigeria and the Syrian Arab Republic

(Unit: Percent)

Type of Tenure	*Thailand*	*Trinidad*	*Nigeria*	*Syrian Arab Republic*
Personally owned	22.5	9	4	-
Husband's land	39	30	23	41
Gift from husband	-	-	30	-
Family land	10	9	12	30
Government land	-	25	-	-
Communal land	-	-	20	-
Squatter land	-	9	-	-
Rented land	29	19	11	22.5

As long as approval is required from a husband or male kin for decision-making, the capacity of women to optimize productivity by managing land and water resources efficiently and effectively will remain constrained. Where women hold land ownership rights and have education equal to that of men, their productivity is significantly higher.

The conditions in India is also not very good.

Membership of Organization

In many countries women are excluded from becoming members of organizations that facilitate access to information and services, and to the benefits of technology transfer because membership is based on land ownership of the "head of household" criterion. Even in countries where membership is open to all, women do not always benefit to the same extent as men, because domestic work leaves little time for women to participate. Table 10 shows households headed by women as a share of total rural households.

Table 10 : Household Headed by Women as A Share of Toral Rural Households

Region	*Percentage*
Asia	9
Asia (excluding China and India)	14
Sub-Saharan Africa	31
Near East and North Africa	17
Latin America and the Caribbean	17
Total developing countries (114)	12
Least developed countries	23

Source: IFAD.

Chapter 3

Fisheries Sector in India : Policies and Programmes

India's Share in World Fish Production

Fish production in India has touched 5.6 million tonnes in 1999-2000. It was a mere 0.75 million tonne in 1950-51. The world production during the same period has gone up from 23.5 million tonnes to around 120 million tonnes in 1999-2000. The trend of fish production in India as compared to the world production during the last 50 years is given in Table 1. The share of India in global fish production has grown gradually from about 2.6 per cent in 1960s and 1970s to 4.7 per cent in 1999-2000. Thus, compared to growth in global fish production, the growth in India has been at a faster rate, mainly due to increasing contribution from inland fish production. This year overall fish production is 6.3 mmt. But of which 3.3 from Indian sector and 3 mmt from marine.

Table 1 : Fish production in India and world during last 50 years

Year	*World (million tonnes)*	*India (million tonnes)*	*India's share(%)*
1950-51	23.50	0.75	3.19
1960-61	43.60	1.16	2.66
1970-71	66.20	1.76	2.66
1980-81	72.30	2.44	3.37
1985-86	85.60	2.88	3.36
1990-91	97.97	3.84	3.92
1999-2000	120.00	5.66	4.72

Source: Hand Book on Fisheries Statistics (2000), Ministry of Agriculture, Government of India.
NCAP 2003. A Profile of People, Technologies and Policies in Fisheries Sector in India (eds Anjani Kumar, Pradeep K. Katiha and P.K. Joshi).

Growth of Fisheries Sector in India

Fisheries sector plays an important role in the Indian economy. It contributes to the national income, exports, food and nutritional security and in employment generation. This sector is also a principal source of livelihood for a large section of economically underprivileged population of the country, especially in the coastal areas. Share of agriculture and allied activities in the GDP is constantly declining. It has been observed that agriculture sector is gradually diversifying towards high value enterprises including fisheries. It is evident from the contribution of fisheries sector to the GDP, which has gone up from 0.46 per cent in 1950-51 to 1.16 per cent in 1999-00 (at current prices) (Table 2). The share of fisheries in Agricultural GDP (Ag. GDP) has increased more impressively during this period from mere 0.84 per cent to 4.19 per cent. This is largely due to a sustained annual growth rate of well over four per cent in the fisheries GDP during the last five decades. The fisheries sector has recorded faster growth as compared to the agricultural sector in all the decades. The growing production of fish suggests that fisheries sector is booming and contributing to the economic growth of the nation. More than 6 million fishermen and fish farmers are totally dependent on fisheries for their livelihood in India.

Table 2 : Contribution and Growth of Fisheries Sector in India

Period	*Percent contribution to*		*Percent annual growth*	
	GDP	*Ag GDP*	*Fisheries GDP*	*AgGDP*
1950-51	0.46	0.84		
1960-61	0.54	1.18	5.63	2.68
1970-71	0.61	1.37	3.92	1.50
1980-81	0.73	1.98	2.86	1.72
1990-91	0.93	3.00	5.11	2.89
1999-2000	1.16	4.19	4.75	3.12

Source: National Accounts Statistics, Central Statistical Organisation, Government of India.

Fish Production : Structure and Trend

The fisheries production in India during 1950s was more pronounced in the marine fisheries and it remained the major contributor till early 1990s (Table 3). Its share in the total fish production was more than 70 per cent in 1960s, but thereafter it started declining and came down to about 62 per cent in 1970s and to 59 per cent in 1980s. In the mid-nineties, the fisheries production witnessed a significant change. The share of inland fish production became almost half of the total fish production in 2000. In seems that marine fisheries production has reached a plateau and at best, it can register only a marginal increase in the near future. On the other hand, inland fish production was on constant rise and its share rose to 38 per cent in 1970s to 41 per cent in 1980s and jumped to over 45 per cent in 1990s. This rise in inland fish production is attributed to the development of aquaculture in our country.

Table 3 : Changes in the Structure of Fish Production in India

(in million tonnes)

Year	*Marine*	*Inland*	*Total*
1950-51	0.53 (71.01)	0.22 (28.99)	0.75
1960-61	0.88 (75.86)	0.28 (24.14)	1.16
1970-71	1.09 (61.85)	0.67 (38.15)	1.76
1980-81	1.5 (59.12)	0.89 (40.88)	2.44
1990-91	2.30 (59.96)	1.54 (40.04)	3.84
1995-96	2.71 (54.70)	2.24 (45.30)	4.95
1999-2000	2.83 (50.09)	2.82 (49.91)	5.66

Figures in parentheses indicate percentage to total.

Source: Hand Book on Fisheries Statistics (2000), Ministry of Agriculture, Government of India.

The growth trends in fisheries production in India during 1980-81 to 1999-2000 is given in Table 4. Since 1980-81 fisheries production in India has been increasing at a rate of 5.12 per cent per year. The inland sector has shown a better performance with an annual growth rate of 6.22 per cent. A disaggregated view of the pattern of growth shows that growth in inland fisheries production has accentuated in the 1990s while marine fish production witnessed deceleration. The latter showed down from 3.73 per cent in the 1980s to 2 per cent in the 1990s. The share of culture fisheries in both freshwater as well as brackish water in the inland sector has increased tremendously in recent years. Its share has risen from 43 per cent in 1984-85 to about 84 per cent in 1994-95. Within the culture fisheries, the major contributor has been the freshwater aquaculture (Krishnan *et al.*, 2000). The policy for fisheries development has also been given a tilt towards inlands fisheries particularly aquaculture in recent years.

Table 4 : Growth trend in fish production in India

Source of fisheries	*Annual Compound Growth rate during different periods(%)*		
	1980-81 to 1989-90	*1990-91 to 1999-2000*	*1980-81 to 1999-2000*
Marine	3.73	2.01	4.23
Inland	5.14	6.34	6.22
Total	4.30	4.03	5.12

Source: Hand Book on Fisheries Statistics (2000), Ministry of Agriculture, Government of India.

Export of Fish and Fish Products

There has been a considerable increase both in the quantum and value of export of fish and fish products since 1960-61. In 1960-61, 0.02 million tonnes worth US$ 10 million were exported (Table 5). It increased to 0.39 million tonnes (about 20 times) worth US$ 1180 million (more than 100 times) in 1999-00. The share of fish and fish products in total exports has increased from 0.74 per cent in 1960-61 to more than 3 per cent in 1999-00.

Table 5 : Development of India's Exports of Fisheries Products

Year	*Quantity (000 tonnes)*	*Value (million US$)*	*%Share in*	
			Ag Export	*Total export*
1960-61	19.9	10	1.68	0.74
1970-71	32.6	40	6.21	1.97
1980-81	69.4	274	10.53	3.23
1990-91	158.9	535	15.19	2.95
1995-96	310.1	1011	16.00	3.18
1999-2000	390.6	1180	20.81	3.14

Source : Monthly Statistics of Foreign Trade of India : Volume Exports and Re-exports (various issues), Ministry of commerce.

The share of fish and fish products in total export was about 2 per cent in 1970-71 and thereafter it has been hovering around 3 per cent. Similarly, the contribution of fish and fish products' exports to agricultural exports also increased from 1.68 per cent in 1960-61 to about 16 per cent in 1990-91 and became about 21 per cent in 1999-00. It seems that the liberalization policies initiated in the 1990s helped the fisheries sector in attaining a higher growth in exports. Four decades ago a humble beginning was made in shrimp export and today the export basket of fisheries includes more than 60 items. Shrimp however, remains the major item of fisheries exports in terms of both quantity and value. In 1998-99 the Share of Shrimp was 26.11 per cent in quantity and 67 per cent in the value of export earnings from fisheries. The share of shrimp has declined and frozen/fresh fish has replaced the shrimp in quantity (Krishnan *et. al.* 2000; Kumar *et. al.* 2002).

DEVELOPMENT PROGRAMMES/POLICIES FOR FISHERIES

Outlays for Fisheries Sector

One of the indicators of development policies and programmes is the allocation of resources for this sector over different periods. The outlay for fisheries sector as per cent of outlay for the agricultural sector over the Five Year Plans has been increasing continuously (Table 6). It increased from 1.74 per cent in the First Five Year Plan to about 6 per cent in the Ninth Five Year Plan. This shows that greater importance in terms of higher allocation of funds to fisheries sub-sector within agriculture has been accorded. Its share in the total outlay during different plans has been hovering between 0.26 and 0.52 per cent.

Table 6 : Outlay for Fisheries Sector During Five Year Plans

Five Year Plan	*Total outlay*	*Outlay for agricultural sector*	*Outlay for fisheries sector*	*Share of fisheries sector(%)*	
				Total outlay	*Agricultural outlay*
First	1960	294	5.13	0.26	1.74
Second	4600	529	12.26	0.27	2.32
Third	7500	1068	28.27	0.38	2.65
Fourth	15902	2728	82.68	0.52	3.03
Fifth	39332	4302	151.24	0.38	3.52
Sixth	97500	6609	371.14	0.38	5.62
Seventh	180000	10524	546.54	0.30	5.19
Eighth	434100	22467	1232.82	0.28	5.49
Ninth	859200	37546	2069.78	0.24	5.51

Source : Hand Book on Fisheries Statistics (2000), Ministry of Agriculture, Government of India.

Development Programmes

The development plans for India's fisheries sector were aimed at increasing the fish production, improving the welfare of fishermen, promoting export and providing food security. The first step towards developing the fishing as an industry was made in 1898, when the then Madras Presidency was advised to strengthen the fishery so that it could fight famine. It took almost 50 years to concretize this idea. After in independence, the first All India Fisheries Conference, held in 1948 in New Delhi, decided to seek foreign co-operation to create necessary infrastructure for modernizing the fisheries sector. In 1952, a tipartite technical co-operation agreement was signed between India, the USA and the United Nations for fisheries development and a year later, the Indo-Norwegian Project (INP) in Kerala was started. From then onwards the modernization of fisheries was initiated in the coastal states in India. Several programmes have been launched for both marine and inland fishery developments in the country, some of which are briefly described below :

Programmes for Development of Inland Fisheries

In recognition of the increasing role of inland fisheries in overall fish production, the Government of India (GOI) has been implementing two important programmes in the inland freshwater sector since the Fifth/Sixth Plans. These are Fish Farmers' Development Agencies (FFDAs) and the National Programme for Fish Seed Development. A network of about 429 FFDAs in functioning today covering all potential districts in the country. The water area brought under the intensive fish culture through the efforts of these FFDAs was 0.46 million hectares (ha) up to 1997-98. The agencies have trained 0.6 million fish farmers in improved practices. Additionally about 0.07 million ha area has been developed for shrimp culture. Some Brackishwater Fish Farmers Development Agencies (BFFDAs) have also been established in the coastal areas of the country; these provide compact package of technical, financial and extension support to shrimp farmers. Under the national programme for fish seed production, more than 50 fish seed hatcheries have been commissioned. It has led to a marked improvement in the production of fish seed. Their production has increased from 409 million fry in 1973-74 to about 20000 million in 1999-2000.

Programmes for Development of Marine Fisheries

The programmes for development of marine fisheries as envisaged in different Five Year Plans include: (*i*) intensive surveys particularly of exclusive economic zone (EEZ), on marine fishery resource assessment, (*ii*) optimum exploitation of marine resources through a judicious mix of traditional country boats, mechanised boats and deep-sea fishing vessels, (*iii*) providing adequate landing and berthing facilities to fishing vessels by completing the ongoing construction of major and minor fishing harbours, (*iv*) intensifying efforts on processing, storage and transportation, (*v*) improving marketing particularly in the co-operative sector, and (*iv*) tapping the vast potential for export of marine products. During the Seventh Plan some selected villages were grouped for setting up "Fisheries Industrial Estates". The major developments include construction of 30 minor fishing harbours and 130 fish landing centres apart from five major fishing harbours *viz.*, Cochin, Chennai, Visakhapatnam, Roychowk and Paradip. They provided landing and berthing facilities to fishing crafts. The Government also provides subsidy to poor fishermen for motorizing their traditional craft which increases the fishing area and the frequency of operation with a consequent increase in catch and earnings of fishermen. About 33,000 traditional crafts were sanctioned for motorization up to 1997-98. Improved beach landing crafts are also being supplied to groups of fishermen. A scheme of re-imbursing Central excise duty on HSD oil used in fishing vessels below 20 m length is also in operation to help the small fishermen to reduce their operational coast.

Welfare Programmes for Traditional Fishermen

There are two important programmes for the welfare of traditional fishermen: (*i*) Group Accident Insurance Scheme for active fishermen, and (*ii*) Development of Model Fishermen Village. Fishermen are insured for Rs 50,000 in case of death or permanent disability and for Rs 25,000 in case or partial disability. About 1.3 million fishermen were insured during 1998-99 under this scheme. Under the programme of Development of Model Fishermen Villages, basic amenities such as housing, drinking water and community hall are provided to fishermen. About 30,000 houses were constructed up to 1998-99 under this programme.

Programmes with International Aid

Several international organizations, including the World Bank, UNDP, DANIDA, NORAD, ODA (UK and Japan) provide aid to India for the development of fisheries sector. Under the Bay of Bengal Programme (BOBP), started in 1979, assistance is provided for the development of small-scale fisheries and enhancing the socio-economic conditions of the fishing communities. ODA (UK) has provided technical aid for the prevention of post-harvest losses in marine fisheries. Recently, FAO launched a scheme for providing technical assistance to implement Hazard Analysis Critical Control Points (HACCP) in seafood processing industries. A Shrimp and Fish Culture Project was started with the assistance of the World Bank in May 1992 and it continued for a period up to December 1999. The states of Andhra Pradesh, Bihar, Orissa, Uttar Pradesh and West Bengal were covered under this project. Six sites covering a brackish water area of 797 ha have been developed for shrimp culture operations. A total of 101 reservoirs and 22 oxbow lakes have been developed for fish culture.

Policies for Fisheries Development

At present, the fisheries sector does not have separate policy of its own like the Science Policy, Technology Policy, Industrial Policy, Telecom Policy or the recently announced Agricultural Policy. Only a passing reference has been made in the Agricultural Policy regarding the fisheries development in the country. However, the successive Five Year Plans of India have set up some broad policies with regard to the production in the fisheries sector and investment in it. Policy makers and planners visualize fisheries as an important sector for agricultural diversification, employment generation, export promotion and food security. The main objectives of fishery development policies through different plans have been: (*a*) enhancing the production of fish and the productivity of fishermen and the fishing industry; (*b*) generating employment and higher income in fisheries sector; (*c*) improving the socioeconomic conditions of traditional fisherfolk and fish farmers; (*d*) augmenting the export of marine, brackish and freshwater fin and shell-fishes and other aquatic species; (*e*) increasing the per capita availability and consumption of fish (present target is 11 kg per annum); (*f*) adopting an integrated approach to fisheries and aquaculture; and (*g*) conservation of aquatic resources and genetic diversity (Planning Commission, GOI). A glimpse at the strategies followed in different Five Year Plans reveals that up to third Five Year Plan the focus was mainly on enhancing the fish production with little attention on issues like marketing, storage, transportation etc. However, in subsequent Plans, measures were initiated to create more facilities for ice-cold storage, processing and canning. Moreover, in 1972, Marine Products Export Development Authority (MPEDA) was established in Cochin with branch offices in all the major centres of seafood production and export in India. In has the responsibility for the promotion and regulation of marine products expot and it is the nodal agency for joint ventures in deep sea fishing. It also promotes brackish water shrimp farming. However, even after 50 years of planning, post-harvest infrastructure is grossly inadequate in India in both the marine and inland fisheries sector (Dehadrai 1996). The marketing, transportation, storage and processing of fin and shellfish are mostly handled by the private sector. This activity has witnessed a relatively slow growth and has lagged behind production trends. It is also a fact that the marine fisheries industry has given more attention towards export and adequate measures have not been made for the development of the domestic market.

Trade Policy and Prospects of Fisheries Exports

The government policies regarding imports and exports play a significant role in influencing the trade structure of a country. Trade policies are in general categorised into two types: (*i*) export promotion oriented policies, and (*ii*) import substitution oriented policies. In the early stages of planned development during the 1950s, Indian development strategy, was heavily oriented towards import substitution. It was only during the midsixties that export promotion explicitly entered the policy frame (Panchmukhi, 1991). However, export orientation for the agricultural sector was not fostered effectively due to various reasons. Firstly, agricultural exports were perceived as a residual and it was generally felt that agricultural production should primarily meet the domestic demand of the Indian population. However, exports of plantation crops, such as tea and coffee, and cash crops, such as tobacco or spices, and later on fish and fish products has been an exception and important source of foreign exchange earnings. For these commodities, the open trade regime has continued from the beginning. Under the new trade policy initiated in 1991, three major changes have been effected in agricultural trade. Firstly, the canalization of agricultural trade has been almost abandoned and the government does not determine now trade value or nature of the exports or imports, except for the export of onion and import of cereals, pulses and edible oils. Secondly, quantitative restrictions on agricultural trade flows have been dismantled completely w.e.f April 1, 2001. Thirdly, reductions in tariffs have been announced. The fish and fish products are exported under the open general license (OGL). As stated earlier, MPEDA is looking after the export promotion and regulation of marine products. The Export Inspection Agency was established in 1969 to ensure quality control of products for the export market. Standards for bacteria, virus, heavy metal contamination etc. are evolved in co-operation with MPEDA and the Indian Institute of Packaging.

The provisions of the World Trade Organization (WTO) include Trade Related Intellectual Property Rights (TRIPS) and the imposition of patent regime, trade related investment measures, reduction in domestic and export subsidies, market access and provision of sanitary and phyto-sanitary measures, and removal of Quantitative Restrictions (QRs) on import. Under TRIPS, the signatories of the General Agreement on Tariffs and Trade (GATT) are obliged to adopt a patent system for microorganisms. However, the patenting higher animal life forms was left unresolved, with countries having the option to use or not to use patents to protect such intellectual property rights. Under Trade Related Investment Measures (TRIMS), countries would have to treat foreign investors at par with the domestic ones. It allows foreign fishing fleets the same access to domestic waters that local people enjoy. This provision has made a deep impact on the global fishing industry, the conservation of fisheries resources and the communities depending upon them.

As per the WTO agreement, developed countries would reduce subsidies and tariffs. Therefore, better overseas market would become available for Indian fish products. It is worth mentioning that the requirement of subsidies reduction under WTO is not applicable to India. Under the provisions of the SPS agreement, all member countries have the right to take sanitary and phyto-sanitary measures necessary for the protection of animal health or life. To challenge any possible threats under SPS measures, the Indian processing industry has to improve quality parameters be accepting Hazard standards. These SPS measures provide protection to Indian industry form the policies of discrimination of developed nations and from disguised restrictions imposed on Indian fisheries exports.

The remove of Quantitative Restrictions (QRs) on the last 714 items by India on April 1, 2001, has developed an atmosphere of anxiety over the entire spectrum of Indian trade, including the fisheries sector. Fish and fish products figured prominently in the list of items on which QR was removed. Fish and fish products figured prominently (60 items) in the list of 714 items on which QR has been removed. This has raised an alarm in the fisheries sector which provides employment to 6 million people directly and indirectly. Perceptions vary among different clientele like fishermen, exporters and consumers. Apprehensions of the fish farmers include crash in prices under large scale import. The exporters are expected to benefit with a regular supply of raw material, which would help processing plants in capacity utilization even during the lean season. The consumers will, by and large be benefited by the import of foreign fish products. At the moment the different stakeholders have conflicting opinions on the removal of QRs. India being a developing country should judiciously use the tariff provision to protect the domestic industry. In the changing global economic scenario it is not possible to prevent imports totally. The only probable solution now is to focus on the changed scenario and gear up to utilize it for benefit. Moreover, India is quite competitive in fisheries export particularly in shrimp (Kumar *et. al.* 2002) and the WTO compulsions can be converted into opportunities by vigorously pursuing the export of fish and fish products particularly the unexplored brackish water segment. This would be in the interest of the coastal fisherfolk also.

Potential of Fisheries Development

India has abundant resources for fish production. In the case of marine fisheries, India has 0.51 million sq. km of continental shelf area and a 8041 km long coastline (Table 7). Based on the available scientific information, exploratory surveys, experimental fishing and other data available, the potential harvestable yield of the Indian economic exclusive zone (EEZ) has been estimated at

3.9 million tonnes. The highest potential (2.3 million tonnes) is in the waters up to 50 meters depts, whereas the potential in water between 50-200 metres depth is 1.3 million tonnes and beyond that only 0.3 million tonnes. The density of fish is highest about 11 tonnes per square km in coastal areas (0-50 m). In waters beyond 50 m, it is less than 1 tonne per sq km. At the moment, most of the catches are from waters less than 50 m deep. There does not seem to be any scope for further expansion in these areas, as many of them have already been over exploited. According to fisheries experts emphasis should be laid on deeper waters and on species which have not been exploited so far like tuna and even anchovies for fishmeal production. However, expansion of fisheries activities in deep waters is highly capital intensive and the local investors have neither the will nor resources to take it up.

Table 7 : Fishery Resources of India

Resource	*Unit*	*Quantity*
Marine		
Continental shelf	'000'sq. km	506
Landing centres	No	2333
Cast line	km	8041
Inland		
Rivers and canals	Million km	0.17
Reservoirs	Million ha	2.05
Tanks and ponds	Million ha	2.86
Beels, oxbow and derelict water	Million ha	0.79
Brackishwater	Million ha	1.42

Source : Hand Book on Fisheries Statistics (2000), Ministry of Agriculture, Government of India.

The inland fisheries resources include a length of 0.17 million kilometres rivers and canals, 2.05 million ha of reservoir area, 2.86 million ha area of ponds and tanks and 0.8 million ha of beels, oxbow and derelict water. The brackish water area for fish production is estimated to be 1.42 million hectare. The inland resources however, have not been tapped fully. Only about 16 per cent of the fresh water area and 10 per cent of the brackish water area are being utilized for fish culture. In the inland sector, the resource potential has been estimated to be 4.5 million tonnes, which takes into account production from both capture and culture fisheries. The productivity however, is low. The average productivity of freshwater aquaculture in 1998-99 was about 2.2 tonnes per ha, while the potential to raise yield was up to 10 tonnes per ha. The realized average productivity of brackish water aquaculture in 1998-99 was 472 kg/ha as against the potential of about 10 tonnes per ha (Krishnan *et. al.* 2000).

It seems that India's marine fisheries production has reached a plateau and at best, only by marginal increase is predicted in the near future. However, inland fish production has exhibited rapid growth and for all future demand, we have to rely on the inland sector, particularly on aquaculture.

SUPPORT SYSTEMS FOR FISHERIES DEVELOPMENT

R&D in Fisheries Sector

India has a huge network of institutions to carry out R&D in fisheries sector. These include: (*i*) Indian Council of Agricultural Research (ICAR) systems; (*ii*) Ministry of Agriculture; (*iii*) Ministry of Commerce; (*iv*) Ministry of Food Processing Industries; (*v*) Council of Scientific and Industrial Research (CSIR); and (*vi*) State Agricultural Universities. Many other organization/agencies also support/conduct R&D in fisheries; these include the Department of Ocean Development (DOD); Department of Science and Technology (DST); Department of Biotechnology (DBT); University Grants Commission (UGC); India Institutes of Technology (IIT); India Institutes of Managements (IIMs) voluntary agencies/private industries. However, the multiplicity of institutes requires a high degree of co-ordination to avoid duplication and diffusion of efforts and paucity of funds. There are overlapping mandates between institutions even within the same system.

Credit Support System

The fisheries sector particularly the aquaculture is on a steady growth path. The fishermen, in general, are poor and practise traditional farming for want of financial resources. The need for credit support for facing the emerging market forces and harnessing the benefits of technological developments has been realized and some measures have been evolved to enhance the flow of credit to the fisheries sector. The National Bank for Agriculture and Rural Development (NABAD), as a refinance agency for commercial banks, co-operative banks and regional rural banks has been the measure facilitator of credit to the fisheries sector. In view of the brackish water aqua boom in the early 1990s, many financial institutions like Industrial Finance Corporation of India (IFCI), Industrial Development Bank of India (IDBI), Shipping Credit and Investment Company of India (SCICI), State Finance Corporations (SFCs) and National Co-operative Development Corporation (NCDC) also entered this sector to lend credit. Credit support from financial institutes is available for almost all the activities of fisheries and for creation of infrastructure. The credit disbursements for the fisheries sector witnessed an increasing trend till 1995-96 but thereafter there has been a decline in disbursements as well as in the number of sanctioned schemes. This could be partly due to the interim order of the Supreme Court of India banning shrimp farming in Coastal Regulation Zone, slow progress in mariculture, systematic changes in refinance policies and the environmental/ disease problems faced by shrimp farming.

Other Infrastructure

The other infrastructure and support system include more than 376 freezing plants, 13 canning plants, 149 ice plants, 15 fish meal plants, 903 shrimp peeling plants, 451 cold storage units and 3 chitison plants.

Training, Extension and Transfer of Technology

Fisheries development is a state subject in India. However, the centre promotes fisheries development through state level programme planning and implementation units. At the national level, the Fisheries Division of the Ministry of agriculture is the planning and policy making body for fisheries development. The training programmes is fisheries are mainly dealt with by the Fish Farmers' Development Agency and Brackishwater Fish Framers' Development Agency. These also provide packages of assistance for popularizing aquaculture technologies. The research institutes and SAUs have also been taking training and extension work as part of their curriculum. The Department of Rural Development promotes fisheries through the Integrated Rural Development Programme. In the states, departments of fisheries have been established at the district level to take care of the fisheries development including training and extension.

The first-line extension system of the ICAR, consisting of demonstration programmes, Lab-to-Land Programme, Operation Research Projects, Krishi Vigyan Kendras and Trainers' Training Centres play an important role in training and extension of fishery development. Technology assessment and refinement through Institution-Village-Linkage programme (IVLP) of the ICAR is a technology integration process fitting the requirements of the farmers suitably in a given farming situation.

Constraints in Fisheries Development

Further enhancement of marine fish production requires diversification of fishing activities not only in the off-shore oceanic regime but also in deep sea fishing which is capital intensive and risk prone. There have already been strong protests in India against foreign equity participation in deep sea fishing and the government had to rescind its Deep Sea Fishing Policy in March 1997. Utilization of marine resources by catch and fishing for unconventional fish species may not be economically viable initially. The conservation of resources and genetic diversity in EEZ would further slow down effort towards higher production from the marine sector.

The story with the inland sector is similar; aquaculture production could be a base but it is beset with varied uncertainties. The aquaculture, particularly intensive and semi-intensive, which has the potential of gaining quantum but it may face a major fish meal trap. Another important intermediary input for aquaculture is seed of culturable fish species. The country is already facing problems with regard to scarcity of breeder stock in the shrimp sector. For diversified aquaculture, various compatible fish species have to be brought under aquaculture operation.

In the case of coastal aquaculture development in India, some social and political conflicts developed at several places. These conflicts were caused largely by disease outbreaks in shrimp farms, environmental pollution due to overcrowding of farms, salination of drinking water wells, conversion of paddy fields into shrimp farms, causing displacement of labour etc. These episodes have already had their effects.

Land and water resources in the country are not available exclusively for fisheries; there is excessive pressure on the resources from several other sectors. Moreover, programs for fisheries management are split between the national and state governments which differ in their policies and approaches. The national policies in India have largely been export oriented, supporting relatively large scale fisheries for shrimp. But for many states, the primary concern is the welfare of local small-scale fishermen.

For the development of fishery and aquaculture, such constraints as well as social, legal and political implications have to be taken into account and innovative strategies and policies have tobe initiated for a balanced and sustainable growth. Now the production has been uncreated 6.7 mmt, 3.4 inland and 3.3 from marine side.

SUBSIDY FOR TUNA FISHING BOATS TO PROMOTE EXPORTS OF 'SASHIMI'

The Marine Products Export Development Authority (MPEDA) is set to introduce a subsidy assistance scheme to promote exports of high value 'Sashimi'–graded chilled tuna fish.

"It is estimated that 2.13 lakh tonnes of oceanic tuna fish exists in the coastal waters that constitute the Exclusive Economic Zone (EEZ). Of this only an average 10,000 tonnes are caught every year owing to inadequate number of tuna long linear."

The subsidy scheme covers the conversion of mechanised vessels (up to 20 metres in length) and deep sea fishing vessels (more than 20 metres) into tuna long liners, thereby equipping them to effectively catch oceanic tuna resources from the EEZ.

As part of the conversion process, decks of the vessels, covered under the scheme, would be modified to accommodate instruments such as 'spooler reel,' 'line hauler setter' : and 'fish hold' compartments. On the salient features of the scheme, the sources said the MPEDA would provide a maximum subsidy of Rs. 7.50 lakh or 50 per cent of the capital cost, whichever is less, to the owners who convert their mechanised vessels into tuna long liners.

In the case of deep sea fishing boats, a maximum subsidy of Rs. 15 lakh or 50 per cent of the capital cost, whichever is less, would be provided to be beneficiaries.

The vessel should be registered with the MPEDA and the age of the vessel should not be more than 12 years. If it exceeds 12 years, the owner should procure a certificate from the Maritime Mercantile Department that the economic life of the vessel was at least eight years from the time of submission of application for the scheme.

It is due to this that MPEDA has suggested an autonomous management system including all the stakeholders.

It was pointed out that through India had the state of the art processing plants, deterioration of the quality of exports happens at the lower ends of the chain *viz* fish landing centers and harbors where quality management is not properly taken care off. The landing centers, which are managed by state govt. or port trust, suffer from infrastructure deficiencies and lack of ideal management. At least 3 to 4 fishing harbors in each state is to be upgraded to the autonomous system.

The EU FVO Mission that visited India recently has pointed out the need of improving the hygienic and sanitary conditions in the fishing vessels. The EIC has proposed a draft notification specifying certain quality requirements specifying certain quality restrictions to be met by fishing vessels. Out of the 50000 mechanized fishing vessels in operation only about 15000 are regulated by MPEDA and this too has been done as means for availing subsidies and to facilitate bank loans.

In fact fisheries sector in India is condemned to perish by man made disasters. It is now like a fish put out of water gasping for oxygen.

In spite of all these adverse circumstances fishermen are doing their best to contribute to the national exchequer. Although they are humiliated by the dwindling catches, eroding returns for the produce, spiraling fuel prices, Govt. : apathy, they contribute substantially to the Rs. 7000 crores worth foreign exchange basket. Even as the country is earning millions of dollars out of the hard work of fishermen, their lives are enmeshed in a web of hostile policies of the central Government and the conflicting policies of the states. Seafood export has always shown an upward trend (Only in Rupee terms) and this was considered as the better performance of the fishing sector. In fact how much sweat the fishermen had to shed and how much they actually benefited for their toils are not brought to limelight.

Nature has blessed our nation with immense volume of water resources, both inland and marine. India has the geographical advantage of having more than 8000 kms of coastline including that of Andamans and Lakshadweep and 2.2 million sq. kms. of seas around it. Although the nation is bestowed with the vital fishery resources which are highly cherished by the nations and people all over the world, the Indian fishermen are still below the poverty line next to the Adivasis. If the prices of diesel continue to rise up again and again as experienced in recent times no chance is left for them to ever cross over the BPL in the near future. May be nothing can be done to bring down the fuel prices but something can be done to ensure better prices for seafood. Fishermen rightly deserve a better price for their produce and adequate returns for their risky hardwork at sea. But who cares!

At present concerned authorities are blaming the international crude prices to justify the price hike in domestic market. In fact the spurt in crude prices is a blessing in disguise for Governments and oil sector companies. By way of royalty, excise, customs duty, Cess, corporate tax and dividend, state sales tax, Octroi, both central and state governments together get more than half of the entire turnover of the oil producing and marketing companies. While all others involved in the "oil trade" make too much gains, the fuel consumers alone have to bear the brunt of the price rise. And fishermen are the worst hit among these consumers.

Indian fisheries sector, in view of its potential contribution to national income, nutritional security, employment opportunities, social objectives and export earnings plays an important role in the socio-economic development of the country. During 2001-02, fisheries sector contributed

Rs. 22,223 crores to the total Gross Domestic Product (GDP), forming 1.4 per cent of the total. Fisheries sector contributes 4.3 per cent to the agricultural GDP and export earnings are presently valued at over Rs. 6,790 crores from a volume of 4.6 lakh tonnes (MPEDA, 2003). Marine products form an important group of primary commodity exported from India accounting for about four per cent of the total export earnings. The important marine products exported from India are frozen shrimp, frozen lobster, frozen fish, frozen squid and frozen cuttlefish. The five major markets contributing to the Indian exports include Japan, USA, European Union, South East Asia and Middle East. In the case of agriculture, including fisheries, India had followed protective trade policies in the past. However, in order to make trade policies consistent with the new economic policies and the provisions of WTO, a number of fisheries products were moved to the Special Import License (SIL). In the recently announced. Exim policy (2002), the import of fisheries commodities was further liberalised and almost all commodities were moved to the list of freely importable commodities. Further, Quantitative Restrictions (*QR*) on agricultural and allied sectors trade was completely removed for the last 714 items with effect from April 2001. More than 120 items of fish and fish products have been affected by these regulations. After complete dismantling of QRs, tariff rates were perceived as the only instrument for restricting imports.

By this one can :

(*a*) estimate the growth of marine products export from India.

(*b*) analyse the instability of marine products export from India.

(*c*) suggest suitable policy options for augmented export earnings for the future in the wake of WTO regime.

Analysis of Growth

Growth rate was used to find out the trend in the export of major marine products during pre and post liberalisation periods. The growth in exports by volume and value and unit value realised from exports, were analysed by using the exponential growth function of the form,

$$Y = \text{abt et} \qquad (1)$$

where, Y = dependent variable for which growth rate was estimated.

a = Intercept $\quad$ b = Regression co-efficient

t = Time variable $\quad$ e = Error term

The compound growth rate was obtained for the logarithmic form of the equation (1) can be written as.

$$\text{In} \quad Y = \text{Ln a} + \text{t Ln b} + e' \qquad (2)$$

Where $\quad e' = \text{In et}$

Then, the compound growth rate (r) was computed using the relationship

$$r = (\text{Anti Ln of b} - 1) \times 100 \qquad (3)$$

The compound growth rates were tested for their significance by the '*t*' statistics given by

$$t = \text{rISE (r)} \qquad (4)$$

(Where, SEe r) = [100 bX SE (Ln b)]/ILn e .... (5)

Export Instability

This analysis was used to find out the fluctuations in export of major marine products during pre and post liberalisation periods. Coppock's instability index was used to estimate the variation in the export of major marine products which, algebraically is expressed in the following estimable form :

$$V - Log = \frac{\sum(\log Xt + 1 - m)^2}{Xt} N - 1 \tag{6}$$

The instability index = (antilog J V log – 1) × 100 (7)

where,

Xt = Value of exports in year t or volume of exports in year t N = Number of years

m = The arithmetic jean of the difference between the logs of Xt and Xt + 1 etc.

V log = Logarithmic variance of the series

Analysis of Growth

Growth patterns in the export of marine products from India during the pre (Liberalisation (1979–1990) and post liberalisation (1991–2002) period in both quantity and values in furnished in Table 8. The commodity wise export of marine products indicated that the post liberalisation period performed better than the pre liberalisation period with respect to quantity and value in rupee terms with compound growth rate of 8.29, and 8.23 per cent per annum respectively.

Table 8 : Export growth of Indian marine products (Commodity wise)

Year	*Pre-liberalisation (1979–1990)*	*Post-liberalisation (1991–2002)*
Total		
Quantity (tonnes)	3.49** (1.53)	8.29** (2.763)
Value (Rs.)	3.33* (1.50)	8.23** (2.58)
Unit Value (Rs.)	– 0.15 (0.10)	– 6.16 (0.04)
Frozen Shrimp		
Quantity (tonnes)	0.83 (0.80)**	5.35** (2.67)
Value (Rs.)	1.95 (0.89)	7.93** (2.36)
Unit Value (Rs.)	11.11 (0.68)	2.45** (1.40)
Frozen Lobster		
Quantity (tonnes)	12.88** (2.94)	2.54 (0.64)
Value (Rs.)	16.05** (2.64)	4.97* (0.83)
Unit Value (Rs.)	2.80 (0.83)	2.36* (0.89)
Frozen Squid		
Quantity (tonnes)	27.26** (2.24)	4.54* (1.02)
Value (Rs.)	26.64** (2.04)	7.37** (1.02)
Unit Value (Rs.)	–0.48 (0.15	2.69* (0.90)
Frozen Cuttlefish		
Quantity (tonnes)	26.03** (3.62)	7.62** (1.58)
Value (Rs.)	126.64** (2.04)	7.04* (1.05)
Unit Value (Rs.)	0.48 (0.06)	0.53 (0.24)

Contd...

1	2	3
Fresh and Frozen Fish		
Quantity (tonnes)	3.49 (0.41)	11.62** (2.29)
Value (Rs.)	8.18** (1.35)	9.59** (1.98)
Unit Value (Rs.)	4.52* (1.14)	– 1.81* (1.66)
Others		
Quantity (tonnes)	– 5.45* (– 0.90)	113.59** (1.80)
Value (Rs.)	– 6.23* (–1.03)	127.44** (1.13)
Unit Value (Rs.)	– 0.83 (0.11)	12.19** (0.77)

Figures in parenthesis indicate the standard errors of the estimates

** one percent level of significance

* five percent level of significance

But it was found that the unit value realization registered a negative growth rate of 6.16 per cent during the post liberalisation period, as compared with the pre liberalisation period, (–0.15 per cent). The export basket during the post liberalization was characterized by the dominance of low value fresh and frozen fishes compared to the high valued species (shrimps and lobster), which resulted in the decreased growth of the unit value realization.

Frozen shrimp, the largest value component registered a 5.35 per cent growth in quantity, 7.93 per cent in value terms and 2.45 per cent in unit value. The important reasons for this significant growth can be attributed to increased landings, higher price realisation and widened markets.

Frozen lobster also registered a positive growth during both the periods, nevertheless the export growth during post liberalisation period lagged behind the pre-liberalisation period mainly due to reduced lobster landings and high domestic prices.

Cephalopods, comprising frozen squid and frozen cuttlefish, performed better during the pre liberalisation period when compared with the post liberalisation period. The reasons for the reduced export growth during post liberalisation periods could be attributed to the higher domestic consumption, reduced landings, EU ban and subsequent rejections on quality grounds due to antibiotic contaminants, microbial and bacterial residues.

Fresh and frozen fish export-the largest quantity component registered higher growth rates during the post liberalisation period on account of the enormous demand by the South East Asian countries like China, Singapore, Malaysia, Hong Kong, Vietnam and Thailand etc. However the decline in unit value posses a threat as the fresh and fish export to the South East Asian countries mainly comprise of low value fin fishes which are being reprocessed with value addition and re-exported.

The most striking feature during the post liberalisation period is the emergence of minor marine products like chilled, dried and live items with the highest value addition generating a commodity diversification and wider export basket.

In general, the growth as a result of increased volume of export can be ascribed to the manifold increase in the marine fish production consequent to the introduction of the mechanisation and deep sea fishing. Also the gradual shift in the preservation systems from dried and salted fish from to frozen items to value added form. Structural changes in the products composition and directional pattern, growth in the number of exporters, growth in the number of ports as the focal points and development in infrastructure, continuous improvement in quality standard and manifold promotional measures by MPEDA and other agencies had resulted in the increased growth.

The analysis of the export growth rate indicated that the export market in United States showed impressive growth rate in terms of quantity, value and unit value during the post liberalisation period with 8.17, 14.79 and 14.73 per cent per annum respectively (Table 9). Even the terrorist attack on September 11, 2001 and subsequent bio-terrorism act did not hamper the export and, incidentally it was during the post liberalisation period (during April-October 2002) that United States emerged as the top most buyer of Indian marine products relegating Japan to the second position after a gap of three decades.

Table 9 : Export growth of Indian marine products (Market wise)

Year	*Pre-liberalisation (1979–1990)*	*Post-liberalisation (1991–2002)*
Total		
Quantity (tonnes)	13.49** (1.53)	8.29** (2.763)
Value (Rs.)	13.33* (1.50)	8.23** (2.58)
Unit Value (Rs.)	- 0.15 (- 0.10)	-6.16 (0.04)
Japan		
Quantity (tonnes)	-0.06 (-0.06)	3.73** (1.00)
Value (Rs.)	10.91** (0.45)	5.03* (1.02)
Unit Value (Rs.)	0.97 (0.52)	1.25 (0.59)
USA		
Quantity (tonnes)	2.62 (0.75)	8.17** (3.57)
Value (Rs.)	3.36* (0.77)	14.79** (3.49)
Unit Value (Rs.)	0.72 (0.51)	14.73** (3.48)
European Union		
Quantity (tonnes)	13.66** (1.61)	-0.66 (-0.17)
Value (Rs.)	11.26** (1.53)	0.97 (0.23)
Unit Value (Rs.)	- 2.11* (- 1.08)	7.35** (2.06)Y
South East Asia		
Quantity (tonnes)	2.14** (0.42)	13.86** (2.04)
Value (Rs.)	4.23 (1.46)	12.54** (1.38)
Unit Value (Rs.)	0.48* (0.31)	-1.51 (0.31)
Middle East		
Quantity (tonnes)	3.42* (1.36)	5.19 (0.85)
Value (Rs.)	2.13 (0.73)	7.84 (0.82)
Unit Value (Rs.)	1.24* (1.19)	12.51 (0.45)
Others		
Quantity (tonnes)	12.84 (3.5)	18.18** (1.52)
Value (Rs.)	16.0** (1.13)	24.39** (1.58)
Unit Value (Rs.)	13.14 (0.63)	5.26** (1.55)

Figures in parenthesis indicate the standard errors of the estimates

** one percent level of significance* five percent level of significance

South East Asia also registered higher growth rated during the post liberalisation period with 13.86 and 12.54 per cent in quantity and value eventhough the growth in unit value realisation reduced by 1.15 per cent. South East Asia happens to be the largest importer in terms of quantity, particularly fresh and frozen fish like pomfret, ribbon fishes, seer, mackerel, tuna and others.

The export to European Union revealed that there had been appreciable growth in value and unit value realisation even though the quantity showed a downward trend. The higher export value, unit value realisation even at a decline in quantity indicated the quality consciousness and higher value for the premium products. It is of utmost importance that European Union ban and rejection of consignments on quality grounds need to be analysed carefully to avoid such recurrance in future.

Japan, as ever continued to be the major importer of Indian marine products. Amidst decline in the export during 2001 due to the depreciation of Yen, the share registered by the Japanese market still shows significant growth in quantity, value and unit value with 3.73, 5.03 and 1.25 per cent respectively. The dependency on Japanese markets as being the highest importer of Indian marine products remains undeterred.

The post liberalisation period scenario of marine products export is heartening in the context of geographic diversification as more number of export markets were destined thereby checking the volatility of exports. In general the analysis of the export growth reveals that there exists considerable improvement on the number of commodities and markets for the Indian marine products exports.

Instability in Exports

The Coppocks instability index results indicated that the degree of instability was more pronounced during the post liberalisation period with 23.4, 27.94 and 16.82 respectively in terms of quantity, value and unit value even though more growth was associated (Table 10).

Frozen shrimp registered higher export quantity variation (11.60 per cent) during post liberalisation period when compared with pre-liberalisation period (7.28 per cent) suggesting that there exist severe competition among the different exporters and the exports are very much responsive to the prices. In addition, the essentiality of a buyers market and lesser number of importers paved the way for higher instability. Contrary to the instability behaviour of frozen shrimp, there existed a lower degree of export quantity variation with respect to frozen squid, frozen cuttlefish and fresh and frozen fish. (Table 10) The widening of European Union domain and trading with South East Asian Countries seemed go generate lesser instability. The category "Others", a new entrant in the export basket exhibited the highest degree of instability cautioning the need for regularising the commodities for export.

Table 10 : Instability Indices of Indian Marine Products Export (Commodity wise)

Year	*Pre-liberalisation (1979–1990)*	*Post-liberalisation (1991–2002)*
Total		
Quantity (tonnes)	12.57	23.40
Value (Rs.)	16.28	27.94
Unit Value (Rs.)	10.23	16.29
Frozen Shrimp		
Quantity (tonnes)	17.28	11.60
Value (Rs.)	17.51	23.67
Unit Value (Rs.)	12.11	15.81

Contd...

1	2	3
Frozen Lobster		
Quantity (tonnes)	128.80	14.18
Value (Rs.)	39.90	40.15
Unit Value (Rs.)	22.54	27.93
Frozen Squid		
Quantity (tonnes)	63.24	19.30
Value (Rs.)	63.00	74.30
Unit Value (Rs.)	127.80	69.64
Frozen Cuttlefish		
Quantity (tonnes)	37.39	15.40
Value (Rs.)	62.96	33.56
Unit Value (Rs.)	58.28	21.77
Fresh and Frozen Fish		
Quantity (tonnes)	61.19	31.64
Value (Rs.)	37.15	35.30
Unit Value (Rs.)	27.57	16.91
Others		
Quantity (tonnes)	42.17	48.33
Value (Rs.)	157.09	235.57
Unit Value (Rs.)	77.60	154.61

The instability indices of major marine products were analysed using the Coppock's Instability Index and the results are furnished in Table 10. The results indicated that the degree of instability was more pronounced during the post liberalisation period with 23.4, 27.94 and 16.82 respectively in terms of quantity, value and unit value even though more growth was associated.

Thus it could be noted that the post liberalisation period generated a higher degree of instability for frozen shrimp, frozen lobster and other whereas a lesser, degree of instability was noticed for frozen squid, frozen cuttlefish and fresh and frozen fish. The analysis suggested the need for diversification of commodities, which would reduce the degree of variability.

Table 11 : Instability Indices of Indian Marine Products Export (Market Wise)

Year	*Pre-liberalisation (1979–1990)*	*Post-liberalisation (1991–2002)*
Total		
Quantity (tonnes)	12.57	23.40
Value (Rs.)	16.28	17.75
Unit Value (Rs.)	10.23	16.29
Japan		
Quantity (tonnes)	7.43	18.04
Value (Rs.)	16.43	24.02
Unit Value (Rs.)	13.73	11.06
USA		
Quantity (tonnes)	27.81	17.96
Value (Rs.)	36.84	29.73
Unit Value (Rs.)	9.79	13.78

Contd...

1	2	3
European Union		
Quantity (tonnes)	36.30	21.09
Value (Rs.)	34.84	22.22
Unit Value (Rs.)	12.79	5.89
South East Asia		
Quantity (tonnes)	18.53	33.35
Value (Rs.)	28.21	46.71
Unit Value (Rs.)	1.056	22.27
Middle East		
Quantity (tonnes)	24.32	36.93
Value (Rs.)	36.52	98.61
Unit Value (Rs.)	17.52	58.59
Others		
Quantity (tonnes)	54.38	60.47
Value (Rs.)	51.02	75.51
Unit Value (Rs.)	32.90	20.30

Analysis of marketwise instability indices indicated that the post liberalisation period produced higher degree of instability as compared to the pre-liberalisation period (Table 11). Japan, a more stable market during the pre-liberalisation period increased its volatility in the post liberalisation period as indicated by the instability index in quantity and value with a marginal decrease in the instability index of unit value. USA on the other hand became less volatile during the post liberalisation period for quantity and value except for the unit value. The major markets that registered considerable reduction in the instability of Indian export during post liberalisation period was that of European Union which had a lesser instability in all the export parameters. This has been possible due to the widening of the European Union and preferential export with respect to quantity and premium products. Though South East Asia and Middle East markets generated a significant growth in export there existed an alarming rate of volatility.

Summary and Conclusions

The export growth estimates for the commodities indicated that there was significant growth in the export parameters for the major marine products with the emergence of new commodities for export, although there was decrease in unit value realization.

The post liberalisation period scenario of marine products export registered considerable geographic diversification with the emergence of export markets like South East Asia and Middle East as compared with the pre liberalisation period.

The post liberalisation period generated a higher degree of instability for frozen shrimp, frozen lobster and other whereas a lesser degree of instability was noticed for froen squid, frozen cuttlefish and fresh and frozen fish. With regard to the markets, Japan, SEA and ME exhibited higher degree of instability when compared with the pre liberalisation period. In general the results indicated that the post liberalisation period produced a higher degree of instability as compared to the pre liberalisation period.

The study suggests the need for the diversification in the geographic concentration and commodity concentration, as these are the important parameters determining the volatility and export instability. The results indicate that there are welcome signs for a quantum jump in fisheries trade. The absence of proper infrastructure to maintain quality standards of marine products is likely to affect Indian exports to a large extent because of rejection of consignments on quality grounds. In order to curb these externalities, the acceptance of the Hazard Analysis and Critical Control Points (HACCP) guidelines to make the Indian seafood to conform to the International quality standards is necessary. Further product diversification tailored to the varied consumption pattern would ensure competitiveness of Indian exports in the world market in the wake of the new WTO order.

SCHEME FOR DEVELOPMENT/STRENGTHENING OF AGRICULTURAL MARKETING INFRASTRUCTURE, GRADING AND STANDARDISATION

Highlights

- **Reform Linked Investment Scheme :** To encourage rapid development of infrastructure projects in agriculture and allied sectors including dairy, meat, fisheries and minor forest produce.
- **Investment subsidy :** 25% of the capital cost up to Rs. 50 lakh on each project providing 'Direct' service delivery to producers farming community in Post-Harvest Management/ Marketing of their produce. In case of NE States, hilly and tribal areas and to SC/ST entrepreneurs and their cooperatives investment subsidy shall be 33.33% **of the capital cost up to Rs. 60 lakh.**
- *No upper ceiling on subsidy in respect of infrastructure projects of State Agencies.*

Conditions

- Applicable only in such States/Union Territories, which undertake reforms in APMC Act to allow 'Direct Marketing' and 'Contract Farming' and to permit agricultural produce markets in private and cooperative sectors.
- Promoter's contribution in project cost to be decided by financing bank with minimum bank loan of 50% in general cases and 46.67% in hilly areas, etc.

Illustrative List of Infrastructure Projects

- Market user common facilities like market yards, platforms for loading, assembling and auctioning of the produce, weighing and mechanical handling equipments etc.
- Functional Infrastructure for assembling, grading, standardization and quality certification, labeling, packaging, value addition facilities (without changing the product form).
- Infrastructure for Direct Marketing from producers to consumer/processing units/bulk buyers etc.
- Infrastructure for E-trading, market extension and market oriented production planning.
- Mobile infrastructure for post harvest operations *viz.*, grading, packaging, quality testing etc., (excluding transport equipment).

1. Background

This scheme has been formulated to develop marketing infrastructure in the country to cater to the post-harvest requirement of production and marketable surplus of various farm product. An Expert Committee set up by the Ministry of Agriculture has estimated that an investment requirement of Rs. 11,172 crore in next 10 years would be necessary for infrastructure development in agricultural marketing. A major portion of this investment is expected to come from private sector, for which an appropriate regulatory and policy environment is necessary. The Department has had several rounds of discussions with the States on restrictive provisions of State Act dealing with agricultural marketing

(APMC Act) and the need to modify and create lawful space for the private sector in the market development and contract farming. This scheme is reform linked and assistance for development of infrastructure projects will be provided in those States/Union Territories which permit setting up of agricultural markets in private and cooperative sectors and allow direct marketing and contract farming.

2. Objectives

The main objectives of the Scheme are :

(*i*) To provide additional agricultural marketing infrastructure to cope up with the large expected marketable surpluses of agricultural and allied commodities including dairy, poultry, fishery, livestock and minor forest produce.

(*ii*) To promote competitive alternative agricultural marketing infrastructure by inducement of private and cooperative sector investments that sustain incentives for quality and enhanced productivity thereby improving farmers' income.

(*iii*) To strengthen existing agricultural marketing infrastructure to enhance efficiency.

(*iv*) To promote direct marketing so as to increase market efficiency through reduction in intermediaries and handling channels thus enhancing farmers' income.

(*v*) To provide infrastructure facilities for grading, standardization and quality certification of agricultural produce so as to ensure price to the farmers commensurate with the quality of the produce.

(*vi*) To promote grading standardization and quality certification system for giving a major thrust for promotion of pledge financing and marketing credit, introduction of negotiable warehousing receipt system and promotion of forward and future markets so as to stabilize market system and increase farmers' income.

(*vii*) To promote direct integration of processing units with producers.

(*viii*) To create general awareness and provide education and training to farmers, entrepreneurs and market functionaries on agricultural marketing including grading, standardization and quality certification.

3. Salient Features of the Scheme

Scheme Linked to Reforms

(*i*) The scheme will be implemented in those States which amend the APMC Act, wherever required, to allow direct marketing and contract farming and to permit setting up of markets in private and cooperative sectors.

(*ii*) Credit linked back-ended subsidy shall be provided on the capital cost of general or commodity specific infrastructure for marketing of agricultural commodities and for strengthening and modernization of existing agricultural markets, wholesale, rural periodic or in tribal areas.

Marketing Infrastructure

(*iii*) 'Marketing Infrastructure' for purpose of the scheme may comprise of any of the following:

(*a*) Functional infrastructure for collection/assembling, drying, cleaning, grading, standardization, SPS (Sanitary & Phytosanitary) measures and quality certification, labeling, packaging, ripening chambers, retailing and wholesaling, value addition facilities (without changing the product form) etc. Transportation facility will not be covered under the scheme;

(*b*) Market user common facilities in the project area like shops/offices, platforms for loading/unloading/assembling and auctioning of the produce, parking sheds, internal roads, garbage disposal arrangements, boundary walls, drinking water, sanitation arrangements, weighing & mechanical handling equipments, etc.;

(*c*) Infrastructure for Direct marketing of agricultural commodities from producers to consumers/processing units/bulk buyers, etc.;

(*d*) Infrastructure for supply of production inputs and need-based services to the farmers;

(*e*) Infrastructure (equipment, hardware, gadgets, etc) for E-trading, market intelligence, extension and market oriented production planning; and

(*f*) Mobile infrastructure for post-harvest operations (excluding transport equipment) will be eligible for assistance under the scheme.

Eligible Persons

(*iv*) The assistance will be available to individuals, Group of farmers/growers/consumes, Partnership/Proprietary firms, Non-Governments Organizations (NGOs), Self Help Groups (SHGs), Companies, Corporations, Cooperatives, Cooperative Marketing Federations, Local Bodies, Agricultural Produce Market Committees & Marketing Boards in the entire country.

(*v*) Bank assisted projects of State agencies, including projects refinanced/co-financed by National Bank for Agriculture and Rural Development (NABARD) for strengthening/ modernization of existing marketing infrastructure would also be eligible for assistance under the scheme.

Credit Linked Assistance

(*ix*) Assistance under the scheme would be credit linked and subject to sanction of the infrastructure project by Commercial/Cooperative/Regional Rural Banks based on economic viability and commercial considerations.

(*x*) Assistance under the scheme shall be available on capital cost of the project only. Banks/National Cooperative Development Corporation (NCDC) will, however, be free to finance other activities/working capital requirement to meet various requirements of the farmers/entrepreneurs.

Subsidy

(*xi*) **Rate of subsidy** shall be 25% of the capital cost of the project. In case of North Eastern States, hilly and tribal areas and to entrepreneurs belonging to Scheduled Caste (SC)/ Scheduled Tribe (ST) and their cooperatives, the rate of subsidy shall be 33.33% of the capital cost of the project.

(*xii*) **Maximum amount of subsidy** shall be restricted to Rs. 50 lakh for each project. In the case of North Eastern States, hilly and tribal areas and to entrepreneurs belonging to SC/ST and their cooperatives, maximum amount of subsidy shall be Rs. 60 lakh for each project.

(*xiii*) In respect, of infrastructure projects of State Agencies, there will be no upper ceiling on subsidy to be provided under the scheme.

(*xiv*) The amount of Central Assistance/Subsidy availed of for the project or any of its components from any other Central Scheme shall be deducted from the amount of subsidy admissible under this scheme.

Release of Subsidy

(*xv*) Subsidy for the projects under this scheme shall be released through NABARD for projects financed by the Commercial, Cooperative and Regional Rural Banks, Agricultural Development Finance Companies (ADFCs), scheduled Primary Cooperative Banks (PCBs), North Eastern Development Financial Corporation (NEDFI) and other institutions eligible for refinance from NABARD and through NCDC for projects financed by NCDC or by Cooperative Banks recognized by NCDC in accordance with its eligibility guidelines.

Implementation Period

(*xviii*) The scheme shall be implemented with effect from 20.10.2004 during 2004-05 and during 2005-06 & 2006-07 in the Tenth Plan with a Central Assistance of Rs. 175 crore for marketing infrastructure projects. In addition, there will be central allocation of Rs. 15 crore for strengthening Agmark laboratory network and for general awareness and training programmes and studies, etc.

Implementing Agency

(*xix*) The scheme shall be implemented by the Directorate of Marketing & Inspection (DMI), an Attached Office of Department of Agriculture and Cooperation. For details of the project please get in touch with secretary SEAI. Fishery can be done with certain modifications in underdeveloped areas.

ROLE OF EIC AND SEA FOOD TRADE

The global consumer with awareness on human health and safety demands more and more stringent hygienic food standards and also protection of food from pollutants and contaminants. As leading seafood importers, EU, USA and Japan have evolved regulations adapting to more stringent standards for the hygiene and safety of food imported to these countries. Even developing nations like Taiwan, Indonesia, Thailand, China etc. are framing their standards in line with those of the developed countries, especially those of the EU. In this context, it is pertinent to analyse the new EU Regulations that are going to be implemented from 1st January 2006 and to study their implications on the Seafood Industry in India.

The role of the Competent Authorities has become very important in the WTO Regime in view of the emphasis given to Equivalence and Mutual Recognition Agreements and therefore, the Export Inspection Council of India (EIC) and Export Inspection Agencies (EIAs), as Official Certifying Bodies and the Competent Authorities, have to play a pivotal role in ensuring the safety and the quality of the food products exported from India and also to safe guard the interest of the importing countries in order to enhance the export potentialities of India. A number of importing countries like USA, Japan, Australia, Italy, Korea, Srilanka, Singapore etc. have already signed equivalence agreement, or entered into Memoranda of Understanding (MOU) with EIC and have recognised EIC's Certification System for a number or commodities including seafood. As an export oriented industry, the seafood industry also needs to adapt itself to the ever-changing international laws and regulations.

New EC Regulations

The European parliament has adopted the following four new Regulations pertaining to the hygiene and safety of food and feed manufactured in the Member Countries and also those imported to the European union.

(*i*) Regulation (EC) No. 852/2004 dated 29.4.2004 on *the hygiene of food stuffs.*

(*ii*) Regulation (EC) No. 853/2004 dated 29.4.2004 laying *down specific hygienic rules for food of animal origin.*

(*iii*) Regulation (EC) No. 854/2004 dated 29.4.2004 *laying down specific rules for the organisation of official controls on products of animal origin intended for human consumption.*

(*iv*) Regulation (EC) No. 882/2004 dated 29.4.2004 on *official controls performed to ensure the verifications of compliance with feed and food law, animal health and animal welfare rules.*

While Regulations 852/2004 and 853/2004 are mainly meant for the business operators, the Regulations 854/2004 and 882/2004 are for the Competent Authorities of the member countries and that of third countries. The requirements for bivalve molluscs and fishery products have been specified in Regulations 853/2004 and 854/2004. However, the Regulation (EC) No. 178/2002, is considered as the mother of all the above-mentioned new EC Regulations and under this Regulation, the European Parliament has framed the "General Food Law" and set up the European Food Safety Authority (EFSA).

General Food Law

The General Food Law is a law governing the food in general and food safety in particular, whether at community or, national level, covering all stages of production, processing and distribution of food and feed. In this context, the food has been defined as "any substance or product, whether processed, or unprocessed, intended to be or reasonably expected to be ingested by humans".

The main objective of the food law is to achieve free and fair movement of food and feed in the community, either manufactured or marketed. Food law envisages "high level of protection of human health" and also the "consumer interest". Here it is interesting to note that the word "consumer interest" has been introduced for the first time by the EC, which covered a vide spectra of objectives like the colour, taste, size, shape, appearance, labelling, packing etc. of the product. Therefore, the EC can reject the fishery products not only on the basis of food safety, but also on the quality, packaging, labelling, etc. of the product.

As per the food law, it is mandatory that scientifically based risk assessment is conducted before framing the standards. However, under the cover of the precautionary principles, a clause is included under they can formulate their own standards without undertaking risk assessment. This precautionary principle, though evolved to safeguard the interest of the consumers, naturally creates trade barriers.

It is also of great significance to note that the EU has fixed the responsibilities for the first time on the business operators for the quality and safety of the food and feed processed by them. The obligations of the business operators have been listed in detail at Article 4 of Regulation (EC) *No. 852/2004*. It is also envisaged that the business operators as well as the importers have to establish traceability of the products/additives at all stages of production.

Another significant aspect is that the primary production has been brought, for the first time, under the purview of the food law, thereby repealing the Council Directive 93/43/EEC. Here the primary production covers all operations of production such as rearing, harvesting, milking, hunting, fishing etc.

Salient Features of New EC Directives

(*i*) The primary responsibility for ensuing the safety and quality of the food and feed lies with the processors. (Article-I of the Regulation (EC) No. 852/2004)

(*ii*) The safety of the food has to be ensured through out the food chain starting from primary production. (Article-I of the Regulation (EC) No. 852/2004)

(*iii*) HACCP has been made mandatory for all business operators other than the primary producers and allied operators (Article-5 of the Regulation (EC) No. 852/2004) (*iv*) Primary producers have to follow good hygienic practices based on HACCP principles and keep record for verification of the Competent Authority (Annex-1-852/2004)

(*v*) Animal health and welfare has to be ensured by the business operators (Annex. 1-852/2004).

(*vi*) It is necessary to establish microbiological criteria and temperature control requirements based on risk assessment (Article 1 of the Regulation (EC) No. 852/2004).

(*vii*) Registration of the facilities of food products of non-animal origin and approval of facilities of food products of animal origin is mandatory. (Article 6-852/2004).

(*viii*) Approval system has been specified in two stages for facilities of food products of animal origin to incorporate approval of processing facilities and a HACCP Audit (Article 3-854/2004).

Implications of the new EC Regulations

The European Commission has not made much changes in the new EC Regulations from those specified in the earlier Directives/Regulations like 91/493/EEC, 91/492/EEC etc. in relation to the existing clauses. However, by bringing the primary productions under the purview of the food law, it has created a lot of concern to the Sea Food Industry in India. In a vast country like India, where the primary production is scattered and mainly confined to un-organised and cottage sectors, the implementation of hygienic standards as specified in the annexure-I of the Regulation (EC) No. 852/2004 may be a difficult task. Moreover, the vast number of primary producers and the lack of formal education of the workforce employed are also stumbling blocks in achieving the targets. Specific areas in which difficulties are expected are given below.

(a) Fishing vessels and landing sites

One of the major recommendations of the FVO Mission that visited India during January 2005 is the implementation of official controls on fishing vessels and landing sites.

At present more than 40,000 mechanised fishing boats, 26,000 motorised boats and over 1,50,000 non mechanised fishing boats are in operation in the coastal wasters of India. In a small State like Kerala, there are nearly 200 major and minor landing sites in operation. The up-gradation of all these fishing vessels and landing sites to bring them at par with the International Standards and monitoring of these is a time consuming task. The Central and State Government Departments are fully engaged in this operation and with the collective efforts of the government bodies and trade associations, this issue would be sorted out in phased manner.

(b) Traceability

More emphasis has been given on traceability of the products and ingredients in the new EC Regulations. Article 18 of the Regulation (EC) No. 178/2002 has defined traceability as "an ability to trace and follow a food, feed, food producing animal or substance intended to be, or expected to be incorporated into a food or feed, through all stages of production, processing and distribution." Therefore, the processor has to establish the traceability of the products and also the ingredients, which include water and ice, at all stages of production starting from the primary production and also through storage and distribution. In Indian condition, establishing the traceability from farm to fork level is not an easy task. However, EIC, MPEDA & SEAI are together working to sort out this issue in the near future. It may also be added that the document on traceability being finalized in relation to inspection and certification systems at Codex level clearly states that traceability is one of the tools, which may be used by a CA for its food inspection and certification systems and a food inspection and certification system without traceability may meet the same objectives as one with traceability. Thus an alternate system to meet the same objectives would need to be acceptable to the EC.

Role of EIC as a Competent Authority

The new EC Regulations has given more important to the role of Competent Authorities of the member countries as well as those of the third countries. The Regulation (EC) *No. 882/2004* has highlighted the responsibilities of the Competent Authorities and their role in maintaining the hygienic standards of the food and feed imported to the European Union. In this connection, it is pertinent to note that the EIC has already taken the following steps to streamline the industry and also to avoid any possible hindrance in the export of seafood from India.

- Revised executive instructions for fish and fishery products have been prepared by EIC in line with the requirements of the new EC Regulations, and these are in force from 1st August 2005. As per the new rules, the approval system has been modified to have a two-level approval for new establishments, a conditional approval once the processing unit is satisfactory and a final approval after carrying out the HACCP audit. Increased emphasis has also been given to the monitoring of the HACCP in the establishments. The responsibilities of the establishment have also been laid down in the revised executive instructions.
- A scheme for the landing sites and auction centres has prepared by EIC in line with the GOI notification and also with those requirements specified in the new EC Regulations and the same will be implemented after due consultation with all stakeholders.
- A draft amendment notification incorporating the conditions of fishing vessels and freezer vessels in line with the new EC Regulations has been prepared which will be submitted to Government for publication after consulting the concerned departments and officials

  Steps have to be strengthened to establish suitable traceability system through out India.

  Laboratories have been strengthened to test for antibiotic levels at sensitivity levels at par with EC requirements and presently all consignments exported to the EU countries are tested for antibitic residues chloramphenicol and nitrofuran metabolites.

Conclusion

EIC, as Competent Authority, recognized by the EC, Australia, Japan and other countries ensures that the processing units meet the most stringent norms of the importing countries. This is done through a system of approval and continuous surveillance of the units. Surveillance is carried out at three levels - routine monitoring visits at a frequency ranging from one month to three months depending on performance of the units, supervisory visits and corporate audits. The fish processing establishments in India have already acquired sophisticated infrastructure facilities and technical know-how at par with the International Standards. Issues as per the new EC regulations, specifically those pertaining to the primary production and traceability, are also being addressed to ensure that Indian exporters continue to supply fish and fishery products to meet importing countries health and safety requirements.

REFERENCES

Handbook on Fisheries Statistics (1996), Ministry of Agriculture, Government of India (GOI), New Delhi

Handbook on Fisheries Statistics (2000), Ministry of Agriculture, Government of India (GOI), New Delhi

Economic Survey, different issues, Ministry of Finance, Government of India (GOI), New Delhi

India, several issues, Ministry of Information and publication, Government of India (GOI), New Delhi

Dehadri, P.V. 1996. Growth in Fisheries and Aquaculture: Resources and Strategies and Policies for Agricultural Development in Ninth Plan. Indian Institute of Management, Ahmedabad

Krishnan, M; Pratap S. Birthal, K. Ponnusamy, M.Kumaran and Harbir Singh 2000. Aquaculture Development in India: Problems and Prospects. Workshop Proceedings held at National Centre for Agricultural Economics and Policy Research, New Delhi, September 6-7, 1999

Kumar, Anjani; P K Joshi and Badruddin. 2002. Export Performance of Indian Fisheries Sector: Strengths and Challenges Ahead. Working Paper 3. National Center for Agricultural Economics and Policy Research, New Delhi

Monthly Statistics of Foreign Trade of India: Volume Exports and Re-exports. (various issues), Ministry of Commerce, Government of India

National Account Statistics (Several Issues) Central Statistical Organisation Government of India, New Delhi

Nayyar, Deepak; and Sen, Abhijit. 1994. *International Trade and Agricultural Sector in India,* In G.S. Bhala (ed.) Economic Liberalization and Indian Agriculture, Institute for Studies in Industrial Development, New Delhi

Panchamukhi, V.R. 1991. *Trade in Agricultural Commodities—Analysis of the Period 1960 to 85,* In M.L. Dantwala (ed.) Indian Agricultural Development Since Independence, Indian Society of Agricultural Economics, Bombay.

Chapter 4
Inland Aquacultural Technologies

Introduction

The inland fisheries in India include both capture and culture fisheries. Capture fisheries have been the major source of inland fish production till mid eighties. But, the fish production from natural waters like rivers, lakes, canals, etc., followed a declining trend, primarily due to proliferation of water control structures, indiscriminate fishing and habitat degradation (Katiha 2000). The depleting resources, energy crisis and resultant high cost of fishing etc. have led to an increased realisation of the potential and versatility of aquaculture as a viable and cost effective alternative to capture fisheries. During past one and half decade, the inland aquaculture production has increased from 0.51 to 2.38 million tonnes, while for inland capture fisheries the same has declined from over 0.59 to 0.40 million tonne (Anonymous 1996a,b; Anonymous 2000; Gopakumar *et al.* 1999). The per centage share of aquaculture has also increased sharply from 46.36 to 85.65 per cent. It is primarily because of tremendous 4.5 fold increase in freshwater aquaculture. Its share in total inland fish production has also increased from 27.95 per cent to 66.4 per cent (Anonymous 1990a,b; Anonymous 2000). Still, its has greater scope for enhancing fish production.

In India, aquaculture witnessed an impressive transformation from highly traditional activity to well developed industry. With rich resource base both in terms of water bodies and fish species, the investments in this sector are following an increasing trend. The recent estimates of freshwater aquacultural production around 2.0 million tonnes contributed over one third of total fish production of India. This outcome in primarily propelled by the appropriate technologies, financial investments and entrepreneurial enthusiasm. The success stories of intensive fish culture started from Kolleru lake basin in Andhra Pradesh in mid-eighties and virtually replicated in states like Punjab, Haryana, Uttar Pradesh and so on (Gopakumer *et. al.* 1999).

By virtue of its geographical situation in monsoon belt, India is endowed with good rainfall and consequently extensive aquacultural water bodies. The inland aquacultural water resources in the form of pond and tanks have been distributed over almost all the states of India. Despite immense efforts for horizontal expansion of this industry, only one third of the area could be brought under scientific fish culture. This untapped production potential can be harnessed through effective and intensive adoptions of available technologies, transfer of technical know how and provision for material inputs. Flexibility in areas of operation and scales of investments and compatibility of freshwater aquaculture practices with other farming systems coupled with high potentials of eco-restoration have provided congenial environment to establish it as a fast growing activity. Considering its potential and impressive annual growth rate of over 6 per cent, Government to India also emphasizing on a pauculture development. The national freshwater aquaculture development plan proposed to increase the area under aquaculture to 1.2 million ha, with average productivity of 2762 kg/ha per year (Gopakumar *et. al.* 1999). To achieve this goal, suitable strategies for enhancement of area coverage and productivity are needed considering components of horizontal and vertical expansion in concurrence with the potential and problems of different states.

Key Freshwater Aquacultural Technologies

The researches on aquaculture technologies got momentum only in seventies at Central Inland Fisheries Research Institute (CIFRI), Barrakpore under All India Coordinated Research Projects on Composite Fish Culture, Air breathing Fish Culture, Riverine Seed Prospecting and Fisheries Management of brackishwater was framed. Aquaculture research in India received the momentum, when a separate centre called Freshwater Aquaculture was established at Dhauli, Bhubaneswar (Orissa). In 1986, it has got the status of independent ICAR institute and named as Central Institute of Freshwater Aquaculture (CIFA). These two institutions are the pioneers in development of aquacultural and culture based fisheries technologies. A brief of aquacultural technologies developed is presented in this chapter.

These technologies may be categorised into technologies for fish seed production and production of table size fish or aquaculture. Both seed production and aquacultural technologies are for different categories of fishes, *i.e.* carps and catfishes including air-breathing fishes, so, described separately. The first part deals with technologies for seed production and fish breeding and second with aquaculture or production of table sized fishes.

FISH BREEDING AND SEED PRODUCTION

Induced Breeding

The development of indigenous technique of hypophysation has revolutionized the seed production of major carps. The eco or circular hatcheries, based on the technology of induced breeding of carps with pituitary gland extract (PGE) are used for commercial fish seed production of Indian and Chinese carps. Under this technology sexually mature fishes which do not breed in captivity are bred in ponds by PGE to spawn then in captivity. Although, the technology was evolved as early as early as 1956-57, it took over decade to popularize in India through All India Co-ordinated project on "Seed Production and Composite Fish Culture". This technique has revolutionized the carp seed production enormously. Nowadays, the synthetic hormone 'Ovaprim' is used as a successful substitute of pituitary hormone.

Intensive Carp Seed Rearing

Availability of adequate quantity of carp seeds of desired species at appropriate time is one of the pre-requisite for success of aquaculture operations. The availability of standard stocking materials in time and space still remains a constraint, despite domestication of induced breeding technology and production of carp seed to the tune of over 16,500 million fry in the country. The raising of seeds in the initial stages is associated with high rates of mortality due to several management problems (Anonymous 2000a). Thus, it is essential to follow standardized package of practices for higher growth and survival in intensive seed raising at higher stocking densities, leading to hypoxic conditions and competition for food and space.

The different standardized package of practices for intensive seed production propose measures to control predatory and weed fish, plant derivatives and soap-oil emulsion to control insects, organic and inorganic fertilizers for fertilization of ponds, stocking densities for carp species for mono-culture, supplementary feed, standard or 'optimum physio-chemical parameters for management of water quality and standard methods for monitoring health-care, etc.

The technology of intensive seed production includes :

- Eco or circular hatchery or collection of spawn from natural abode;
- Raising the spawn to fry in nursery ponds; and
- Rearing of fry to fingerlings in ponds.

Eco Circular Hatchery

The essential features of the eco-hatchery are :

(*i*) Tube well or a dependable source of potable waste.

(*ii*) Overhead tank (25000 to 30000 liter capacity) with arrangement of continuous water supply to various hatchery components.

(*iii*) Circular spawning pool (8 m diameter) capable of holding the spawners and male population.

(*iv*) Incubation pool is a circular double walled chamber of 3m diameters. The eggs are released in the outer chamber. The water intake through floor mini pipes prevents eggs to settle down. The hatchlings are kept in the outer chambers for 72 hours.

(*v*) Spawn collection pool is rectangular in shape. The spawn is collected in a rectangular sac like cloth piece called *hapa.*

From the earthen pits to double walled *hapa* hatcheries and associated modifications, carp hatcheries have come a long way in terms of running water glass jar or circular hatcheries (Bhowmick 1978; Dwivedi and Rabindranathan 1982; Dwivedi and Zaidi 1983; Jhingran and Pullin 1985 and Rath and Gupta 1997). These eco-hatcheries have not only provided the scope to produce and handle mass quantities of eggs during hatching but also to greater extent reduced the requirement of water and manpower.

Raising Fry from Spawn

Generally the size of nursery is 0.04-0.1 ha. The pond preparation includes treatment of the ponds with Mahua oil cake (MOC) atleast 15 days prior to stocking for eradication of unwanted fishes and application of lime. The fertilization includes application of groundnut oil/mustard oil cake @ 700 kg, cow dung 200 kg and 50 kg single super phosphate per ha, after making a thick paste of the three ingredients. These are applied in three doses i.e. 50 per cent of the paste 3 days before stocking, 25 per cent 5 days after stocking and remaining 25 per cent 10 days after stocking.

Stocking density is 3-5 million spawn per ha is usually followed by fish farmers in earthen nurseries, however the intensity can be as high as 10-20 million spawn in cemented nurseries (Jena *et. al.* 1998a). Generally monoculture is done for raising the fry.

The supplementary feed applied is the mixture of rice and oil cake at 1:1 ratio. The feeding is done @ 6kg per day per million spawn for first 5 days followed by 12 kg per day million spawn for next 10 days, in spilt doses during early morning and evening hours (Jena *et. al.* 1998b).

The rearing period is usually for a period of 15 days during which the fry attain a size of 25-30 mm.

In the beginning, during 1950s, the survival rate was 10-20 per cent, but at present it is 50-60 per cent.

Rearing of Fry to Fingerlings

Generally the size of rearing pond is 0.1-0.2 ha. The pond preparation is almost same as that of nursery for raising fry to fingerlings. It includes MOC treatment of the ponds atleast 15 days prior to stocking, eradication of unwanted fishes and application of lime. Fertilization includes application of both organic and inorganic fertilizers at conventional doses (Jena *et al.* 1998b).

The stocking density is 0.1-0.2 million fry per ha. The rearing may be done as poly-culture for raising the fingerlings.

Rice bran and oil cake in ratio of 1:1 per provided as the supplementary feed. The doses of feed over the rearing period are 8-10 per cent of fish biomass per day in first month, 6-8 per cent of fish biomass per day in second month and 3-5 per cent of fish biomass per day in third month.

The rearing period is two to three months till the fingerlings attain a mean size of 100 mm in length and 10 g in weight.

Initially the survival rate was very low, but now it is 60-80 per cent.

Cost and Returns of Seed Production

The process of fish seed production has three stages namely, spawn, fry and fingerlings. Therefore, its economics has been worked out for raising fry from spawn, 3-4 crops may be taken in a year leading to production of 3.6 to 4.8 million fry. The major components of the operating cost were value of seed and lease value. The benefit cost ratio was 1.5 for nursery management. At the other stage of rearing of fry to fingerling the costs incurred on feed, lease value and seed cut the major share in cost. The average number of fingerlings produced were 0.15 million per ha. The benefit cost ratio was 1.32.

Table 1 : Economics of Seed Production

Item Area	*Nursery 1 ha*	*Rearing 1 ha*
Lease value (Rs./crop for nursery & Rs./year for rearing)	5000	15000
Pond preparation		
Predatory and weed fish clearance	7500	7500
Insect control	1000	
Fertilisation	7500	4000
Seed (Spawn 3 million, fry 2 lakh)	15000	12000
Supplementary feed	4500	24000
Labour charges	5000	12000
Miscellaneous	2000	3000
Total cost	47500	79500
Returns (Survival rate 40%)	72000	
Returns (Survival rate 75%)		105000
Profits	24500	25500

Breeding and Seed Production of Catfishes, Magur and Singhi

The air-breathing catfishes *Clarias batrachus* (magur) and *Heteropneustes fossilis* (singhi) are adapted to adverse ecological conditions, *viz.*, water bodies with low oxygen and pH, high CO_2, H_2S, CH_4 and heavy silt with decaying vegetation, organic load, etc. These can be stocked @ 20,000-50,000 fingerlings per ha, which attain 100-200 g in 6-8 months (Anonymous 2000a).

Management of Brood Stock

Proper care and maintenance and provision of balanced supplemented feed play a key role in achieving successful spawning. The broad fishes are stocked generally in flow though (21 pen min) cement cisterns (3 m × 1m × 1m) with 10-15 cm thick soil base, an inlet at the top of cistern and outlet at about 20 cm from the bottom.

Induced Breeding Technique

Following standardized induced breeding technique, using different inducing agents like carp or catfish pituitary extract, ovaprim, HCG, LHRHA + Domperidone, etc. the species can be bred from March to September.

The incubation time in singhi is less than magur. Proper flow through system is used for incubating eggs to make seed available over a longer period of a year.

Larval Rearing

The newly hatched larve for a period of 15 days in indoor conditions are stocked at density 2,000-4,000 per sq. m in well aerated water till air-breathing habit commences. They are fed with mixed zooplankters or *Artemia* larva or *Tubifex* spp. with replenishment of water at least twice a day initially for a few days, followed by compounded supplementary feed. The laboratory reared fry are ready to be reared in earthen nursery ponds or outdoor cement cisterns (4m × 1 m × 0.5m) with soil base, after 10-12 days at 200-300 per sq. m to raise fingerlings of 6-8 cm in a month. Survival levels over 60 per cent may be obtained during raising of fingerlings under optimum rearing conditions.

Grow out Culture for Table size Fish Production

India's aquaculture is basically carp-oriented and the contribution of other species is marginal. The carps both India and exotic contributed over 90 per cent of freshwater aquaculture production. The major freshwater culture technologies (Table 2) may be classified into the following types :

1. Polyculture of Indian carps or Indian and exotic carps together (Composite carp culture)
2. Mono- and polyculture of air-breathing fishes
3. Mono- and polyculture of freshwater prawns
4. Integrated fish farming
5. Cage culture
6. Pen culture
7. Pearl culture

The prevalent freshwater aquacultural technologies along with their cultural practices are summarised in Table 2.

(*a*) Low input of fertiliser based system

(*b*) Medium input or fertiliser and feed based system

(*c*) High input or intensive feed and aeration based system

(*d*) Sewage fed water based system

(*e*) Aquatic weed-based system

(*f*) Livestock based or Integrated fish farming

Species Mix

The species mix of 3-6 carps has been most prevalent including three indigenous (*Catla catla, Labeo rohita* and *Cirrhinus mrigala*) and three exotic (*Hypophthalmichthys molitrix, Ctenopharyngodon idella* and *Cyprinus carpio*) carps. Some of the other fish species and species combination adopted for freshwater aquaculture technologies are *Labeo bata* and *C. reba* in sewage fed. *Ctenopharyngodon idella* (grass carp more than 50 per cent) in weed based, medium and minor carps, viz. *Labeo calbasu, L. gonius, L. bata, Puntius pulchellus, P. sarana* and *Cirrhinus cirrhosa* in integrated fish farming with paddy-cum-fish culture.

The low input or fertiliser based, medium input or fertilizers and feed based, sewage fed and weed-based are the culture practices having fish yield from 1-3 tonnes per ha, are considered as extensive technologies. The most important carp culture technology is composite culture or intensive carp culture therefore, considered for discussion in the present chapter.

Carp Polyculture or Composite Carp Culture

The research and development efforts during last five decades have greatly enhanced average fish yields in the country making carp culture an important economic enterprise. It has grown in geographical coverage, diversification of culture species and methods, besides intensification of farming systems. The three Indian major carps, *viz.*, catla (*Catla catla*), rohu (*Labeo rohita*) and mrigal (*Cirrhinus mrigala*) were the principal species cultured by the farmers in ponds since ages and production from these systems remained significantly low (at 600kg/ha/year) till the introduction of carp polyculture technology. The introduction of exotic species like silver carp (*Hypophthalmichthys molitrix*), grass carp (*Ctenopharyngodon idella*) and common carp (*Cyprinus carpio*) into the carp polyculture system during early sixties also added new dimension to the aquaculture development of the country. With the adoption of technology of carp polyculture or composite carp culture production level of 3-5 tonnes/ha/year could be demonstrated in different regions of the country. Probably it is the technology of carp polyculture that has virtually revolutionized the freshwater aquaculture sector and brought the country from a level of backyard activity to that of a fast growing and well organized industry and placed the country on the threshold of blue revolution. The average national production from still-water ponds has gone up from 600 kg/ha/year to over 2 tonnes/ha/year, with several farmers even demonstrating higher production level of 8-12 tonnes/ha/year.

The standard recommended carp culture in India involves three species of Indian major carps or combination of three Indian major carps and three exotic carps, though adoption has been with several modifications depending on the market demand and resource availability. Standardized package of practices for carp poly culture include predatory and weed fish control by use of certain chemicals or plant derivatives: stocking of Indian major carps and exotic carps at densities of 4,000-10,000 fingerlings/ha: pond fertilization with application of organic manure like cattle dung or poultry droppings and inorganic fertilizers: provision supplementary feed and water quality management. The researches over the years (Lakshmanan *et. al.* 1971); Sinda *et. al.* 1973; Chaudhuri *et. al.* 1974, 1975; Chakrabarty *et. al.* 1979; Saha *et. al.* 1979; Sinha and Saha 1980; Tripathi and Mishra 1986; Rao and Raju 1989; Tripathi *et al.* 2000; Ayyappan and Jena 2001 and Jena *et al.* 2002 a,b) have led to the development, refinement and standardization of host of technologies with varied production levels depending on the input use and finally resulted in technology of intensive carp culture.

Intensive Carp Culture Technology

The average production of the country from still water ponds is about 2 tonnes/ha/year, with the packages of practices developed in the ICAR (Anonymous 2000a). There are possibilities of producing 10 tonnes and 15 tonnes/ha/year. Standardized package of practices for intensive carp polyculture include:

(*i*) predatory and weed fish control by use of certain chemicals or plant derivatives;

(*ii*) pond fertilization with application of Azolla at 40 tonnes/ha/year at weekly split doses as bio-fertilizer, substituting traditional organic and inorganic fertilization;

(*iii*) stocking of Indian major carps and exotic carps of 25-50 g size at densities of 15,000-25,000 fingerlings ha;

(*iv*) provision of balanced formulated supplementary feed comprising rice-bran, ground nut oil-cake, soybean flour, fish meal and vitamin mineral premix;

(*v*) provision of 4-6 paddle-wheel aspirator aerators per hectare of water to keep dissolved oxygen within desirable limits especially during night maintenance of water column of 1.5-2 m;

(*vi*) water replenishment depending on the water quality; and

(*vii*) fish health management through prophylactic and curative measures depending on the necessity.

Though harvesting of the table-size fish is done usually at the end of 10-11 months, partial harvesting of bigger size fishes is done during monthly samplings, after a growth period of 6-7 months, which provides congenial environment for remaining fishes and also reduces amount of supplementary feed provision. Stocking of large size of seed, preferable 25-50 g, minimizes mortality during initial months and thus leads to higher survival at harvest. Supplementary feed being major input, contributing over 60-70 per cent of input cost, needs judicious application and the quantities are divided based on the fish biomass present at any gives point of time. Supplementary feed in the form of dry pellet, provided at 2-3 rations per day, helps in its effective utilization and minimal wastage.

Mono/Polyculture of Air-breathing Fishes

The air-breathing fishes are distinguished by possession of an accessory respiratory organ, which enables them to survive for hours outside water or indefinitely in water with low oxygen content. These are extremely hardy for environmental stresses and adaptable for the waters unsuitable for conventional cultivable species.

Magur (*Clarias Batrachus*), singhi (*Heteropneustes fossilis*), koi (*Anabas testidineus*) murrels, giant murrel (*Channa marulius*), stripped murrel (*C. striatus*) and spotted murrel (*C.punctatus*) are the most important culturable species in India.

The air-breathing fish culture is particularly oriented to shallow waters (2-3 ft depth). The material inputs needed are only the fingerlings (6-10 gm) and feed. Replenishment of water becomes an essential input in case of very heavy stocking and multiple cropping to obtain high yields. The pond size should be 0.1-0.2 ha, for effective management. Growth of magur and singhi goes very well upto water temperature 32°C. Fishes are stressed around 35°C and mortality starts at 38°C. Collection of their seed from nature continues to be dependable source for stocking material. The peak season for collection of seed is pre-winter period. Availability of air-breathing fish seed is in plenty in part of Assam, Andhra Pradesh, Bihar, West Bengal and Karnataka. The fry rearing phase in murrels is complex due to cannibalism. It can be reduced with supplementary feeding.

The mono-culture of magur and singhi permits high stocking density (40-50 thousand), while for poly-culture with varies between 20-30 thousand. For mono-culture of murrels the stocking density ranged between 15-25 thousand, with the lowest for giant and highest for spotted murrels. Feed for singhi and magur includes fish offal, slaughter house waste, dried silkworm pupae mixed

Table 2 : Cultural practices under different aquacultural technologies

System	*Species*	*Stocking (In 000 fingerling ha^{-1})*	*Fertilization / Liming ha^{-1}*	*Feed day^{-1}*	*Management Practices*	*Duration of rearing (Month)*	*Average yield (t ha^{-1} yr^{-1})*
Low input	3-6 species	3-5	Cow dung 10-15t/ Poultry droppings 3-5t, Urea 2q, SSP 3q	No feed	Fertiliser use, maintenance of water depth at 1.5-2.5 m	10-12	1-2
Medium input	3-6 species	5-10	Cow dung 10-15t/Poultry droppings 3-5t, Urea 2q, SSP 3q	Rice bran and oil cake, @2-3% of fish biomass	Maintenance of water depth at 1.5-2 m, Intermediate liming at 3 month interval @100 kg ha^{-1}	10-12	3-6
High input	3-6 species	15-25	Less use of organic manure, Bio-fertilization with *Azolla*, SSP	Rice bran, oil cake, fish meal, Vitamin and mineral mix, @ 2–3% of fish biomass	Aeration, water exchange towards later part, inter-mittent liming at every quarter @ 100kg ha^{-1}, maintenance of water depth at 2–2.5 m.	10-12 Periodical harvest	10-15
Sewage fed	3-6 species + *L. bata, C. reba*	30-50 (total in intermittent stocking)	Domestic Sewage water	No feed With feed	Multiple stocking and multiple harvesting (Size 100-200 gm), Maintenance of water depth at 0.7-1.5 m	8-10	2-5 3-7
Weed based	50% Grass Carp and 50% other species	4-5	SSP 3q for one crop to be applied at 15 days interval Liming @ 100 kg/quarter	Aquatic weed (*Hydrilla Najas Ceratophyllum*, Duck weeds like *Spirodella, Lemna, Wolffia*, etc.	Maintenance of water depth at 1.5-2 m	10-12	3-4
Integrated : Cattle (3–4 ha^{-1}) Duck (300 ha^{-1}) Poultry (500ha^{-1}) Pig (50 ha^{-1})	3-6 species	5-10	No fertilised use, liming	Rice bran and oil cake, 2-3% of fish biomass	Maintenance of water depth at 1.5-2 m	8-10	3-5
Paddy-cum-Fish	3-6 species and Medium & Minor carp	5-10	Cow dung 10-15 t	Rice bran and oil cake, 2-3% of fish biomass	Maintenance of water depth at 1.5-2 m in pond	6	0.5-2.0 of fish 3-6 of paddy
Pen	3-6 species	5-10	Liming	Rice bran and oil cake, 2-3% of fish biomass	Maintenance of water depth at 1.5-2 m	8-10	3-5*
Cage	Single species			Experimental stage			10-15*
Running water	Single species			Experimental stage			20-50*
Air-breathing	Mono-culture	20-50	Cow dung 10-15t/Poultry droppings 3-5t, Urea 2q/ha, SSP 3q	Rice bran, oil cake and Fish meal	Maintenance of water depth at 1-1.5 m	8-10	3-6
Freshwater Prawn	Mono-culture	20-50	Cow dung 10-15t/Poultry droppings 3-5t, Urea 2q, SSP 3q	Palletted feed	Maintenance of water depth at 1-1.5 m	6-8	1-1.5
Polyculture of Carp with Prawn	2-3 species of carp + Prawn	Fish 5+ Prawn 10-15	Cow dung 10-15t/poultry droppings, 3-5t, Urea 2q/ha, SSP 3q	Rice bran and oil cake, 2-3% of fish biomass	Maintenance of water depth at 1-1.5 m	10-12	Fish 3-4 Prawn 0.3-0.5

* kg m^{-2} yr^{-1}

with rice bran and oil cake in the ratio of 1:1:1. The mixture of rice bran, oil cake and bio-gas slurry in the ratio of 1:1:1 also proved successful. The feeding schedule varies over the culture period and for different species. Feeding may be done either by broadcasting the feed in small quantity fishes or by lowering feed basket near banks in addition to broadcasting. The culture period for these may vary between 8-10 months with an average yield of 3-6 tonnes/ha.

Mono/Poly-Culture of Freshwater Prawn

In India, freshwater prawn culture is becoming popular. Mono-culture of *Macrobachium rosenbergii* and *malcolmsonii* and their polyculture with carps are common (Reddy *et al.* 1985; Jhingran 1991; Tripathi 1992). They are available in freshwater resources like rivers, streams, canals, beels, swamps, lakes, *etc.* The prawn seeds can be collected from natural resources or produced at government/private prawn hatcheries. Freshwater prawn culture can also be taken up in pens or cages. They feed on algae, insect larvae, molluscs, worms, smaller weed fishes, cereals, slaughter house wastes, oil cakes, etc. Fresh water prawns can tolerate very high range of salinity (upto 28 per cent), but salt concentration upto 5 to 6 per cent is preferred. Rectangular ponds of 0.1-2.0 ha size having unpolluted freshwater, with high concentration of oxygen are considered ideal.

Other culture methodologies are similar to carp farming. Liming and pond fertilization help freshwater prawn in attaining quicker and the healthy growth. Normally stocking density ranges from 20-50 thousand per ha. Male grow bigger than females and attain about 70 gm average weight in 6-8 months.

Periodic sampling to monitor the growth, survival an also to decide the feeding dosages, etc, is essential. The prawn so grown can attain marketable size in 6-8 months. The production ranging from 1-1.5 tonne/ha can be achieved in scientifically managed system. Fresh water prawn farming is assuming greater importance due to very high demand, good price and high returns.

Integrated Fish Farming

Integrated fish farming is the link of two or more normally separate farming systems, which become sub systems of a whole farming system (Anonymous 2000a). Such farming systems can broadly be categorized into two types:

(*i*) systems with no direct byproduct utilization from one to other subsystem, but optimal utilization of farming space and time *e.g.* paddy-cum-fish culture; and

(*ii*) systems where byproduct *i.e.*, waste from one subsystem is being utilized for sustenance of other *e.g.*, fish-pig/poultry/duck farming.

Paddy-cum-fish Culture

The practice is undertaken in deep water bodies with fairly strong dykes to prevent escape of cultivated fishes during floods. Presence of channels, small ponds of sump near to the field is essential to give shelter to fish against heat and predators. In India, fish species like catla, rohu, mrigal, common carp, murrel, magure etc. at 5,000-10,000 per ha are used in paddy-cum-fish culture. The excreta of fish and leftover supplementary feed help in fertilizing soil thereby increasing paddy production. Some fishes eat harmful insects and their larvae, which otherwise can cause problems to paddy. A production level of 0.5-1 tonne fish per ha and 3-6 tonnes paddy per ha can be achieved in a well managed system.

Fish-cum-cattle Farming

The pond embankments can be used for cattle shed and their washings drained directly into pond. A better way to utilize dung is in the form of slurry. About 30-60 tonnes slurry per ha could be applied to pond. It has been estimated that dung and urine obtained from 3 to 4 cattle is sufficient to fertilize a pond of 1 hectare. The production levels of 0.5-2 tonnes/ha year can be achieved from this system without addition of any supplementary feed.

Pig-cum-fish Farming

The excreta from 30-50 pigs have been found adequate to fertilize 1 hectare pound. The pig west acts as an excellent fertilizer and for some fish species this acts as feed. Production level as much as 4 tonnes/ha/year of fish have been achieved along with 16 tonnes pig meat per ha (live weight) from such an integrated farming.

Duck-cum-fish Culture

Ducks feed on tadpoles, snails, insects, etc. A total of 200-300 ducks are sufficient to fertilize 1 ha pond. The embankments are used for night shelter. During day when they are in search of food they also aerate pond water, in addition to helping in pond bottom raking effect. The fish yield from duck-cum-fish farming system ranges from 3-5 tonnes/ha/year, in addition to 4,000-8,000 duck eggs and 2 tonnes duck meat per ha from the unit.

Poultry-cum-fish Farming

In this system 500 country birds are adequate to fertilize 1 ha pond. The dosage of application of poultry manure is about one third the rate of cow dung. A production of 3-5 tonnes fish, 28,000 eggs and 5 tonnes meat per ha is expected from this farming system in a year.

Cage Culture

Intensive culture of fishes through non-conventional system like cage culture is gaining importance owing to higher productivity potentials of the systems and possibilities of higher revenue generation from unit water area (Anonymous 2000a). Culture of fish in cages is largely accepted all over the would because of its usefulness in exploitation of large water bodies, which otherwise are under-utilised for fish production, employment and income generation. Various lakes, tanks, and coastal waters can be brought under cage culture technique and can be practiced at various management levels. Cage cultures have many advantages, viz., large extent of large water bodies can be utilized for aquaculture, which otherwise are not fully exploited for fisheries; high production per unit area can be obtained with high stocking density and intensive feeding; and monitoring of stock for growth and well being is easy; harvesting is simple and cages can be dismantled and reused in other locations as per requirement.

Cages are circular, cubic or basket like. The may be floating at the surface, just submerged or set at the bottom and enclosed at the bottom as well as sides by bamboo mesh, metal screens or netting (webbing) material.

Seed Production and Rearing in Cages

In nursery phase of cage culture, spawn or early fry are reared to fingerlings within 2-3 months for stocking in grow out cages or other systems, by adopting high density stocking with supplementation of protein rich diet.

The fingerlings of carps can be raised in commercial scale in cages of 5 sq. m. with a depth of 1.5 m. In situations where nursing of fry is not feasible, cages can be conveniently used.

Grow-out Production Systems

The fish production level obtained in grow out cages largely depend on the stocking density, species, provisions of inputs like supplementary feeding and overall management. The number of fishes that can be stocked in a cage depends on the productively of water body, rate of circulation, fish species, quality and quantity of feed supplied. The initial size of the fish to be stocked depends primarily on the length of the growing season and the desired size at harvest.

Carp fingerlings for stocking in 16-26 mm mesh cages should be of 10-15 g to expect of final size of over 500 g in a rearing period of 6 months. When natural fish food organisms are limited for high density rearing in cages, supplementary feeding from the vital component of production. In carps, feeding is provided at 4-5 per cent of fish biomass per day until 100 g size and reduced thereafter to only 2-3 per cent.

Pen Culture Technology for Foodplain Wetlands

The floodplain wetlands are commonly known as beels, mauns, chaurs or pats in various states of India. Though there is a great potential of more than 1,000 kg per ha production from floodplain wetlands, an average of 120-300 kg per ha is recorded. Most of these systems are weed choked and are under productive. Efficient use of popular gears is not practical in such water bodies. Till recently, the mainstay of fish production from these waters was through capture fishery. To boost fish production, shift of operation from mere extraction to capture-cum-culture fishery has yielded better results.

The pen may be square, rectangular, oval, elongated or horseshore shaped depending upon nature of banks, land and water depth. For better management pen area should be 0.1-0.2 ha. The pen consists of thick bamboo frame, splite bamboo or cane screen covered with nylon net lining. Most of the wetlands are infested with unwanted flora and fauna, so, deweeding, eradication of unwanted fauna and liming is essential prior to fixation of the pen(s). The selection of fish species depends upon the productivity and group of flora and fauna. The species combination of indigenous and exotic carps with giant freshwater prawn (*Macrobrachium rosenbergii*) is proved to be successful, although, culture of *M. rosenbergii* alone is more profitable. The stocking density varies according to species combination, *e.g.* monoculture of carps - 4000-5000 per ha, carp + prawn culture 3,000-4,000 carps and 10,000-20,000 prawn per ha and for monoculture of prawn, it may be as high as 30,000-40,000 per ha. Most of the floodplains are rich in natural food. So, supplementary feed is required is special cases like monoculture of prawn. Pen culture can be done round the year, avoiding monsoon months. The culture period may vary between 4-6 months. Therefore, it is possible to take two crops in a year. The range of fish yield for carp culture is 4-5 tonnes/ha/year, for carp + prawn is 2-2.5 tonnes/ha/year carps and 0.3-0.5 tonne/ha/year prawn and for monoculture of prawn 1.3 tonnes/ha/year.

Freshwater Pearl Culture Technology

Freshwater pearl culture is akin to cash crops of land based agriculture system and the technology is a privy to a very few countries in the world, viz., Japan and China. Cultured pearls are produced both in marine and freshwater environments. Freshwater pearl culture is more advantageous is terms of commercial scale availability of nature stocks of pearl mussels is easily accessible habitats, wider area of farming, even in non-maritime regions, operational easiness in management of freshwater culture environment, absence of natural fouling, boring and predatory organisms and overall cost effectiveness of operations. Realizing the potential and scope of inland pearl culture, a package of practices for producing cultured pearl from common freshwater mussels *Lammelidens marginalis, L.corrianus* and *Parreysia corrugata* has been developed (Anonymous 2000a).

The process include :

(*i*) Collection and conditioning of native pearl mussels

(*ii*) Surgical implantation of mantle grafts and appropriate nuclei in internal organs of the mussels (*iii*) Post-operation care of implanted mussels

(*iv*) Pond culture of implanted mussels in specially designed culture units in natural pond environment for 12 months

The pearl products developed at the CIFA includes:

(*i*) Shell attached half-round and designed pearls

(*ii*) Unattached non-nucleated oval to round pearls and nucleated large round pearls and alternate nuclear material.

In addition to producing regular, free, round cultured pearls, irregular non-nucleated pearls and pearl images (up to 1.0 cm) have also been produced successfully, which are drawing attention of several entrepreneurs.

Cost and Returns

The cost structure returns and benefit cost rations for different aquacultural technologies are presented in Table 3. Cost structure has primarily the items of leas value of the water body, cost of organic manure and inorganic fertiliser, seed, management harvesting. The specific costs related to particular technology included expenses on bird/animals in integrated fish culture, cost of paddy cultivation in paddy-cum-fish culture, construction of pens in pen culture, etc.

Need for Capital Formation : We have already discussed capital formation is a previous chapter and also the measures for promoting it to break the vicious circle of poverty. Here we discuss it from the point of view of economic growth. Capital formation is the very core of economic development. It may be a predominantly private enterprise system like the American, or a communist economy like the Soviet, economic development cannot take place without capital accumulation. No economic development is possible without the construction of irrigation works, the production of agricultural tools and implements, land reclamation, building of dams, bridges and factories with machines installed in them, roads, railways, and airports, ships and harbours—all the "produced means of further production" associated with high level of productivity. It seems unquestionable that the insufficiency of capital accumplation is the most serious limiting factor in under-developed countries. In the view of many economists, capital occupies the central and strategic position in the process of economic development.

Capital formation indeed plays a decisive role in determining the level and growth of national income, hence economic development. This is due to the fact that of all sectors of production capital has unlimited expansibility. It is man-made and is capable of increasing in quantity and improving in quality. There is no doubt that productive capacity of an economy can be increased only by increasing the quantity and improving the quality of its capital equipment.

Thus, in any programme of planned economic development capital formation must be assigned a significant role on account of a very close connection between economic growth and capital growth. It enables the adoption of more productive methods of production. Capital widening makes the economy diversified and broad-based. It exerts an interacting and cumulative effect on the whole economy. It facilitates technical progress. In all these and several other ways, capital formation promotes economic growth.

It could of course be argued that without the presence of other factors favourable to development, the supply or creation of capital alone would not be of much avail, indeed, as pointed out above, it has often been argued that economic development is a matter of changing social attitudes and economic institutions rather than a simple process of increasing capital per head. Yet the history of economic development shows that widespread changes in attitudes, values and institutions came about in the very process of economic development and not prior to development.

Table 3 : Cost and returns for different freshwater aquaculture technologies

Cost of Cultivation/	*Carp Polyculture*			*Sewage fed*		*Weed Based*	*Integrated*				*Pen culture*	*Air Breathing*	*Prawn culture*	*Carp-Culture*
	Low input	*Medium input*	*High input*	*Without feed*	*With feed*		*Duck*	*Poultry*	*Pig*	*Paddy*				*Prawn*
Cost														
Lease value (year[1])	10000	10000	10000	10000	10000	10000	10000	10000	10000	5000	2000	10000	10000	10000
Pond preparation	7500	7500	7500	7500	7500	7500	7500	7500	7500	2000	7500	7500	7500	7500
Fertilizers & Lime	10000	7500	7500	2500	2500	2500	2500	2500	2500	2500	7500	7500	7500	7500
Fingerlings (Seed)	3500	7000	20000	7000	7000	3500	3500	3500	3500	3500	7000	20000	30000	15000
Bird/Animal							3600	4000	450					
Paddy										7500				
Pen											30000			
Feed (Bird/Animals		60000	20000		30000		10000	50000	7500					
Fish Feed				7500	7500						20000	80000	60000	50000
Sewage cost														
Harvesting	5000	15000	30000	10000	15000	20000	15000	15000	15000	15000	15000	30000	30000	15000
Miscellaneous	3000	5000	10000	5000	5000	5000	5000	5000	5000	5000	5000	5000	5000	5000
Interest	2925	8400	21375	3713	6338	3638	4238	7313	4163	3038	7050	12000	11250	8250
Total cost	41925	120400	306375	53213	90838	52138	61383	104813	59663	43538	101050	172000	161250	118250
Fish yield (tonnes)	2.5	6	12.5	3	5	3	3	3	3	1	4	4	1.5	3
Gross returns	75000	180000	375000	90000	150000	90000	110000	148000	96400	60000	120000	240000	300000	190000
Profits	33075	59600	68625	36788	59163	37863	48618	43188	36738	16463	18950	68000	138750	71750
B.C ratio	1.79	1.50	1.22	1.69	1.65	1.73	1.79	1.41	1.62	1.38	1.19	1.40	1.86	1.61

Note: The obtained from Duck, Poultry and Pig is 2,5 & 16 quaintals per ha respectively and eggs obtained are 8000 & 28000 nos. from ducks and poultry respectively. The values of these outputs were included in gross returns wherever applicable.

Process of Capital Formation : The process of building up the necessary stock of capital equipment requires huge resources for financing it. Either a part of national income must be saved for the production of capital goods of the necessary funds for the purpose must be borrowed from abroad. The various methods of financing economic development, will be discussed in detail in a separate section. Here we may only emphasize that domestic saving is a sine qua non of capital formation. In fact, Professor Arthur Lewis has transforming a country from a 5 per cent to 15 per cent saver. But saving though necessary are not sufficient for the purpose of capital formation, which involves the following two independent activities:

(*a*) a finance and credit mechanism, so that the available resources may be availed of by private investors or government for capital formation; and

(*b*) the act of investment itself, so that resources are used for the production of capital goods.

Although Schumpeter showed that investment can and does exceed by the banking system, yet the requirements of capital accumulation cannot be simply met by monetary expansion. Without additional real savings, monetary expansion may merely generate inflation. The basic point is that the cost of development must be measured in real terms and not in monetary terms. The real costs are those of the resources that must be mobilized to carry out the development programme: the foreign and domestic services, materials, and equipment directly required for its execution; and the additional goods and servives for which more demand will indirectly be created through development expenditures.

Can capital accumulation take place without technological progress? A community could just go on building more transport facilities, more sources of power, more factories of the existing type. This process of duplication the existing technique is sometimes called "widening of capital," in contrast with "deepening of capital", which implies use of more capital-intensive techniques. In fact, capital accumulation and technological go hand in hand. Technological improvement is virtually impossible without prior capital accumulation. This is because the most efficient techniques require heavy investment for their introduction, even if they reduce capital costs per unit of output, once they are installed and are operating. Thus, no nation, that is not willing either to save and pay taxes to save and pay taxes or to borrow abroad, will enjoy the fruits of the advanced techniques.

Capital-output Ratio

Meaning : Apart from the ratio of capital formation to the aggregate national income, the growth of output depends upon the capital-output ratio. "The capital-output ratio may be defined as the relationship of investment in a given economy or industry for a given time period to the output of that economy or industry for a similar time period." The capital-out-put ratio thus determines the rate at which output grows as a result of given volume of capital investment than a higher capital-output ratio. For example, a capital-output ratio of 3:1 would mean, in Indian rupees, that a capital investment of Rs 3 results in the addition of output worth Re. 1. Hence, given the output, smaller capital in investment would be needed if the capital-output ratio is lower than when it is higher.

Factors Determining Capital-output Ratio : It is difficult to estimate the capital-output ratio for an economy. The productivity of capital depends upon many factors such as the degree of technological development associated with capital investment, the efficiency of handling new types of equipment, the quality of managerial and organizational skill, the existence and the extent of the utilization of economic overheads of investment. For instance, the higher the proportion of investment devoted to the production of direct commodities, the lower the capital-output ratio; and higher the proportion of investment devoted to public utilities, *i.e.*, economic and social overheads, the higher shall be the capital-output ratio, and vice versa. Higher the investment devoted to heavy industry, the higher will be the capital-output ratio, and vice versa. Higher the rate of investment and greater the technological progress, the lower will be the capital-output ratio. The capital-output ratio also varies with the prices of inputs.

Why High in Under-developed Countries : It is agreed that capital-output ratio in under-developed countries is generally higher, *i.e.,* the capital is less productive in them than in developed countries. This is so because there is a relative inefficiency of the industries which produce capital goods. There is the greater wastage of capital in the process of production due to low level of technical knowledge and there is the scarcity of economic overheads. Besides, owing to indivisibilities, certain kinds of investment are bound to be initially under-outilized. As development proceeds, naturally the pattern of demand will shift towards the more capital intensive industries.

Various estimates have been made of capital output ratios in poor countries. A group of experts appointed by the United Nations used a ratio ranging from 2 : 1 to 5 : 1. The Second Five-Year Plan of India assumed an average capital-output ratio of 23 : 1. It was 2 : 1 in the First Five Year Plan. Kurihara has assumed that in most under developed countries the ratio is of the order of 5 : 1. Singer in his model of economic development assumed a ratio of 6 : 1 in the non-agricultural sector and 4 : 1 in the agricultural sector and Rosentein-Rodan estimates that the ratio is at least 3 : 1.

Importance : Thus, the objective of capital accumulation, howsoever important, should not be over stressed. For to gain the most from capital formation, a country must also undergo technological and organizational progress, so that the capital may be used more productivity. The growth of the rate of output depends not only on the amount of capital accumulated but also on how much capital is required per unit increase in output (*i.e.,* capital-output ratio). A low capital-output ratio is, thus, as significant as capital accumulation. But it must also be pointed out that a low ratio requires technological and organizational progress. So that capital becomes more productive.

Thus, capital-output ratio plays a vital role in accelerating economic growth. The lower the capital-output ratio, more accelerated is the economic growth. The capital-output ratio can be reduced by means of technological progress and administrative improvements.

Stages of Economic Development

Prof. Rostow, an eminent economic historian and a specialist in economic development, has divided the historical process of economic growth into three states: (1) the preparatory stage, (2) the 'take-off' period and (3) the period of self-sustained growth. Now a word about each of these.

Preparatory Stage covers a long period a century or more during which the preconditions for take-off are established. These conditions mainly comprise fundamental changes in the social, political and economic fields; for example (*a*) a change in society's attitudes towards science, risk-taking and profit-earning; (*b*) the adaptability of the labour force; (*c*) political sovereignty; (*d*) development of a centralised tax system and financial institutions; and (*e*) the construction of certain economic and social overheads like rail-roads and educational institutions.

The "Take-off" : This is the crucial stage which covers a relatively brief period of two or three decades in which the economy transforms itself in such a why that economic growth subsequently takes place more or less automatically. "The take-off" is defined as "the interval during which the rate of investment increases in such a way that real output per capita rises and this initial increases carries with it radical changes in the techniques of production and the disposition of income flows which perpetuate the new scale of investment and perpetuate thereby the rising trend in per capita output."

The term 'take-off' implies three things, firstly the proportion of investment to national income must rise from 12 to 15 per cent definitely outstripping the likely population increase; secondly, the period must be relatively short so that it should show the characteristics of an economic revolution; and thirdly, it must culminate in self-sustaining and self-generating economic growth.

Period of Self-sustained Growth : This is, of course, a long period of self-generating and self-propelling economic growth. The rates of savings and investment are of such magnitude that economic development becomes automatic. Overall capital per head increases as the economy matures. The structure of the economy changes increasingly. The initial key industries which sparked the take-off decelerate as diminishing set in. But the average rate of growth is maintained by a succession of new rapidly growing sectors with a new set of pioneering leaders; the proportion of the population engaged in rural pursuits declines, and the structure of the country's foreign trade undergoes a radical change.

It is both with the problems and the cyclical movements of national income in such growing economies in the third stage that the bulk of modern theoretical economics in concerned. The students of contemporary under-developed countries and also of economic history are more likely to be concerned with the economics of the first two stages, that is, the economics of the preparatory and the 'take-off' stages. If we are to have a useful and adequate theory of economic growth, it must, obviously, be comprehensive enough to embrace these two stages as well, especially the economics of the "take-off".

On y-axis capital, investment and savings are measured on x-axis income is measured. The 45 degree line measured $I = S$ (investment = savings) and $S_{0,}$ S_1, S_2 .... is higher rate of economic growth.

(*i*) Initially the economy is at 'Y_0' which brings about 'S_0' savings which leads to increase is 'I_0' investment, this increases the rate of income in the economy from 'Y_0' to 'Y_1'. (Due to K_1Y_1 Capital-out-put ratio)

(*ii*) When economy settles at Y_1' income, which brings 'S_1' savings and 'I_1' investment. This shifts the economy from Y_1 to Y_2 (Due to K_2 Y_2 capital output ratio) through "$I_1S_1Y_2$ route.

(*iii*) As economy settles at 'Y_2' this brings about "S_2" saving to and 'I_2' investment. Hence through the '$I_2S_2Y_2$' route that means on due to I investment the income increases to 'OY_3' and further the capital output ratio increases to 'K_3Y_3.

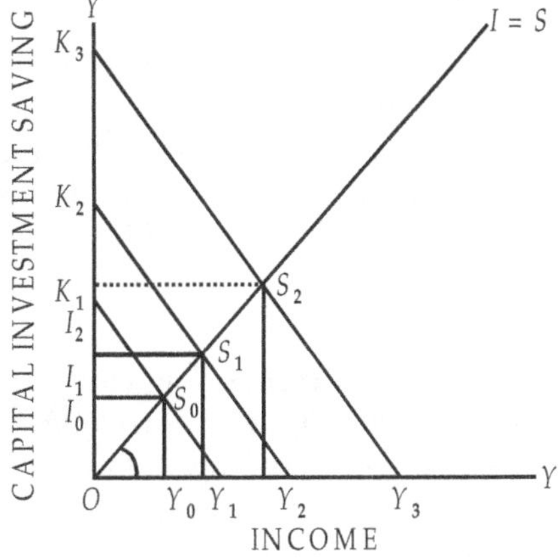

Self sustained im mulative process

In this way the economy reaches into a self-sustained cumulative process, which automatically increases the growth rate from S_1,S_2,S_3.........and so on.

Determinants of Economic Growth

We have said that economic development means the transformation from low income to high income society. Let us see now the conditions which facilitate this transformation of maintain a sustained and steady rate of growth. The process of economic development is a highly complex phenomenon and is influenced by numerous and varied factors, such as political, social and cultural factors. As such, economic analysis can provide only a partial explanation of this process. To repeat here the remark of Prof. Ragnar Nurkse in this connection, "Economic development has much to do with human endowments, social attitudes, political condition and historical accidents. Capital is a necessary but not a sufficient condition of progress." The supply of natural resources, the growth of scientific and technological knowledge—all these too have a strong bearing on the process of economic growth. We shall briefly notice some of these factors one by one. Form the stand-point of economic analysis, the most important factors determining that rate of economic development are :

(*i*) Availability of natural resources;
(*ii*) The rate of capital formation;
(*iii*) Capital-output ratio;
(*iv*) Technological progress;
(*v*) Dynamic Entrepreneurship;
(*vi*) Rate of growth of population;
(*vii*) Social overheads like education and health;
(*viii*) Non-economic factors.

Availability of Natural Resources

The quantity and quality of natural resources vitally affect the economic growth of a country. Among the natural resources, we generally include the land area and the quality of the soil, forest wealth, good river system, minerals and oil-resources, good and bracing climate, *etc.* A country's productive capacity largely depends on the natural resources available. Without a minimum availability of natural resources it is idle to expect any sizable economic growth. But it may be noted that the existence of natural resources is not a sufficient condition of economic growth. For instance, India is blessed by natural with good and sufficient resources, yet it is poor and under-developed. This is due to the fact that the natural resources have not been properly harnessed and fully exploited. Hence, availability of natural resources by itself cannot bring about economic development. Ability to utilisation is also required.

The supply of natural resources can be increased by research and technological progress. Technological progress helps in the discovery of new resources, *e.g.*, oil resources in India and putting to economical use resources which have been lying useless hitherto. Also, the shortage of some natural resources can be made good by syntherials. For instance, in the advanced countries, synthetic rubber is being used more and more in place of natural rubber and nylon is being largely used for natural silk.

The use of natural resources and their contribution to economic development depends on the type of technology. The resource use has a close connection with the type and level of technology. To know this one need not go fat in to history. For instance, petroleum which is considered so valuable today was not considered so important a short while ago. Now on account of scientific discoveries and technological development petroleum is regarded vary useful. Besides, just now radioactive substances are considered very valuable. There is no doubt that there exists in the under-developed countries abundant mineral resources which are not being used owing to the lack of technological progress.

Capital Formation

According to classical economists, the main factor, which helped capital formation, was the accumulation of capital. Profits made by the business community constituted the major part of the savings of the community and what was saved was assumed to be invested. Adam Smith too emphasized the virtues of savings. He said: "Capitals are increased by parsimony and adminished by prodigality and misconduct." Keynes also ascribed the economic development of Europe to the accumulation of capital. He said: "Europe was so organised socially and economically as to secure the maximum accumulation of capital." Later, Schumpeter showed that increased investment made possible a rise in gross output in money terms.

Thus, the crux of the problem of economic development in an under-developed economy lies in a rapid expansion of the rate of its capital investment so that it attains a rate growth of output which exceeds the rate of growth of population by a significant margin. Only with such a rate of capital investment will the living standards begin to improve in a developing country.

Joan Robinson was perhaps the first economist who used the term 'disguised unemployment'. But she used this term for people taking to occupation with comparatively low productivity and income instead of occupations of high productivity and large income during periods of depression in the developed and advanced countries. But the term 'disguised unemployment' is used in a different sense in the under-developed countries.

In the under-developed countries, 'disguised unemployment' refers to a situation where too many people are engaged in agriculture. A common characteristic of the over-populated under-developed countries is that a large majority of population draw their livelihood from agriculture. In a situation of rapidly increasing population and owing to slow rate of industrialisation, naturally a large number of people gravitate to land, because sufficient employment opportunities are not available in the non-agricultural sector to absorb the growing population. The result is that more people are apparently engaged in agriculture than are warranted by the size of the land holdings and capital available and the techniques of cultivation. If some of them are withdrawn, it will no reduce agricultural output and may perhaps increase it, because as it is said too many cooks spoil the broth. This disguised unemployment is found in the self employed agricultural population. The term 'disguised unemployment' is used to such a situation because such people are only apparently employed. In fact they are unemployed or only partly employed and their unemployment is concealed. Since more people seem to be working in agriculture than it is necessary, some of them can be withdrawn without reducing the total output. In other words, their marginal productivity is zero.

In Nurkse's words, "These is disguised unemployment in the sense that even with unchanged techniques of agriculture, a large part of the population engaged in agriculture could be removed without reducing agricultural output...... The same from output could be got with a smaller labour force."

Some economists are of the view that the term 'disguised unemployment' refers to seasonal unemployment, because all workers are able to get full employment during the harvesting season. This is true to some extent, but even in the harvesting season, work can be so arranged as to be able to manage is with smaller number or people. Even when employment is seasonal, there is still the question of making a productive use of this labour. The seasonal unemployment too has an important role to play in capital formation in under-developed countries. Economists like Nurkse think that disguised unemployment is not merely seasonal in the under-developed countries but is to be throughout the year.

Difference between Disguised Unemployment and Open Industrial Unemployment

The disguised unemployment of under-developed countries in agriculture is different from the open industrial unemployment to be found in the developed countries. The cause of open unemployment in the industrial countries is the deficiency of effective demand during depression. Owing to a reduction in aggregate demand, output is reduced in some factories and other factories are altogether closed on account of lack of demand for their goods. As a result, labour employed in such countries is retrenched. Thus, there is open unemployment of industrial labour, in spite of the availability of capital. The cause of this unemployment, is the reduction in aggregate demand. This type of unemployment can be removed by increasing aggregate demand by creating new money or by deficit financing, *i.e.*, by putting new purchasing power in the hands of the people.

On the contrary, the disguised unemployment to be found in the agricultural sector in the under-developed countries is due not to the deficiency of demand, as in the case of open unemployment in the industrial sector of the developed countries, but to the deficiency of capital equipment, *i.e.*, a low rate of capital formation as compared with a high rate of population growth.

In other words, the disguised unemployment in the under-developed countries is caused by a lack of capital formation, industrialisation and economic development commensurate with the rapid in crease in their population. That is why it cannot be cured by deficit financing and by creating new money. Deficit financing would merely raise prices in such countries and there would be inflation because owing to deficiency of capital, output of goods cannot be increased in these countries as fast. Hence, deficit financing will have no effect in removing this unemployment.

Another important difference between agricultural disguised unemployment in the under-developed countries and the open industrial unemployment in the developed countries is that, in the developed minor jobs for a temporary period during depression the rights and privileges and conveniently ignores their duties and responsibilities.

(*v*) **Productive Employment of Surplus Labour** in Disguised Unemployment. In the under-developed countries, there is lot of surplus labour to the found in the farm of disguised unemployment. Here it may suffice to say that in the agricultural sector, in the under-developed but over-populated countries, more people are apparently employed than there is need for them. This surplus labour can be withdrawn from agriculture without in any way diminishing the agricultural output (since in agricultural their marginal productivity is zero) and they can be employed elsewhere more productively, *e.g.*, in road making, irrigation works which are labour-intensive. But the full effect of capital contribution from the transfer of surplus from agriculture would follow only if their consumption level does not rise. That is, the labour left behind does not consume more than before nor does the labour transferred to more remunerative employment, start consuming more, otherwise the saving and investment potential will be reduced. The level of consumption can be prevented from rising by means of direct or indirect taxation.

(*vi*) **Encouraging Investment :** So far we have tried to tackle the problem of capital formation from the supply side, *i.e.*, side of savings. Now let us see what can be done to break the vicious circle of poverty on the side of demand, *i.e.*, investment side. We said that the under-developed countries are poor because there is not much inducement to invest. Obviously, if active steps are taken to encourage investment, the level of output and income will rise. Through wise monetary and fiscal policies, the Government can encourage investment. The Government may follow cheap money policy and give tax concessions and rebates on new investment. For instance, there are provided in India tax holiday for new enterprises, liberal depreciation allowance in corporation tax, *etc.* Protection is granted to domestic industries from foreign competition. Infrastructure (*i.e.*, economic and social overhead) are built up to promote trade and industry. Industrial estates are set up and so on.

If the flow of finance into investment is obstructed by institutional factors, it can be facilitated by making institutional changes and by setting up financial institutions, *e.g.*, in India were set up Industrial Finance Corporation of India. State Financial Corporations, Industrial, Credit and Investment Corporation, Industrial Development Bank of India, *etc.*

(*vii*) **Strategy of Balanced Growth :** Again, to break the vicious circle of poverty of the demand side of capital formation, Nurkse recommends the strategy of balanced growth. According to him, if investment is made in one particular industry, it is likely to fail owing to low income and low purchasing power of the people. That is why private investors are discouraged from investment in a particular industry. But Nurkse says investment is made in several industries simultaneously, then the persons employed in different industries become consumers of the goods produced by one another since they have all acquired more purchasing power. That is, the industries in which investment has been made create demand for one another. In this way, balanced growth, in which investment is made simultaneously in a number of industries, creates its own demand. This is how in Nurkes's opinion, the vicious circle of poverty can be broken on the demand side by means of balanced growth.

Industry	Agriculture
Construction	Transport

Balanced growth calls for simultaneous in investment in various sectors.

The low level of income reflecting low productivity is the crucial point both in the demand circle and the supply circle. Of these, the supply end in the more difficult to break than the demand end. It is obviously ease to create or increase demand for capital but it is not so easy to make up the deficiency of capital. The country may also suffer from lack of natural resources like water and mineral resources or the poverty of the soil. But in the matter of economic development, the things of crucial importance are the small capacity to save and small inducement to invest. Other deficiencies can be made up and the handicap of the natural factor removed, if the problem of capital formation is successfully tackled.

Foreign Aid and Its Role in Economic Development

We have already referred to a low rate of capital formation as one of the primary causes of the vicious circle of poverty in the under-developed countries. The domestic saving rate being very low in such countries. Foreign and assumes great significance if a poor country wants to come out of the vicious circle. Let us therefore consider at some length first the concept of foreign aid and then its role in economic development.

Role in Economic Development

The objective of foreign aid is the achievement of sustained economic growth by the recipient country *i.e.*, achieving target rate of growth which can be sustained without further external assistance.

We may notice three basic approaches to foreign aid requirements for a developing country: (*i*) The Savings-investment gap approach, (*ii*) Foreign exchange earnings and expenditure gap; and (*iii*) The capital absorption approach. The first two approaches *viz.*, the Saving-investment gap and foreign exchange earnings and expenditure gap yield identical results. Foreign aid is equal to both the gap between imports and exports and the gap between domestic investment expenditure and domestic savings.

The third approach *viz.*, the capital absorption approach assesses the capital requirements of a developing country on the basis of the ability of an economy to utilise both domestic and foreign capital efficiently *i.e.* it should yield a minimum rate of return. In other words, it has to be seen that foreign aid is not just frittered away in senseless and useless plans. Foreign aid is regarded as a means of overcoming internal obstacles to growth and as a catalyst for mobilising domestic resources for economic development. According to H.M. Chenery and H.M. Strout, foreign aid should make a contribution to the transformation of a poor stagnant economy by raising the levels of skills and improving economic organisation through removing resource bottle-necks and encouraging self-help measures in the administration of foreign aid. This is a more comprehensive view than merely focussing attention of investment-saving gap or import requirement and foreign exchange earnings gap.

Thus, foreign aid makes a significant contribution to the acceleration of the pace of economic growth (*a*) by overcoming shortages and (*b*) by supplementing domestic resources.

As per the Census 2001 report, out of the country's total 402.51 million workers, 235.08 million are directly dependent on agriculture. A majority of them consists of small and marginal farmers and agricultural labourers, and a large numbers of them are either unemployed or under-employed. Although the per centage of workforce dependent on agriculture has declined from 64.8 per cent in 1991 to 58.40 per cent in 2001, the absolute number has significantly increased. If we look at the contribution of this sector to the total GDP (about 22 per cent in 2003-04), this per centage dependence on agriculture seems to be very high and indicates to the intensity of disguised unemployment and under-employment in the sector. The employment absorption capacity of the sector may further decline, due to the application of labour-saving technology; declined public investment; and lifting of quantitative restrictions (QRS) on imports of agricultural products. Similarly, urban organised manufacturing sector does not appear to be promising in absorbing growing workforce. This is evident from decline in the absolute number of jobs in public sector enterprises due to retrenchment and disinvestment policy and a slow growth in employment in private manufacturing sector due to application of more capital intensive technology. Although, urban informal sector is growing very fast and a large number of workers being absorbed by it, but the exodus of rural workers in urban areas has been creating serious problem, to the carrying capacity of urban sector. Under these constraints, the viable option for providing gainful employment, to the growing rural workforce and to lessen the burden of manpower on agriculture seems to be in generating more employment, in the Rural Non-Farm Sector (RNFS), which constitutes mainly three components: services such as hotels and *dhabas,* transport and communication, repairs shops, tailoring and hair cutting, cobblers etc.; trade and commerce such as wood, iron cloth, groceries shops, medical stores, and other business; and manufacturing such as shoe making, farm implements, bakeries, pottery, basket making, soap making, rice mills, sugar and Khandsari, fruits and vegetables processing units, dairy and meat processing units and other small and cottage industries.

Review of Development Initiatives

During the planning period, rural development strategy has shifted from growth oriented to welfare oriented and further to empowerment oriented. Up to the Fourth Plan, our developmental strategy was based no the 'trickle down theory' in which emphasis was laid on acceleration of growth on the presumption, that its benefits would percolate down to the lowest strata of society and would spread evenly among the different regions. This 'top down approach' of development could not prove effective in alleviating rural poverty and reducing regional imbalances. Therefore, from the Fifth Plan, Government of India tilted its development strategy towards the welfare of downtrodden and underprivileged sections of rural society and also towards the development of backward areas to reduce regional imbalances. Since then, a number of targets oriented wage-employment and self-employment rural development programmes have been implementing by the government. But the major weakness of these programmes was that the people, for whom these programmes were made, were involved in their conception, planning, execution and monitoring. However, with the amendment of 73rd Constitution Act, this weakness has been removed to some extent and the participatory development process can become a reality in rural area. Under this new dispensation, rural industrialization is also the subject of *Panchayati Raj Institutions (PRIs).*

Another drawback of the self-employment programmes was that the group entrepreneurship approach was not followed in them. DWACRA was the only programme that was based on the group approach, covering 10-15 poor rural women in a group. The programme aimed at developing income-generating skills and promoting activities among them. However, due to lack of literacy and entrepreneurial skills among the rural poor women, the programme could not succeed in achieving the objective for which it was made.

TRYSEM, another self-employment programme, started with the objective to developing technical and entrepreneurial skills among poor rural youth to enable them to take, up income-generating activities, also could not make much impact on the generation of productive employment. The mid term appraisal of the 9th Plan observed that the trainees under the programme were only interested in the stipends; they did not apply the training inputs for furthering self-employment prospects. Due to rapid changes in rural market, small entrepreneurs under TRYSEM find it difficult to have a niche in the market.

More recently, Government of India has launched a new self-employment scheme- Swarnjayanti Gram Swarozgar Yojana (SGSY), which is based on group approach. This scheme replaces the earlier self-employment and allied programmes, such as, IRDP, TRYSEM, DWCRA, SITRA, GKY, and MWS. It aims at establishing a large number of rural micro enterprises. It is rooted in the belief that the rural poor in India have competencies and given the right support, they can be successful producers of valuable goods, services. This scheme is conceived as a holistic programme covering all aspects of self employment, *viz.,* organisation of the rural poor into self-help groups (SHGs) and their capacity building, planning of activity clusters, infrastructure build up, technology, credit and marketing. It is a credit-cum-subsidy programme. It envisages a greater involvement of the banks in the planning and preparation of projects, identification of activity clusters, infrastructure planning as well as capacity building and choice of activity of the SHGs. The programme also provides for promotion of marketing of the goods produced by the Swarozgaris. This would involve market intelligence, development of markets, consultancy services as well as institutional arrangement for marketing of the goods including exports.

Review of these self-employment schemes evince that the policy focus has largely been on poverty alleviation rather than on creation of income and wealth on sustainable basis through investment in the productive employment generating activities. As beneficiaries of most of these lack necessary skills to run effectively their activities chosen under the schemes. A big or medium enterprise can easily hire the workers with different skills. However, it is not feasible for a micro enterprise to have access to all these skills, and therefore, due to lack of necessary skills, many of them could not survive in the rapidly changing markets. Therefore, development of rural industries under group entrepreneurship may be a good option in this regard.

Potential for Investment in Rural Industries

Rural industry constitutes only 7.4 per cent of total rural workforce. Expansion of rural industrialization can play a vital role in abating distress migration from rural area; stopping skill drain; reducing the pressure on urban civic services; raising local demand for building rural infrastructure; and boosting rural income and employment. India produces annually over 200 million tonnes of food grains and about 130 million tonnes of fruits and vegetables. She ranks second in the production of fruits and vegetables and first in milk production and fourth in fisheries. The country also has the largest cattle population in the world and holds a huge size of other livestock. However, this huge supplying primary agricultural products has yet partially been utilised as raw material by the food-processing industries. For instance, in India only two per cent of fruits and vegetables are processed compared to 70 per cent in Brazil, 30 per cent in Thailand, 78 per cent in Philippines and 80 per cent in Malaysia (Government of India, 2001). Due to lack of proper transport, storage, and processing facilities, annual post-harvest losses in India are as much as Rs. 50000 crores. These losses could be minimized through investment in transport, storage, and communication infrastructure and investment in the processing of primary agricultural products.

In Uttar Pradesh, Panchayat Udyogs (PUs) have been set up, as rural micro enterprises, to provide employment to rural artisans and other weaker sections, and to ensure proper utilization of locally available resources. Currently, over 900 PUs are working in the state. They represent a unique model, where micro enterprise is promoted through local bodies. Over 1000 products, such as, wooden and iron furniture, cotton mats, printing press, boats, blankets, candle, slate, leather products stationery, stone work, horse and buffalo carts etc. are being manufactured by them. Thus, PRIs have already some experience to run micro enterprises. These enterprises need to be revitalized and upgraded through private partnership.

For promoting group entrepreneurship among the rural enterprising youths. PRIs should identify the educated unemployed youths having diploma in engineering, ITIs other management courses in their areas and organize them in groups for jointly taking up some enterprising projects. Each group man have 5-10 rural youths trained in different skills. Such groups may also be integrated with SGSY by involving the beneficiaries of the programme. Agro and non-agro based rural small-scale industries should be established by PRIs in joint venture by involving these identified engineering diploma holders, ITI certificate holders and other skilled rural unemployed youths. These industries should be based on the locally available resources. The funds of such industries may be mobilised through bank loans and subsidy given to the target groups. The role of financial institutions and NGOs because very vital in this regard. Gram Panchayat should be entrusted the responsibility of identifying the different type of unemployed workers willing to join the group and formation of groups for taking up the different projects. In forming the group, more than one GP may be involved. Kshetra Panchayat should conduct survey of natural, physical and human resources of the area and identify the various productive activities based on the locally available resources. Block Development Office should create information database that may have, among others, the details of unemployed rural youths having training in different skills. The database should be well connected through internet, with other development block, district statistical office, training institutes, and the district panchayat office. The overall promotion and development of rural industrialization through group entrepreneurship should be left of the district Panchayat.

There will be several advantages from such rural industries. At present, gram panchayats do not have adequate financial resources of their own Grant-in-aid from state and central governments, is the only source of their income for the implementation, of various rural development activities. These rural industries, in which gram panchayats will have their share capital, may be an important source of their income. The employees of such industry will also be able to generate sustained income for their families because over and above their capital. Since, the owners of firm will also be its employees; they will work with complete devotion and sincerity to enhance productivity, profitability and income. In this way, the twin objective–economic growth and social justice may be achieved in a more meaningful way. To study the effectiveness and feasibility of the scheme, pilot projects man initially be launched in a few selected blocks.

ESSENTIAL FOR GROUP ENTREPRENEURSHIP DEVELOPMENT

Skills Formation

Improved technical and others skills among youths associated with rural enterprises are of prime importance, for enhancing the productivity and marketability, of their products in the competitive market environment. We have a huge infrastructure, for vocational training in the country that needs to be upgraded and made demand-driven, so that trainees may have access to those skill, which are required in the changing labour market. There are 4.274 Industrial Training Institutes (ITIs) with an intake of 6.2 lakh students and six Advanced Training Centres (ATCs). Apart from these, there are also private institutes providing various types of training in areas ranging from computer applications to catering. The KVIC also provides training through its 51 training center and 10 training-cum-production centres. There is also a scheme of Community Polytechnics (CPs) run by polytechnics to provide short term training to youths. NGOs also provide training to skill development with financial assistance from CAPART.

It has been observed that the training provided in most of these institutes is for the skills and have little demand. The curriculum is not attuned to current market requirements due to lack of revision. The training is oriented towards individual job seeking and not towards setting up self-owned enterprises. Trainees lack multi-skills, and are not generally trained in skills to having access to information, credit or market knowledge. The training is a one-time affair, and not a continuous process.

Strengthening the employability of the youth should be the main concern of training institutes. Training modules, should be attuned with the changing needs of the labour market, minimizing mismatch between demand and supply of different skills. After formation of the group of young entrepreneurs, each group should be provided training for developing necessary skills, related to the activity identified by the group for micro or small enterprise. Knowledge about availability of technology, machines and equipments, raw materials and other inputs, finance, and marketing of the products should be provided to them during the training. Apart from these, it is also expected that the trainees would also be imparted training in learning, new skills, required in the changing world of work. However, skills formation among the rural youth is the necessary, not the sufficient condition for rural industrialization. Post-training follow-up assistance in the form of other support services, such as assistance in the procurement of technology, inputs, and finance and marketing support is equally important.

Up-gradation of Training Institutions

We have large number of training institutes in the country. For promoting entrepreneurship development and strengthening of the human capital base, training infrastructure should be modernized and upgraded. The existing infrastructure of DIC, ITIs, rural polytechnics and other institutions supporting small industries should be strengthened. District Panchayat Office should have a training department, that may work as a coordinating agency to create linkages, with all training institutes and technical colleges under the jurisdiction of the district.

Developing Effective Market Mechanism

Experience shows that many of micro and small enterprises in the country, could not survive, due to lack of market support. Their growth or decline, largely depends on markets, links with larger firms, and access to technology, credit, skills and marketing. There is, therefore, a need to develop a suitable institutional market mechanism for the products of these enterprises. In case of certain products, the existing marketing infrastructure of KVIC and other government outlets could be used. Cooperative marketing institutions could also be involved. What is more required is to build brand equity for their products. This could be done by developing a market mechanism under the decentralized framework of cooperative marketing.

Policy Support

1. Well coordinate efforts from NABARD, SIDBI, other commercial banks, rural local government, and NGOs are required to promote group entrepreneur ship. To study the effectiveness and feasibility of the scheme, a few pilot projects can be initiated by them in some selected development blocks.
2. Government should provide subsidy to the group in the same way as the self-help group under SGSY receives.
3. Micro and small enterprises, to a greater extent, suffers from the technological inefficiencies and lack of effective demand. Therefore, transfer of cost-effective technology and its constant upgradation is necessary for raising productivity, improving product quantity and competitiveness, and increasing profitability of the enterprise. B.Tech students from local engineering institutions, can be motivated to take up-projects, on rural technology up-gradation. There is need to build an effective linkage between the technical institutions and these enterprises.
4. Currently, development blocks in the states, like Uttar Pradesh do not have a sound mechanism, for information and data management. They do not have internet link with each other and with the district statistical office. State planning department of Uttar Pradesh has done a commendable job by providing on line data related to districts and in some cases even to block. However, specific information related to rural industrialization is not available. Central government should devolve adequate funds to block development offices, for computerization and creating a database, especially having information on unemployed youths trained in different skills.

Conclusion

Agriculture cannot provide gainful employment to the entire rural workforce. The organized sector has been shrinking due to retrenchment and investment policy in public sector and application of more capital intensive technology by the private sector. The fast growing urban informal sector is putting more pressure on the carrying capacity of urban sector due to influx of workforce from rural areas. Our past experiences with the rural self-employment schemes show that the policy focus has largely been in poverty alleviation, rather on creation of income and wealth on sustainable basis, through investment in workers' education and skill in this backdrop, development of group entrepreneurship, among rural youths may be a good alternative strategy for rural industrialization. A group of 5-10 youth having trained in different skills can jointly run micro or small enterprises with the involvement of PRIs and assistance from NGOs, financial institutions, and the government. Initially pilot projects can be started in a few selected blocks to assess the effectiveness of the approach. To create enabling and conducive environment for the promotion group entrepreneurship in the rural areas, several policy interventions, such as the development of institutional market mechanism, arrangement of necessary inputs and finance, revitalizing the existing training infrastructure and its orientation towards integrated training focussing on group entrepreneurship, and transfer of modern technology are required.

Chapter 5

Emerging Trends of Coastal Aquaculture in India

Among the various food producing sectors of the world, aquaculture has been the recognized as the fastest growing in the recent times. For the past decade it has grown at an average annual rate of 9%. As per latest FAO estimates for the year 2003, the contribution from aquaculture stood at 42.30 million metric tons out of the 132.52 million metric tons of global fish production (32%). China continues to be the world leader in aquaculture production with 28.89 million metric tons. Although India is placed second, the production is much lower at 2.22 million metric tons from aquaculture sources.

In the recently held FAO/NACA workshop at Ramsar, Iran, while reviewing the status and prospective development of Asian aquaculture it was observed that six major trends are seen in Asia Aquaculture of late.

- Increasing intensification driven by restrictions on and limits to aquaculture expansion
- Continued diversification of production systems
- Increasing influence of markets, trade and consumption patterns
- Enhanced regulation and better governance
- Drive for better management

India, being part of Asia the trend and situation in aquaculture is not an exception to this and appears to be more or less similar to other countries in the region and the following chapters will highlight some of these trends and challenges observed in coastal aquaculture.

From time immemorial, coastal shrimp/scampi culture is an important component of aquaculture in India, and we cherish a long history of traditional brackish water farming in some of the coastal States. However, the development and progress of this sector on scientific lines has taken place only during the past two decades. The combined production of shrimp and scampi through aquaculture in India during 2004-05 was 0.16 million metric tons. Although coastal aquaculture production of shrimps appears to be much lower in terms of quantity compared to inland fish production (3.5 million metric tons), shrimp aquaculture is more significant due to the high unit value realization for shrimps. Coastal shrimp farming is contributing substantially to the rural economy of the country besides earning valuable foreign exchange. Therefore sustainability of this sector has become a key factor for rural development, social progress and also for sustaining export production.

For the past decade or so, coastal shrimp farming has been playing a leading role in the seafood exports of the country. Shrimps are the most important seafood commodities exported and about 60% of the raw material supply is from aquaculture sources.

As per the latest estimate, an area of about 2,09,710 ha. is brought under coastal shrimp/scampi culture by 2004-05 and there is an increase in production to 1,64,390 metric tons. This includes 1,25,670 metric tons of shrimp from 1,36,390 ha. and 38,720 metric tons of scampi from 36,990 ha. of farm area. It is observed that shrimp culture had developed at a fast pace during 1990's and this sector subsequently stabilized during the current decade with sustained growth. However, as compared to estimated area available for brackish water aquaculture of 1.2 million ha, the development has been just around 13.8% of the available area. Hence, huge resources in the form of suitable land are available in the country for judicious exploitation and sustainable production. The progress of shrimp and scampi farming in various coastal states during the past three years is furnished in Table 1. It could be seen that the east coast states are leading in coastal aquaculture production, with Andhra Pradesh, West Bengal and Orissa contributing significantly. Among the west coast states, Kerala is leading in culture shrimp production, while there is tremendous potential for development in other states like Gujarat and Maharashtra. The present thrust is to augment the production through increasing the productivity of existing farms as well as bringing more area under farming.

Table 1 : State-wise details of Coastal Aquaculture (Shrimp & Scampi) Production (2002–03 to 2004–05)

		2004–05		*2003–04*		*2002–03*	
Sl.No.	*State*	*Area under Culture (Ha)*	*Production (MT)*	*Area under Culture (Ha)*	*Production (MT)*	*Area under Culture (Ha)*	*Production (MT)*
1.	West Bengal	54610	38625	54375	32149	53150	30410
2.	Orissa	10218	10366	12586	12840	11995	10690
3.	Andhra Pradesh	89525	96150	96924	85209	93000	86210
4.	Tamil Nadu	3876	6818	3373	6203	3800	5120
5.	Kerala	11708	7841	14915	6699	14510	7770
6.	Karnataka	1693	1426	3291	1943	3205	2800
7.	Goa	295	534	963	700	930	710
8.	Maharashtra	567	1129	7596	1287	4880	930
9.	Gujarat	895	1502	2443	1616	1240	1130
	Total	173380	164390	196470	148650	186710	145770

Statutory powers have been bestowed with the aquaculture sector through the enactment of the Coastal Aquaculture Authority Act by the Parliament in July 2005. Infact, coastal aquaculture had been facing several legal issues, perhaps due to multi user conflicts of the available resources. Since aqua farms are scattered around in remote coastal villages, they are yet to be well organized to defend their causes. The Coastal Aquaculture Authority being constituted as per the provisions of this Act is expected to streamline the procedure for licensing of shrimp farms in the country, and provide a legal/statutory identity to shrimp farms.

MAJOR HURDLES IN THE DEVELOPMENT OF AQUACULTURE

Despite the potentials and the scope for development, the progress of development is confronted with many issues. Some of these challenges affecting the aquaculture growth are discussed below :

(*a*) Land Leasing Policies of State Governments

It is estimated that about 1.19 million ha. of brackish water area is available in the country for development of shrimp farming. Out of this, the maximum area seems to be available in West Bengal (4.05 lakh ha.) Gujarat (3.76 lakh ha.) Andhra Pradesh (1.50 lakh ha.), Maharashtra (0.80 lakh ha.) etc. The area available in various maritime states and its utilization for coastal shrimp culture is furnished in Table 2. It is observed that the area utilization in various states is low because much of the available area falls under Govt. land. Although, the state Govts. have formulated land leasing policies, the allotment is at a slow phase, resulting in poor progress.

Table 2

State	*Estimated Potential Brackish Water*	*Area Developed (Ha)*	*% of available Potential*
West Bengal	4,05,000 (34.01%)	50,405	12.44
Orissa	31,600 (2.65%)	12,877	40.75
Andhra Pradesh	1,50,000 (12.60%)	76,687	51.12
Tamil Nadu	56,000 (4.70%)	5,286	9.44
Pondicherry	65,000 (0.7%)	14,106	16.25
Kerala	1,693 (5.46%)	1,426	21.70
Karnataka	8,000 (0.67%)	1,910	23.87
Goa	18,500 (1.55%)	310	1.68
Maharashtra	80,000 (6.72%)	1,281	1.60
Gujarat	3,76,000 (31.57%)	2,271	0.60
Total	11,90,900 (100.00%)	165,263	13.882

Figures in bracket shows % of total potential of the country.

Expeditious allotment of Government land could accelerate the developments in coastal shrimp farming sector.

(*b*) Financial Support

Aquaculture is a capital-intensive project, and institutional support is essential to mobilize resources. It takes several years for repayment of loan. Unfortunately, if the Project is some times affected by unforeseen losses due to disease attack and crop loss, the farmer's repayment capacity is affected and the project ends up in doldrums. In such cases, rescheduling the repayment package is essential for breathing life in to sick projects. On the other hand, many of the corporate companies have closed down their projects, and the financing agencies are finding it extremely difficult to retrieve their money. This conflicting situation calls for redrafting norms for financial assistance to aquaculture projects. As financial institutions are taking a back seat, insurance agencies are also not coming forward to insure aquaculture operations.

(c) Infrastructure Developments

Coastal aquaculture cannot develop in isolation, as it requires strong backward and forward integration. It is observed that shrimp farming has concentrated in some areas, where the conditions are conducive. Infrastructure facilities such as hatcheries, feed manufacture/supply, processing factories, ice plants etc. have also come up in such areas. However, the development appears to be lower in other areas due to the absence of supporting industries. Therefore, it is essential to formulate areawise master plans for potential sites, for effectively planning sustained growth. Unless, the infrastructure facilities are developed simultaneously, shrimp farming in new areas cannot achieve nurtured growth.

(d) Increasing Competition in International Market

The developing countries from the South East Asian, region especially Indonesia, Thailand, Vietnam etc. have substantially increased their share of aquaculture production of shrimp, as it has become a major source of their export revenue. Compared to the progress achieved by these countries, the performance of the coastal aquaculture sector in India in the recent years has been rather slow. Therefore, unless we increase the production of quality shrimps to substantial levels, there is a chance to lose international markets.

Due to increasing competition among producing countries, the international price of shrimps has come down. Therefore, apart from boosting the production of quality shrimps, more efforts should also be focused on reduction of production costs along with widening our markets besides developing 'niche' markets for specific variety/sizes.

(e) Sustainability of Aquaculture Sector

Coastal aquaculture is an age-old profession and the traditional culture practices are still continuing in some of the Maritime States. In view of the possible threat of intensive farming on the environment, semi intensive and intensive farming practices are prohibited in India. Therefore, the development is more on extensive system of culture, which is considered as environmentally friendly. In the extensive system of culture, the stocking density is low and the risks are also less. In a country like India, where the land resources are abundant, the thrust could be on increased coverage of area under shrimp farming and not on the intensity of operation for augmenting production.

In order to ensure sustainability of the aquaculture operation, MPEDA has been implementing several programmes mainly focusing on shrimp health management. These include the testing of shrimp seeds for viral disease through PCR laboratories, registration of shrimp hatcheries which are following MPEDA's Code of Practice, introduction of effluent treatment system, re-circulation facility etc., to prevent horizontal transmission of diseases. Besides assisting the preparation of area-wise Master Plan, MPEDA has also been promoting, mangrove afforestation programme around the shrimp farms in order to ensure ecofriendly farming. Regular surveillance is also conducted by MPEDA on aqua farms to ensure scientific management. The average productivity of shrimp farms in the country is around 920 kg per hectare and as compared to many aquaculture-producing countries in the region, the culture practices followed here are more sustainable and ecofriendly.

A brief account of MPEDA's efforts towards sustainability in coastal aquaculture is furnished below :

1. Registration of Shrimp Hatcheries

Quality of shrimp seed is very critical for successful aquaculture operation. Therefore, MPEDA has prescribed certain "Code of Practices" to be followed for responsible hatchery operation. MPEDA is implementing a scheme to register the hatcheries, which are voluntarily adopting the "Code of Practices", and have the basic infrastructure facilities prescribed under this scheme.

2. Registration of Aquaculture Consultants

The country is estimated to have more than one lakh farmers and extending technical guidance and timely advice to such a vast population is a challenging task. The Govt. agencies like MPEDA and BFDAs or State Governments have limited manpower and extension machinery to provide timely technical assistance to farmers. Therefore, private consultants have entered the field and rendering technical assistance to needy farmers. In order to ensure that only qualified and experienced consultants are guiding the farmers, MPEDA has started a scheme to register aquaculture consultants.

3. Standardization of Aquaculture Inputs

Inputs used in the operation of shrimp farms are expected to affect the quality of the harvested materials. Therefore, these inputs should adhere to certain quality standards, for which stipulations have been formulated by MPEDA, and issued to trade.

4. Study on Shrimp Disease Control

The White Spot Syndrome Viral disease has been ravaging the shrimp culture activities for more than a decade now. Although, MPEDA has been taking some active steps to contain the disease, the outbreaks continue to affect in various parts with moderate intensities. Hence, every year about 10,000-15,000 metric tons of the products is lost on account of disease incidences. MPEDA has been taking several initiatives to control the situation. One of these efforts is to conduct a detailed study on "Shrimp disease control and coastal management" in shrimp farms, for which MPEDA has coordinated a research project with Network of Aquaculture Centers in Asia-Pacific (NACA). This study was carried out in major producing centers of Andhra Pradesh, the leading state in Aquaculture. The findings of the study and the interventions so evolved for successful farm operation and shrimp health management have been brought out in a Manual and are circulated to farmers for adopting better management measures in their farms. Further, the study has highlighted the role of "Self Help Group" and participatory approach by farmers in aquaculture, and as a result, aqua clubs are now taking roots with an initiative for a cooperative movement in this sector. This study and approach for farmers' empowerment is presently being expanded in to other States for initiating a successful co-operative movement in coastal aquaculture. Besides this, the participatory approach has helped farmers to get higher yields and returns and better quality products, enjoying rarer incidence of diseases on farms while causing less impact on the environment.

5. PCR Laboratories for Disease Diagnosis

Shrimp farming is besieged with a multitude of problems related to health management and disease control. In order to provide proper guidance to farmers, MPEDA has set up PCR Laboratories in various coastal states, and these labs are assisting the farmers to screen shrimp seeds for the presence of pathogens. Besides operating labs, MPEDA is financially assisting the establishment of PCR laboratories as in house facility in private shrimp hatcheries as well as in service aqua laboratories. Presently international assistance is being availed from Australian Centre for International Agriculture Research (ACIAR) and Commonwealth Scientific and Industrial Research Organisation (CSIRO) for the accreditation of PCR laboratories in the country for increasing their credibility.

6. Financial Incentives to Shrimp Farmers

MPEDA is extending subsidy assistance schemes for the benefit of shrimp/scampi farmers and hatchery operators, for meeting part of the capital expenditure. Apart from the development-oriented schemes, some innovative schemes have been drawn up towards focusing sustainability of this sector. This includes subsidy packages for setting up effluent treatment systems or re-circulation system in shrimp farms. Farmers are also given marginal financial assistance to purchase water quality testing equipments like Dissolved Oxygen meter, pH meter, salinity refractometer etc. so that optimum culture conditions could be maintained in the farm.

7. Preparation of Master Plans for Aquaculture

The long-term sustainability of aquaculture ventures is presently focused, to ensure sustained production from the farms, without causing detrimental effects on the environment. Therefore, planned development is essential to avoid adverse impacts on the environment. Area-wise master plan development is proposed with the help of remote sensing to identify potential areas and to develop aquaculture estates. Protection of agriculture lands, mangrove forests, etc. will be given due priority in such master plan which should have provision for main in take canal, effluent treatment system, reservoirs and other common facilities. Therefore, identification of suitable areas and avoiding over concentration of farms in such areas are of primary concern for sustainable development.

8. Development of Specific Pathogen Free (SPF) Brood Stock

The impact of virus attack has not remained confined to the shrimp farms, but has spread over to the wild resources. In the natural belt, the virus has infected both the targeted as well as non-targeted organisms. It is found that even the wild brooders are contaminated with various pathogens and these brooders when used for production of seed in the hatchery, results in multiplication of the pathogens, through the progeny. To prevent this, it is to be ensured that the brooders used in the hatchery are healthy and disease free. In order to produce Specific Pathogen Free brood stock, MPEDA has identified a suitable location in Andaman & Nicobar Island, for setting up a "SPF Brood stock Project" and "Domestication Center" for tiger shrimp. Initial surveys have proved that the Andaman Coast is relatively free from infection of brood stock and highly biosecure from disease risks. The project is being implemented through the Rajiv Gandhi Center for Aquaculture (RGCA), availing overseas expertise.

9. Quality Issues in Cultured Shrimp

India is the 12th largest exporter of fish and fishery products in the world. The seafood exports from the country stood at 1.48 billion US dollar during 2004-05. A product-wise analysis of the export reveals that shrimp constitute 63% in terms of value and 30% in terms of volume of the marine products exports. The share of cultured shrimp in the shrimp exports from the country has been steadily rising, and currently, it is claiming about 88% of the value.

In the international market the competition is very high and quality of the seafood plays a major role for the market edge. Therefore, aqua cultured shrimp is expected to maintain the highest quality and should be free from contaminants. The importing countries have stipulated standards for the imports and these have to be strictly adhered for avoiding setbacks. For example, the EU countries are vigilant on the residues of antibiotics or steroids. The Japanese market have complaints on muddy-mouldy smell is cultured shrimps imported from Bhimavaram and nearby areas. Infact, there has been a drastic reduction in the export of shrimps to Japan, over the past three years, due to the off flavour problems. Therefore MPEDA is actively conducting village level campaign to create awareness among farmers on the repressions of the quality issues in overseas markets. Further, a "National Residue Control Plan" is also designed and implemented by MPEDA to monitor the residues of undesired chemicals/contaminants at the tissue level. According to this plan, MPEDA is collecting samples from hatcheries, farm, processing plants, feed mills etc., and monitor them for various parameters.

These efforts are expected to result in maintaining the quality of cultured shrimps exported from the country.

10. Aquaculture Diversification Programme

Presently, our commercial aquaculture sector in the maritime States is targeting only a single candidate, *viz*, the shrimp. However, concentration of efforts on any one species is not advisable, in view of the growing seafood market. Therefore, MPEDA has been conducting demonstration programmes for alternate species to prove their commercial feasibility.

Besides, a separate society has been established for promotion of aquaculture diversification programmes of MPEDA. The Rajiv Gandhi Centre for Aquaculture (RGCA) has been piloting some efforts for introducing commercially important species such as Asian Sea bass, Groupers, Tilapia, Mud crabs, lobsters, Artemia etc. The hatchery technology as well as grow out techniques of various species is being perfected by this Centres with technology sourced from abroad.

Chapter 6

Socio-Economic Issues in Fisheries Sector

Introduction

Indian fisheries have evolved from the stage of a domestic activity during the 1950s and 60s to a status of an industry by 1990s. During the process of this transformation many changes in the socioeconomic status of fishers have taken place. This paper traces some of such changes that have taken place in the process of fisheries development. The paper addresses the profile of the fish producers in India and the changes over time, the pattern of commercialization across regions, salt and freshwater, capture and culture, between larger and smaller scales of production and between higher and lower value species. The paper also addresses the question of increasing commercialization of the fisheries sector in India and its impact on poor in terms of direct and indirect income gains, employment and the acquisition of skills that contribute to income-earning capacity.

India is endowed with 2.02 million sq. km of EEZ (Exclusive economic Zone) along with a coastline of 8129 km and 0.5 million sq. km continental shelf with a catchable annual fishery potential of 3.93 million tonnes occupying a very important strategic position in the Indian Ocean. The aquaculture resources in the country comprise 2.25 million hectres of ponds and tanks, 1.3 million hectares of bheels and derelict waters, 2.09 million hectares of lakes and reservoirs and also 0.12 million kilometers of irrigation canals. Among the Asian countries India ranks second in the culture and third in capture fish production and one of the top leading exporters of sea foods (Sampath 1998). The marine environment of India consists of unique ecosystems known for their aesthetic beauty and provide habitat for numerous biological species. The ecosystem is divided into three basic categories as estuarine, inter-tidal and coral reef. The estuarine ecosystem is a fresh water ecosystem comprising estuaries, mangroves and other wetlands rich in microscopic plant life and abundant in vegetation. They are the rich breeding grounds for larvae of some commercial species, a broad range of algae, fungi and lichens among others. More than 75 per cent of the commercial fish catch in India is dependent on estuaries for part of their life cycle. India ranks 14th in the list of world's major mangrove areas and 5th in the Indo-Pacific regions—the major mangrove areas in India include the northern Bay of Bengal and the Sunder bans (~690 sq. km) (B Sahai 1993).

Marine Fish Production

The marine fish production in India consists of a large number of species using different crafts and gears mostly in the depth range of 0-50 meters. The annual average landings during 1995-99 period was 2.5 million tonnes principally constituted by the Indian mackerel (8.5 per cent), penaeid prawns (7.7 per cent), croakers (6.8 per cent), oil sardine (6.7 per cent), carangids (6.1 per cent), perches (6.1 per cent), non-penaeid prawns (5.2 per cent), ribbon fishes (4.9 per cent), cephalopods (4.1 per cent) and others (CMFRI 1997).

The marine fish production in India is characterized by its annual fluctuations. This phenomenon has led to considerable uncertainties about investment in the production process. Marine fisheries still remains open access and suffer from overcapitalization. The near shore area within 40-80 meter depth range covering an area of 0.45 million sq. km is subjected to heavy fishing pressure (Kurup and Devaraj 2000). About 2,43,000 fishing vessels (1,82,096 artisanal crafts, 26171 motorized crafts and 3471 mechanized crafts) exploit this area, where the estimated annual potential is 2.2 million tonnes. A conservative estimation made by Kurup and Devaraj (2000) shows that the capital investment in fishing technologies (crafts and gears) at current price is about 33.4 billion, but the return per unit of investment seems to be economically not viable. The estimation of optimum size of fishing fleets which would allow sustainable yields become very important for the better utilization of scarce resources of the society.

Socio-economic Status of Marine Fishers

The coastal communities in India follow multiple fishing and non-fishing activities and most of their income in generated from open access/common property resources. The coastal poor are not confined to any one sector and change occupations as and when necessary. Most coastal people in rural areas also work as seasonal labourers in agriculture or as part-time farmers or occasional wage earners in order to supplement their family incomes. Working as labour in tourism, industries, ports, mining and other industries is a relatively new occupation and it is mostly confined to specific areas from where these industries have come up. As pointed out by many authors, the employment generation potential of many of these industries is often much less than the livelihoods that are adversely affected by them. The issue of some of the social and environmental costs of economic reforms and growth has received considerable attention from the policy makers and researchers. Many studies have shown that during the process of liberalization and structural adjustments the vulnerable groups suffer more than the others. There are ample evidences to believe that the common pool resources of coastal regions. Which provide substantial part of the income of the coastal poor communities is declining and degraded. The industrialization of the one hand and developmental projects on the other such as ports, tourism, aquaculture have led to decline of coastal biodiversity and thereby deprived the poor people of the common benefits which they used to get from such resources otherwise. According to Central Water Commission (1996) 16,935 hectares of fertile land was lost and 51,105 people have been displaced in three coastal districts of Karnataka. The CRZ notification relating to coastal protection explicity states that all estuaries, fish-breeding centres, mangroves etc. are to be declared CRZ-I areas. The coastal zone management plans are yet to be considered as an approved document by the state authorities. The decline of traditional community management institutions and the absence of a strong legal framework are some of the other reasons, which made the poor stakeholders more vulnerable.

Dominance of the Informal Sector

The marine fishing units are classified into three broad categories depending on their scale of operation namely mechanized, motorized and traditional gears. Each category could be further into different classes. The mechanized boats capable of fishing in deeper waters (500 meters depth and above) and for 10-15 days are called multi-day fishing vessels. They generally have all electronic fish finding devices and on-board ice storage facilities. The mechanized boats fishing in near shore waters (50-100 meter depth) normally return in the evening but they are gradually converting themselves into multi-day vessels. The motorized fishing units such as gill nets, harvest valuable fishes and their size is also increasing and currently it ranges between 40-50 feet in overall length, fitted with engines of 25-40 BHP. Theoretically, the main difference between the mechanized and motorized boates is that the motorized boats use the motor power only for reaching the fishing grounds. The motors only help in reducing the risks and time required to reach the fishing grounds.

The traditional fishing units normally restrict the fishing to inshore and estuaries and they are competed out by the motorized units.

In India, the crewmen are always paid a share of the value of the catch and not a fixed wage rate. In addition to a share they also get bonus for every fishing trip if the sales revenue is above a specified limit. The onboard food expense is paid by the owner and in some cases this food expense is deducted from the gross income to arrive at share value. The expense is deducted from the gross income to arrive at share value. The crew share depends on the degree of mechanization. In the case of motorized sector, the sharing pattern is different. The owner receives 50 per cent of the share of the net income (after deducting the operating expenses such as diesel, oil, food and marketing commission) and the rest is shared among the crewmen. The sharing system among traditional fishing category is relatively simple. The entire revenue is divided into three parts and crewmen share one and a half part. Although there is no uniformity in the sharing system, generally with higher mechanization, the per centage share of crewmen decreases, though the absolute amount could be higher than their counter parts in the other sectors. The owners are expected to give advance credit to crewmen to the extent of Rs. 5000-6000 depending on the experience and relationship. The field observations in Andhra Pradesh show that the variability of income among the crewmen is relatively higher than the owners.

In coastal fishing 99 per cent of the workforce in fishing and post harvest activities work in the unorganised sector where they do not get any social security benefits. The main economic activity of men is fishing in estuarine and coastal waters on a cooperative basis using motorized and outboard motorboats. These motorized boats are smaller than rampanies and fitted with 20-50 HP motors and go up to 20-50 km of distance in the sea. On the other hand, the traditional rampanies are operated manually from the shore. They work as crewmen earning a share of the total value of the catch. The quality of fish they land is fresher than fish harvested by large mechanized boats, which reach the market only after three or four days. There is a virtual absence of alternative employment, unemployment insurance and other elements of a social wage. Though most members of the local communities are employed in fisheries-related activities, a large proportion of them earn an extremely low income. Thus, though the unemployment rate is negligible, the level of poverty is high and insecurity of income is a characteristic feature of many post harvest activities. Younger men suffer from higher rates of unemployment as with higher education they then to move away from fisheries and look for alternative employment, which is not available.

Post harvest activities are very heterogeneous containing both high return activities such as trade in export varieties and employment in shrimp processing plants and low return and low wage work such as fresh fish vending and sun drying of fish. Fresh fish, cured and dried fish for domestic consumption are distributed and marketed in the unorganised sector. The organised sector is concerned with freezing, canning, fish oil and fishmeal production mainly to meet export demand. The utilisation of marine fish varies from state to state depending on catch composition and availability of facilities for preservation, processing and storage. Field studies in Karnataka have shown that some of the important harvest activities in which the poorest are to be found are shown in Table 1.

As soon as the catch arrives at the landing centre, it is auctioned through the commission agents with whom the fishermen have a financial relationship. The four main types of traders/buyers who market the fish are :

- Bulk buyers who buy in large quantities and transport fish to interior parts of Karnataka and other coastal states;
- Wholesale commission agents who buy exportable varieties such as shrimp, cuttlefish, squids, etc. at reduced rates and supply to processors–cum–exporters;

- Cooperative societies;
- Men and women retailers who buy fish regularly in the landing centres and sell to consumers through shops and vendors who either walk (mostly women) or transport fish on cycles/auto rickshaws usually men.

Table 1 : Categories of poor in the fisheries sector

Class of Worker	*Activities*
Fishermen	They work as crewmen during the fishing season and are partners in small-scale fishing activities during the monsoon season.
Head loaders	Men and women are hired by marketing or commission agents to unload fish from the boats to the marketing yard. They get wages paid on the basis of baskets lifted.
Processors	Small-scale processors (both men and women) undertake processing activities such as drying and curing in small scale units and export—oriented peeling sheds and canning companies; they also help in transporting and retailing on behalf of traders. Their livelihood is fast eroding due to the development of the commercial organised sector, which depends on the use of ice, improved transportation and centralisation of landing catches of fish. They are paid at piece rates or per sack of dried fish.
Commission agents and petty traders	Small-scale village level operators (women and men) who trade and lease out their facilities to big merchants.
Workers in fish processing units	Contract workers in peeling sheds, surimi plants, ice plants, fish curing and drying yards.

The most important factor that influence the web of relationships that makes up the marking structure is the mode of sale or more precisely the organisation of transaction between fishermen and buyers. Negotiations are generally conducted through intermediaries who facilitate the sale. Often, middlemen financiers advance loans to fishermen to by craft and gear and corner the right to sell the fish to buyers of their choice at a 'fair price'. Sales take place in three ways: through bargaining which is the rule when buyers are few; by auction when there when there are numerous buyers, when supplies fluctuate or when there is a wide variety of fish and at fixed prices, which is the only way exportable fish are sold. The first two ways of selling are most common in disposing the catch on the front itself.

Women's Participation in the Marine Fisheries Sector

There has been little research on the role of women in coastal fishing, marketing of fish and their contribution to family income. Their role as 'facilitators' of fish distribution, particularly in states like Kerala is indeed significant. Although women retailers form only a small segment of the total fish trade, they are a vital link between wholesalers and consumers. Women fish traders in Kerala compete with the local traders for fish and if one goes by external appearances—dress, type of fish basket, manner of selling and the like—there seems to be very little difference among women fish traders. Field observations however show that differences exist between them in terms of working capital, sources of fish supply the mode of transport used and the points of disposal of the harvest of fish.

In the dry fish trade, traditionally in north Kerala, women produce and sell directly to consumers or supply merchants. A few self-help groups have promoted production and marketing. But hundreds of others, wives of fishermen, work for low wages as cheap labour to sort and dry fish for large establishments. Women in Kerala work for wages as processors and sorters in landing centers in the unorganized sectors well as in the organized sector where they dominate in prawn/shrimp processing and specialize in peeling work. In recent years, highly developed peeling facilities have led to the decrease in the demand for such workers. They are also employed in modern surumi plants established in the 1990s to cater to Southeast Asian markets and in processing factories. These workers are not organized despite the fact that their contract with the contractors violates existing laws.

In Karnataka, small-scale women fish distributors, generally from the traditional fishing community, Mogaveera as well as a few Muslim women, are in the business primarily for subsistence. They buy small quantities of fish, transport it over short distances and serve more or less a regular clientele and make nominal earnings. They usually participate in auctions, purchasing a few baskets depending on their capacity to sell in the retail market to regular customers, their relative bargaining position in the market being backed and also limited by the level of consumer demand. While wholesale merchants buy large quantities to transport over distances to make high profits, small scale women retailers spent proportionately higher amount on transportation and ice, make low profits, expend more physical labour and work long hours. Although mainly involved in vending fresh and dried fish, they also work for wages as head-loaders, processors and sorters at the landing centres. In the export factories in Karnataka peelers from Kerala-usually young girls and women in their twenties—are preferred as they are considered highly skilled and those migration is encouraged.

Education Among Coastal People

Very little information is available on the educational status of the coastal communities and is scattered in internal documents and reports of NGOs. Tietze (1987 and 1996) publications mention about the poor educational standards of the coastal fishers. Large family size and poor quality of life characterize coastal families since even small children can participate in income generating activities and it is argued that with decline in resource base, the coastal poor feel a need to have large families that can extract enough for survival.

Vivekanandan found that educational standards amongst the fisherfolk caste of Pattapu in southern Andhra Pradesh are abysmally low. He attributed this to the low age of entry (around 12 years) into Catamaran fisheries, in case of males and in case of girls, the need for taking care of younger children in the absence of their mother who goes to work. Salagrama (1990) in a study of nomadic fishing community found that children were almost as productive as the elders and had no inclination to study. In another paper, Salagrama (2000) suggests that educational standards of the people who migrated out of the fisheries sector are better.

Impact of Fisheries Management Regulations

The ultimate objectives of regulations aim at increasing the productivity of the stock and the net economic yield. The fishery manager would be interested in expanding fishing effort up to the point of maximum rent, which lies before the maximum sustainable yield point. The maximum economic yield (MEY) is preferable to the maximum sustainable yield (MSY) both from the economic point of view. The regulatory mechanisms usually adopted are:

- Regulations such as gear selectivity and seasons are closures
- Regulations that control the fishing effort and catching.

The first fishery regulation was enacted in 1897 to control destructive fishing activities in both marine and inland waters. It explicitly banned the use of explosives and poisons in harvesting fish. However, until 1970s the state governments did not find the need for controlling fishing effort, as the fishing was mostly artisanal in character and mechanized fishing was negligible. With the rising foreign and domestic demand for fishery products private firms apart from traditional communities saw a good opportunity for financial profits in the exploitation of marine resources. Modern mechanized trawlers were subsequently introduced in India with state and central subsidy programmes through cooperatives and commercial bank. During the initial years most of these fishing units operated by domestic firms from within and outside the local communities. Thus, the influx of large number of big trawlers in the early eighties increased the fishing pressure on the marine fishery resources tremendously. This leads to stagnation of catch and decline of average profitability.

Panayotou (1982) identifies broadly two parameters, which the fishery administrators have tried to manipulate: (*a*) the age or size of fish at first capture (*b*) the total amount of fishing effort. The ultimate objective of both the approach is to increase the productivity of the stock and net economic yield. However, the productivity and sustainability of measuring the marine fish production cannot be done in the same way as the productivity of the land or forests. The fishery catch depends on the stock of fish in the fishery grounds well as on inputs in terms of fishing efforts and quantity of fishing gear used.

The ban on fishing is one of the methods of fish conservation by prohibiting fish harvesting during the breeding (monsoon). The ban period varies from state to state. Goa observes fishing ban from 1st June to 24th July, every year. In Kerala the ban is from 15th June to 29th July. The ban in Karnataka is from 1st August. In Maharashtra, it appears that recently a decision has been taken to ban fishing from 10th June to 7th August each year. Gujarat on the western coast does not have any fishing ban. Recently the Goa High Court gave a judgment (The High Court of Bombay at Goa, 2002) that the State Government of Goa should strictly implement the above fishing ban against all kinds of mechanized vessels, including country crafts and boats/canoes fitted with inboard or outboard motor and other mechanized both using nets for the purpose of fishing within the territorial waters of the State Goa, *i.e.* 22 kms. from the sea coast. It is made clear that the traditional fishing by boats, without any mechanized motors, etc., are permissible and this order will not come in the way of the "ramponkars" earning their day-to-day livelihood by traditional fishing.

However, such a measure may affect the livelihood opportunities of the small-scale fishermen who are dependent on fishing during the monsoon period. Thus, the fisheries management measures need to be implemented cautiously without harming the interests of some of the stakeholders in the resources.

Socio-economic Status of Inland Fishers and Fish Farmers

Fresh water aquaculture resources in the country consists of 2.25 million hectares of ponds and tanks, 1.3 million hectares of bheels and derelict waters, 2.09 million hectares of lakes and reservoirs as also 0.12 million kilometres of irrigation canals and channels.

Indian aquaculture production consists mainly of Indian major carps and common carps. Indian fresh water aquaculture is mainly based on carps such as Indian Major carps (catla, rohu and mrigal), kalbasu carp, grass carp and common carps etc. The factors such as culture practices breeding and seeds production and socio economic factors play a very important role in productivity enhancement. The socio-demographic characteristics of the freshwater fish producers in the six selected states of India based on a comprehensive survey (Bhatta 2001) are presented in Table 2. The age of carp farmers ranges between 38 years in Andhra Pradesh to 58 years in Haryana with an all India average of 47 years. The number of years of formal school education is the lowest for Haryana and Andhra Pradesh farmers had highest educational attainment with 10 years of schooling. The number of working days indicating the generation of employment was found to be the highest for West Bengal (75 man days) and Karnataka producers spent only 17 man-days on an average.

Table 2 : Profile of Fish Farmers

Variable	*Overall*	*States*					
		Andhra Pradesh	*Haryana*	*Karnataka*	*Orissa*	*Uttar Pradesh*	*West Bengal*
Total Respondents	417	66	56	64	62	123	46
Household size	8	5	23	5	8	6	7
Age (Years)	47	38	58	47	46	47	51
Education (years)	6	10	2	9	6	6	6
No. of working days available for aquaculture in a year (in % age)	68.66	60.94	47.96	17.48	50.68	44.94	75.07
Average farming duration (number of months in a year)	8	7	9	8	9	8	-
Experience in carp culture	10.21	9.75	3.67	3.14	5.17	7.01	44.3

Source : Bhatta (2001).

The farming durations on the other hand extended up to 8 months in UP with a minimum of 7 month in Andhra Pradesh. The results show that Andhra Pradesh farmers took less farming duration as they have stated stocking stunted fingerlings. The size of family varies from 5 in Karnataka to as high as 23 in Haryana. In terms of experience, West Bengal has uniqueness since the back yard pond system has been in existence for many years.

The share of different sources of income of carp farmers presented in Table 3 shows that 80 per cent of the income is generated from carp farming followed by agricultural crops and others. However, there is wide variation between the states. In Andhra Pradesh the producers are basically fish farmers getting 95 per cent of their income from carp farming, while in Karnataka, Orissa and West Bengal substantial part of the income is generated from agricultural crops. Secondly in Orissa the carp producers received 25 per cent of their income from salaries and wages.

Table 4 shows the land use pattern and allocation of land and water spread area of the sample and basic infrastructure facilities available for farmers in selected states. The holding size of carp farms is highest in Andhra Pradesh (8.42 ha) followed by Haryana (7.88 ha) and Karnataka (5.63 ha). The farm size is smallest in West Bengal (0.85 ha) and highest in Andhra Pradesh. In Haryana commercial crops occupied 75 per cent of their land while in Karnataka it was only 25 per cent. The maximum utilization of water-spread area for carp farming was observed in Andhra Pradesh (97 per cent). The minimum water retention level during dry and wet season is one of the major factors, which affects the growth.

Table 3 : Share of Income of the Carp Producers (in Per centage)

Variable	*Overall*	*States*					
		Andhra Pradesh	*Haryana*	*Karnataka*	*Orissa*	*Uttar Pradesh*	*West Bengal*
Mean household gross income (in Rs.)	410818	1822701	231304	337711	79444	58178	33009
Total Respondents	417	66	56	64	62	123	46

Contd...

1	2	3	4	5	6	7	8
% share of income from :							
Culture	79.66	95.26	54.89	30.1	14.98	58.86	49.26
Fish capture	0.16	-	-	0.61	0.49	1.68	0.02
Harchery	0.06	-	-	-	0.06	1.30	-
Paddy	9.32	-	22.25	49.7	28.04	4.8	32.89
Other crops	3.76	0.02	22.09	14.19	0.79	4.88	3.01
Live stock	0.03	-	-	-	-	0.8	-
Business	5.18	4.72	0.77	0.75	31.12	15.33	12.05
Salaries & Wages	1.54	-	-	2.91	24.52	10.86	1.9
Others	0.29	-	-	1.74	-	1.47	0.87
Total	100	100	100	100	100	100	100

Source : Bhatta (2001)

West Bengal and Andhra Pradesh farmers maintain maximum level of water, with highest stocking density. Whereas, in Karnataka and Haryana farmers suffer from lack of water during summer season, because many of the ponds are village where the water during the dry season is being utilized for other household and irrigation activities. In most of the states the farmers are residing within the radius of 1-2 kms except in Andhra Pradesh where the farms are located 8 km away from their house.

Table 4 : Profile of Fish Farms

Variable	*Overall*	*States*					
		Andhra Pradesh	*Haryana*	*Karnataka*	*Orissa*	*Uttar Pradesh*	*West Bengal*
1	2	3	4	5	6	7	8
Total area (in ha.)	4.24	8.42	7.88	5.63	1.74	2.15	0.85
Per centage of the total area under Home stead (%)	1.18	-	1.14	-	11.49	-	7.06
Commercial crops (%)	34.91	2.14	74.75	25.22	66.09	41.86	9.41
Food crops (%)	13.0	-	-	59.15	-	-	38.82
Water spread area including fish pond (%)	50.9	97.86	24.11	15.63	22.14	5.86	44.71
Total owned	3.53	6.68	7.23	5.22	1.77	1.18	0.81
Leased out	0.04	0.15	-	-	0.07	0.02	0.04
State owned	0.11	-	0.43	0.31	-	-	0.06
Minimum water depth (in mts)							
Dry season	2.89	4.38	1.73	1.7	3.84	1.61	5.84
West season	4.78	6.18	2.16	2.58	6.79	4.02	8.33
Distance of the farm from (in km)							
District head quarters	66.3	54.18	81.98	64.56	26.04	24.08	231.52
Main road of the district	4.68	2.7	7.58	3.43	4.63	3.42	9.07
Main river of the district	10.3	1	9.17	17.44	8.9	13.01	9.8
Nearest village Market	5.43	-	11.85	10.09	2.77	2.47	2.12
Nearest fish seed supply	17.0	8.28	38.24	19.92	17.71	15.52	2.25
Home	2.52	8.12	1.6	2.11	1.27	1.27	1

Source : Bhatta (2001).

Table 5 shows the various problems faced by the aquaculture farmers and their ranking. The table indicates that poaching is the most severe problem (4.31) faced by the farmers followed by disease (4.73). It is important to note that the farmers faced a variety of problems in their farming operations.

Table 5 : Mean Ranking of Problems Encountered by the Producers (1 = most severe)

Variable	*N*	*Mean*	*Std dev*	*Minimum*	*Maximum*
None	45	6.91	3.03	1	18
Poaching	180	4.31	3.45	1	24
Bad weather	57	7.39	2.45	3	14
Flood	125	4.67	3.77	1	19
Drought	126	5.42	4.30	1	22
Unreliable water supply	93	6.66	4.81	1	24
High cost of water	105	6.41	4.75	1	24
Polluted water	85	7.27	4.77	1	28
Sulphur upwelling	57	8.98	4.91	3	27
Net/Pond destruction	55	8.64	4.48	2	26
Poor/slow growth of fry	68	7.72	4.03	2	23
High fingerling mortality	69	7.42	4.19	1	24
Small size of fish at harvest	90	6.21	3.84	1	27
Uncertainty of access to present location	56	7.50	4.86	3	29
Proliferation of carp farms	56	7.16	4.38	1	22
High price of figerlings	56	6.61	4.26	1	23
Increasing cost of inputs	82	5.55	4.20	1	25
Difficulty in obtaining credit	108	4.81	4.17	1	28
Lack of technical assistance	80	5.71	4.11	1	24
Limited management expertise	57	6.58	4.32	1	21
No skilled workers	70	6.23	4.30	1	20
High capital requirement	70	5.83	4.75	1	22
High marketing cost	61	6.90	6.23	2	28
Disease	103	4.73	3.57	1	26
Cold	55	6.98	5.12	1	29
No buyers at market	91	5.18	4.97	1	28
Others	80	3.56	2.10	1	10

Source : Bhatta (2001).

The intensity of these problems differs between regions, size and intensity of the farming methods.

Fish Consumption Pattern

The studies on fish consumption in India are very few. The study of National Council for Applied Economic Research (NCAER 1980) is a benchmark exhaustive study of fish consumption pattern in some of the metropolitan cities of India. It revealed that 45-88 per cent of households consume fish in big cities like Bangalore, Kolkata and Delhi. The per capita monthly consumption was in the range of 0.56 kg at Bangalore to 1.01 kg at Kolkata. The proportion of expenditure on fish in total expenditure on food ranged between 6.3 to 14.6 per cent in Delhi and Kolkata, respectively. Another study by Sekar *et. al.* (1996) in south India indicated that on an average the urban consumers by around 4.5 kg of fish per month and spend around 7.5 per cent of their total food expenditure on fish. Birthal and Singh (1997) estimated demand for livestock and fish products in Uttar Pradesh. The results showed that live stock products such as milk, mutton eggs and fish together accounted for 18 per cent of the total consumption expenditure. The average expenditure share of fish in the total food expenditure increased with increase in income initially but marginally declined with higher income classes. This was explained by socio-cultural factors than economic ones. The study further revealed that at the commodity level, meat, fish, eggs together shared 4.83 per cent of the total expenditure. Average expenditure share of fish across different income groups varied from 0.29 per cent in the lowest class to 0.33 per cent in the highest income class. The income elasticity of demand for fish found to be 0.37 irrespective of the income class. The National Sample Survey of India also gathered information on fish consumption all over India. The findings of this section are based on the results of a comprehensive survey of urban and rural fish consumers in five Indian states namely, Haryana, Karnataka, Orissa, Uttar Pradesh and West Bengal (Bhatta 2001). Primary data were collected through a food consumption survey of randomly selected 890 (421 rural producer-consumer and the rest from urban areas) sample households ten districts of six states. Some of the important results of the survey are presented in this section.

Species wise monthly consumption of carps other species across income classes and regions of urban households are presented in Table 6. The all India monthly average household consumption of fish was 3.17 kg/month, among 'very poor' income classes. In the very poor category and also in all other income classes West Bengal had the highest monthly consumption. The household monthly fish consumption was found to increase as income increases except for 'rich' income class. Rohu and catla constituted higher per centage of consumption among all the income classes and states. In Karnataka the share of common carp varies between 14.78 per cent among very poor class to 11.59 per cent among medium income classes. Even the rich income classes consumed 17.65 per cent of their total monthly household consumption of 0.68 kg.

Table 6 : Species-wise Household Inland Fish Consumption in Urban Area (kg/month)

Income Group	*Overall*	*States*				
		Haryana	*Karnataka*	*Orissa*	*Uttar Pradesh*	*West Bengal*
Very Poor						
Total (kg)	3.17	2.66	3.18	2.62	2.51	4.95
Per centage of						
Rohu	48.58	92.86	15.72	59.16	44.62	41.82
Catla	36.91	4.14	69.50	37.02	28.29	37.58
Mrigal	8.83	1.50	0.00	3.82	13.55	18.79
Common carp	2.84	0.00	14.78	0.00	0.00	0.00
Others	2.84	1.50	0.00	0.00	13.55	1.82
Poor						
Toral (kg)	3.17	2.87	2.45	3.28	2.89	4.39
Per centage of						
Rohu	53.00	87.80	35.92	44.51	52.25	46.79
Catla	35.02	3.14	51.02	50.30	24.22	42.43

Contd...

1	2	3	4	5	6	7
Mrigal	7.26	9.06	0.00	5.18	10.38	9.40
Common carp	1.89	0.00	13.06	0.00	0.00	0.00
Others	2.84	0.00	0.00	0.00	13.15	1.38
Medium						
Total (kg)	3.60	2.74	2.76	4.80	2.94	4.76
Per centage of						
Rohu	62.78	97.81	43.12	71.25	62.24	46.43
Catla	29.72	2.19	44.57	25.21	28.57	42.02
Mrigal	4.44	0.00	0.72	3.54	4.76	9.45
Common carp	1.67	0.00	11.59	0.00	0.00	0.00
Others	1.39	0.00	0.00	0.00	4.42	2.10
Rich						
Total (kg)	2.91	2.36	0.68	3.73	3.47	4.30
Per centage of						
Rohu	54.64	90.68	35.29	53.62	54.18	39.30
Catla	30.24	8.47	47.06	29.22	31.41	39.30
Mrigal	12.71	0.85	0.00	17.16	7.49	21.40
Common carp	0.69	0.00	17.65	0.00	0.00	0.00
Others	1.72	0.00	0.00	0.00	6.92	0.00

Source : Bhatta (2001).

Species wise monthly household consumption among rural household are presented in Table 7. Since the consumption by rural households represent only producer—consumers, the figures are slightly higher than the average. The average household monthly consumption of 'very poor' class was 7.25 kg and West Bengal had highest consumption of 7.87 kg per month. In the very poor income class Karnataka, Orissa and Uttar Pradesh had least consumption of fish of 2.20, 2.26 and 3.33 kg per month respectively.

Table 7 : Species wise household fish consumption in rural areas (kg/month)

Income Group	*India*	*States*				
		Haryana	*Karnataka*	*Orissa*	*Uttar Pradesh*	*West Bengal*
Very Poor						
Total (kg)	4.49	6.81	2.20	2.25	3.33	7.87
Per centage of						
Rohu	44.97	68.87	41.69	55.56	44.74	28.34
Catla	40.97	26.58	58.31	44.44	23.42	24.65
Mrigal	6.48	4.55	0.00	0.00	11.71	20.84
Common carp	0.00	0.00	0.00	0.00	0.00	0.00
Others	7.59	0.00	0.00	0.00	20.12	26.18
Poor						
Total (kg)	7.55	11	3.74	6.55	9.02	7.43
Per centage of						

Contd...

1	2	3	4	5	6	7
Rohu	46.36	56.82	24.33	66.26	41.24	30.55
Catla	31.92	34.09	52.67	26.56	30.60	24.36
Mrigal	10.33	0.00	1.60	7.18	18.29	23.28
Common carps	2.12	0.00	21.39	0.00	0.00	0.00
Others	9.27	9.09	0.00	0.00	9.87	24.80
Medium						
Total (kg)	10.55	12.51	4.29	13.55	11.45	11
Per centage of						
Rohu	46.45	43.96	38.96	82.88	30.74	23.64
Catla	25.12	35.97	41.49	12.77	28.47	18.18
Mrigal	12.42	15.03	1.17	4.35	22.97	12.73
Common carp	1.52	0.00	18.65	0.00	0.00	0.00
Others	14.50	5.04	0.00	0.00	17.82	45.45
Rich						
Total (kg)	17.21	12.37	3.94	44.71	11.4	13.66
Per centage of						
Rohu	71.93	60.95	44.92	95.39	40.35	39.02
Catla	18.71	33.87	43.15	3.15	30.44	39.02
Mrigal	5.46	2.59	0.00	1.45	15.18	14.64
Common carps	0.52	0.00	11.93	0.00	0.00	0.00
Others	3.37	2.59	0.00	0.00	14.04	7.32

Source : Bhatta (2001).

Table 8 and 9 shows the annual per capita consumption of fish among different income classes and across states on urban and rural. The household overall annual per capita fish consumption among rural households was 8.52 kg for India, which varies between 4.98 kg in Karnataka and 12.61 kg in West Bengal. Among rural producer-consumer households the overall consumption was double than urban households. Another interesting result has that the rural- producer-consumer consumption level of Haryana was 29.70 kg per annum indicating the impact of increased production and accessibility of fish on consumption pattern. Further, variation in consumption between rural income classes was more than the urban income classes. The overall per capita annual consumption of fish among producer-consumer increased from 7.55 kg by 'very poor' to 31.06 kg by 'rich' income class. In West Bengal there is more uniformity across income classes compared to other states.

Table 8 : Annual per Capita Consumption in Urban Areas (Units in Kilograms)

Income Group	*India*	*Haryana*	*Karnataka*	*Orissa*	*Uttar Pradesh*	*West Bengal*
Very Poor	6.42	7.31	5.33	4.73	5.03	10.59
Poor	8.16	10.15	5.85	7.61	6.25	12.53
Medium	10.33	8.87	7.71	12.32	7.90	15.19
Rich	9.16	8.09	2.06	11.44	10.53	14.00
Over all	8.52	8.33	4.93	7.02	6.91	12.61

Source : Bhatta (2001).

Table 9 : Annual per capita consumption in rural areas

Income Group	*India*	*Haryana*	*Karnataka*	*Orissa*	*Uttar Pradesh*	*West Bengal*
Very Poor	7.55	5.13	10.67	2.70	5.57	14.55
Poor	11.29	8.66	7.90	10.78	16.17	17.16
Medium	18.10	10.00	11.22	27.30	28.31	28.70
Rich	31.06	39.78	13.64	32.75	23.86	49.20
Over all	16.99	29.70	11.32	18.61	13.62	18.40

Source : Bhatta (2001)

Purchasing Power and Price Behaviour of Aquaculture Products

Economists suggest several methods to measure changes in degree of economic access to food. One way to measure such changes is to examine the trends in the proportion of per capita required to buy a unit of food (Tyagi 1988; 1990). Using the fish retail prices, it was observed that the average per capita income required to buy a kg of fish declined by 50 per cent in 1994-95 in the of rohu and 20 per cent in the case of pomfret which is a marine fish (Table 10).

Table 10 : Average per Capita Income and Retail Prices of Fish

Year	*Per capita income at current prices (Rs.)*	*Average retail prices (Rs./kg)*	
		Rohu	*Pomfret*
1986-87	2303	28(1.22)	20(0.87)
1990-91	2837	36(0.72)	30(0.60)
1993-94	7060	47(0.67)	44(0.62)
1994-95	9983	50(0.61)	50(0.61)
1996-97	10771	60(0.55)	62(0.57)

Figures in parenthesis indicate per centage to per capita income.
Source : Bhata (2001).

The economic access to fish has increased in these years. If certain section of the population is not able to have due to very low purchasing power, the solution lies in creating more employment and income opportunities for them rather than solely relying on keeping the product price stable. It may not encourage farmers to adopt new technology and making investment in fisheries and aquaculture.

Fish Marketing

The commercialisation of fish harvesting technologies has led to many changes in the total fish landings as well as on composition of the landings. Improved market infrastructure and centralization of landings has no doubt benefited the fisher folk through remunerative prices since such centralized port markets are effectively linked to urban wholesale markers and thus increased marketing efficiency. However, such developments also affected other income generating activities derived from the catches in traditional fishing committees. Thus, many scale fisher-women processors depending on the landings of smaller ports of their processing activities are finding less fish in the nearer landing centers and also have to face tough competition from the industrial buyers. For the fish varieties, which are frozen and exported (particularly shrimp and cuttlesfish) there has been a significant increase in the price received by the fishermen. Considerable competition exists among the processor-exporters to obtain supply for their under-utilized processing plants. Moreover, the high value of such products enables the fishermen to spend on ice better handling system, which increases their withholding capacity. Many associations of commercial fishermen were organized in order to get a larger share of the export price for their member-fishermen.

The scenario of the domestic marketing is totally different. Although over the years the proportion of the total landings exported is increasing, still 80-85 per cent of the total quantity is traded and consumed in the domestic market. Hence, major part of the income received by the fishermen still depends on the unit value realized from domestic marketing. A case study on marine fish marketing in Tamil Nadu indicated that the average retail price recorded manifold increase between 1974-75 and 1997-98 (Sathiadhas 1998). The price of seer fish increased about 10 times from Rs. 9.00/kg in 1973-74 to Rs. 100/kg in 1997-98. The price of pomfrets increased from Rs. 2.50/kg to Rs. 120/kg during the corresponding period. Similarly the price of sharks increased from Rs. 2.50/ kg in 73-74 to Rs. 60/kg in 1997-98. However, fishermen frequently feel that they receive unfair price for their catch, particularly when there are a few buyers or when there are gluts in the landings. The seasonality of catches has also changed significantly. For example in Karnataka due to the development of fishing technology capable of fishing during monsoon season the monsoon landings have increased from a mere 7 per cent of the total during the pre mechanization period (1956-78) to 20 per cent during the post-mechanization period (1985-1993) (Bhatta. R. 1996). This has resulted in round the year availability of fresh to the consumers, this led to decline in the demand for dried fish in the domestic market. Further, the proportion of dried items in the seafood export declined from 41 per cent in 1966 to less than 1 per cent in 1996 (Sathiadhas 1998). These changes in the utilization pattern many socio-economic implications on the poor income groups living in rural and urban areas.

The cultured fishes such as Indian major carps and freshwater prawns are sold in urban markets and transported to West Bengal and other North-Eastern states. For instance, in Vellore a town in the northern part of Tamil Nadu about 5000 hectares of area is under freshwater fish culture with an estimated fish production of 12,000 tonnes. Most of the cultured fish and fresh water prawns are exported to North-Eastern states and West Bengal. The farmers some times enter into buy back arrangements with the merchants. Under this arrangement the merchants supply inputs such as nets, crafts and feed in advance with an understanding that the entire fish is sold to him by the farmer at the prevailing market price. The total revenue realized by the sale of fish is paid to fishermen after deducting the cost of input advances made.

Role of Fisher Groups and Cooperatives

Some of the fishermen groups attempted to market their own fish by forming community level associations and cooperatives. Pritchard *et al.* (1997) documented some of the success stories of the fisher-women groups in marketing fresh in Tamil Nadu. However, in general only a few of these initiatives were successful. Some of the important reasons for the failure of the group initiatives were the competition by the merchants by offering higher prices temporarily to attract the sales away from the group, lack of quick and right sale points to get better price and paucity of working capital.

Fishing cooperative were also expected to play an important role in marine and inland fish marketing. There were about 9500 primary societies with a membership of about one million linked to 108 district cooperative federations. Under the National Federation of Fishermen Cooperatives in New Delhi. An examination of the working of these cooperative showed that these cooperatives were successful when markets has following characteristic features:

- The production of fish was centralized
- Fishing operations were highly mechanized and capital intensive with higher share of inputs imported from outside the region.

The Employment in the Export Oriented Enterprises

The processing units in different parts of the country employ largely women from Kerala mainly on short-term contractual arrangements. The problems of the unorganized sector have become more significant today when liberalization has become center-stage. The work in the processing plants have been divided into two parts; the pre-processing part that involves peeling and cleaning of raw materials and the next stage grading and packing of the product. The companies mainly employ two types of workers casual daily workers drawn from the local areas and the contract workers who are mainly migrant women workers. It is very difficult to estimate the number of actual work force employed in these processing units. The number of permanent workers is relatively insignificant compared to total number of casual workers employed in these unit. Further, no systematic data is being maintained by the government organizations such as Marine Products Export Development Authority (MPEDA) and labour commissioners' office.

Socio-economic Impact of Aquatic Biodiversity Conservation

The Ministry of Environments and Forests, Government of India, considering the over harvesting of some of the rare species such as whale sharks included it in the Schedule I of the Wildlife Protection Act. The whale shark is the first species to get protection under the Act. After including the whale shark in the negative list, it is expected that the official exports are likely to come down. However, underground exports are likely to flourish unless the whale fishermen are properly rehabilitated.

According to the reports of the TRAFFIC there has been large-scale fishing of whale sharks for their meat, fins liver, skin and cartilage (Hanfee 2001). The whale sharks are found largely in the west coast. TRAFFIC India's survey revealed that between 1999-2000, 600 whale skarks were caught, smallest catch was two meters long and 0.5 tonne and the largest 14.5 meters and 12 tonnes. While the fresh and frozen meat of whale shark is sold at Rs. 40 and Rs. 70 per kg in India. It sells for US$ 15,00 (Rs. 750) in Taiwan. Export of many biologically sensitive marine species such as crabs, mussels, snail, seaweed (agar) etc. in different product foms is going on. The export of total increased from 1844 metric tonnes worth Rs. 2283 lakhs in 1994-95 to 2586 metric tonnes worth Rs. 3350 lakhs in 1999-2000. The export of shrimp in different product form has increased from 74563 metric tonnes in 1992-93 to 110564 metric tonnes in 1999-2000. The export of fresh water prawn (scampi) has doubled from 102 metric tonnes in 1995-96 to 217 metric tonnes in 1999-2000. With the result freshwater fish biodiversity issues have been completely ignored. Similarly the export of cultured tiger shrimp *P. Monodon* from coastal wetlands has increased significantly. Many local fishes such as mullets and pearl spot have to be eliminated before stocking the shrimp in costal shrimp farms. These fishes were the staple food of the local communities. The tremendous growth of exports of the cultured shrimp affected biodiversity adversely left less availability of local fishes for the local communities. International trade in many marine species is prohibited under various Acts and notifications. The export of some of the species such as marine turtles, shells, gastropods except the giant clams are banned under the Wild life protection Act 1972 and CITES declaration. The sea cucumber (Beche-de-mer) is another commercially important marine species, which has very high export value. In 1982 Government of India put a ban on the export of Beche-de-mer below the size of 7.5 cm. In Andaman and Nicobar Islands fishing for sea cucumber is totally banned. Corals and associated species like sea fans, sea-sponges are heavily exploited for their known sources of bioactive substances with wide application in the pharmaceutical industry. Especially sea fans (Gorgonids), which constitute only source of prostaglandins and terpenoids (Hanfee 2001). Black corals were listed in CITES Appendix II in 1981 to protect the highly exploited stony corals. However, control of coral trade is difficult since they are often collected in offshore areas not directly controlled by the coastal nations.

Conclusion and Policy Implications

The marine production in India is reaching maximum sustainable yield levels and in the case of some commercially important species the symptoms of over harvesting such as stagnation of total production, decline in the catch per unit of fishing effort are observed. This has negative socio-economic implications in terms of lack of fish availability to local community and nutritional insecurity.

India initiated the aquaculture development programs during 1980s and 1990s. There were restraints to the fish farming initially but the growth in production was accelerated in some regions such as Andhra Pradesh, Punjab, Haryana and Bihar. Hopes have been raised about the increasing domestic consumption, inter state movements and exports with the changing food preferences and urban growth. The composition of species is responding to market changes. It is becoming concentrated in rohu, common carps and fresh water prawns.

However, the current level of fish consumption is very low compared to other countries. The studies on demand projections indicate that there will be very rapid growth in the demand for fish. The socio-economic conditions of fisher folk in terms of education, employment, income, food and nutrition security are not encouraging.

REFERENCES

Ayyappan S., 2000. Technological potential for the development of freshwater Aquaculture. In M. Krishnan and Pratap Birthal (Ed.) Aquaculture Development in India: Problems and Prospects Workshop Proceedings 7, National Centre for Agricultural Economics and Policy Research, New Delhi, pp. 78-86.

Bhatta, R. 1996. Role of cooperatives in fisheries management and development in coastal Karnataka, R.Rajgopala (Ed.) rediscovering Cooperation, Vol. II: Strategies for the Models of Tomorrow, Institute of Rural Management, Anand.

Bhatta R. 2001. Carp Production and Consumption in India: Socioeconomic. Analysis (unpublished), report submitted to ICLARM, Penang 76p.

Birthal, P.S., & Singh, M.K. 1997. Demand for livestock products by rural households in Western Utter Pradesh. *Agricultural Economics Research Review,* 10(1): 95-105.

Central Marine Fisheries Research Institute, 1998. Status of Research in Marine Fisheries and Mariculture. CMFRI Spl. Publn., No. 67:35.

Hanfee F., 2001. General giants of the sea TRAFFIC — India/WWF. India, New Delhi. 56p.

High Court of Bombay at Goa 2002. Goa Environmental Federation Versus State of Goa. PIL Writ Petition No. 212 of 2000, Panaji.

Kurup and Devaraj, 2000, Estimates of Optimum Fleet Size for the Exploited Indian Shelf Fisheries, Central Marine Fisheries Research Institute, Cochin. *Mar. Fish. Infor. Serv. T & E Ser., No.165– July–August–September 2000 2-11 pp.*

National Council of Applied Economic Research 1980. Demand for fish and its transportation and storage in selected cities. NCAER, New Delhi. p. 121.

Panayotou, T. 1982. Management Concepts for Small-Scale Fisheries: Economic and Social Aspects. *FAO Fish. Tech. Pap.,* (228):53.

Pritchard M, A. Gorden, G. Patterson and A Gray, 1997. Utilization of small pelagic fish species in Asia. Natural Resources Institute, The University of Greenwich, p. 24.

Sahai, B. 1993. Applications of remote sensing for environmental management in India Space and Environment report of the International astronomical Federation, 44th Congress Graz, Austria.

Salagrama, V. 1990. Our Boats are Our Homes. News Article in the Bay of Bengal News, Issue. No. 39. Bay of Bengal Programme, Madras.

Salagrama, V. 2000. Small-scale Fisheries: Does it Exist Anymore? in Bay of Bengal News, Issue No. 16, April 2000.

Sampath, V. 1998. "Living Resources of India's Exclusive Economic Zone" In Quasim S. Z, Roonwal G S. (Eds.).

Sathiadhas, R. 1998. Marine Fisheries in Indian Economy, *Paper presented at Symposium on Advances and Priorities in Fisheries Technology,* 11-13 February 1998, Kochi, India.

Sekar, C., Randhir, O.T., & Meenarakshisundaram, V. 1996 Urban Fish Marketing - A Consumption Behaviour Analysis. *Agril. Marketing* 39(2).

Tietze. U. 1987. Bank Credit for Artisanal Marine Fisherfolk of Orissa, India. BOBP/REP/32. Bay of Bengal Programme, Madras.

Tietze, U. & Shreshta, B.P. 1996. Report of the Regional Consultation on Institutional Credit for Sustainable Fish Marketing, Capture and Management in Asia and the Pacific. Manila, Philippines, 3-7 July 1995. FAO Fisheries Report. No. 540. Rome: FAO.

Tyagi, D.S. 1988. Increasing economic and physical access to food: A case study of two macro-policies in India. Asia Regional Seminar, China, October 4-8.

Tyagi, D.S. 1990. *Managing India's Food Economy,* New Delhi: Sage Publications.

Chapter 7

Fisheries Economics

Concept of Economics : What is Economics?

Concept : Economics is mainly concerned with choices at all levels of society, choices made by individuals, firm or by the government. Economics is the study of why choice are necessary and how they are made. A study generally under taken with the aim to improve in some way or other the outcome of the choices.

Why we study economics. The fundamental facts are :

(*i*) Human wants are unlimited.

(*ii*) The different wants have different importance.

(*iii*) Resources are limited.

(*iv*) There are many alternative uses of Resources.

Indian population : Over a billion total land area of India : 328 million hectares (ha.)

Definitions of Economics

1. First definition of economics was given by Adam Smith known as the "Father of Economics." "Economics is the science of wealth".
2. Marshal describes economics is a study of mankind in the ordinary business of life, it examines that part of individual and social action which is most closely connected with the attainment and with the use of the material requisites of well being.
3. Robbins said economics is the science which studies human behaviour as a relationship between ends and scarce means which have alternative uses :

Robbins emphasised four important points.

(*i*) Study of human behaviour.

(*ii*) Objective you want to satisfy.

(*iii*) Scarce means (resources) through which you can satisfy your want.

(*iv*) These scarce means have alternative uses.

Factors of Production

(*i*) land (rent)

(*ii*) labour (wages)

(*iii*) capital (interest)

(*iv*) management

(*v*) organisation or entrepreneur

(*vi*) profit and loss

The modern economic period 1990 onwards.

All the economics activities can be divided into five categories :

1. Production
2. Consumption
3. Distribution
4. Exchange
5. Public Finance

1. **Production :** Can be defined as "creation of utility".
2. **Consumption :** "Destruction of utility".
3. **Distribution :** Distribution or sharing of wealth among the community of owners which have been responsible for the production process.

Example :	land	rent
	labour	wages
	capital	interest
	management organisation and entrepreneurs	profit

4. **Exchange :** Change of ownership of the right of the goods and services from one individual or community to another group or individual.
5. **Public Finance :** In this we study
   (*i*) public income
   (*ii*) expenditure
   (*iii*) department
   (*iv*) tax policy of govt.

Basic Terms and Concepts used in Economics

1. **Goods and Services :** Anything which is capable of satisfying human want *i.e.* possesses a quality or qualities by virtue of which it satisfies human want (one or more).
   (*a*) ***Free Good :*** Exist in nature in super abundance.
   (*b*) ***Economic Good :*** Scarce in nature and one has to pay a price to get it.
2. **Utility :** Want satisfying power of any commodity or good or service.
3. **Value and Price :** Value of a commodity simply means its power to get other commodities in exchange for itself.
4. **Wealth and Income :** Total stalk (store) of goods and services at a particular time.

Difference Between Want, Desire and Demand

1. **Want :** An effective desire or a desire backed by willingness and ability to pay or of purchasing power.
2. **Desire :** It may be defined as a psychological feeling of having something.
3. **Demand**

Modern concept of economics : It can be divided into 3 parts :

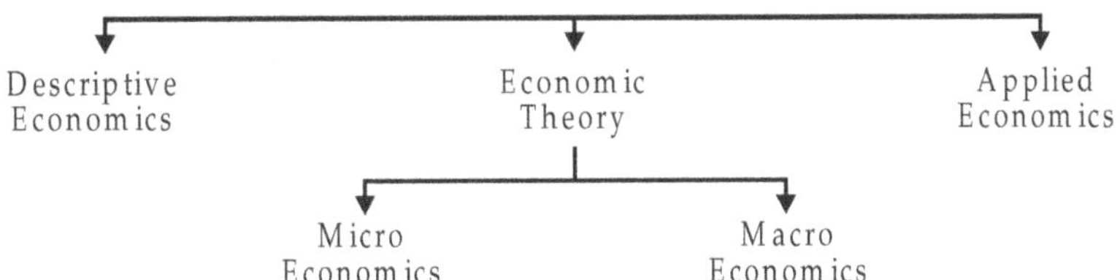

1. **Descriptive Economics :** Under this we collect all the relevant facts about a particular topic; concerned.
2. **Economic Theory :** It gives a simplified explanation of the way in which an economic system works and the important features of such a system. It is divided into two parts.
   (*a*) Micro Economics
   (*b*) Macro Economics

(*a*) ***Micro Economics :*** In this we study the behaviour of individual decision making units. Examples :
   (*i*) individual consumer
   (*ii*) resource owner
   (*iii*) produces and etc.

(*b*) ***Macro Economics :*** It is concerned with the aggregate and average of the entire economy *e.g.* national income, aggregate or total output, total employment, total saving, total investment, total demand supply etc.

3. ***Applied Economics :*** It takes the framework of analysis provided by economics theory and tries either to use this analysis to explain the cause and significance of events reported by descriptive economics or it tries to test the economic theories :

$$\text{Demand} \times \frac{1}{\text{Price}} \quad \left[D \times \frac{1}{P}\right]$$

Basic Assumptions Applied in Economics

1. **Ceteris Peribus :** It means that other things remain the same.
2. **Rationality :** It means rational behaviour of individual decision makers *e.g.* Rational behaviour of producer and consumer. For producer to get max. profit and for consumer to get max. satisfaction and comfort.

   Example : If a person prefers banana over apple and apple over mango.

   i.e. B $\leftarrow$ A

   A $\leftarrow$ M

Now if that person is given banana and mango, then his rational behaviour shows that he should prefer banana over mango :

Consumer Behaviour

Demand : By demand we mean that the various quantities of a given commodity or service which consumer would buy in a market in a given period of time at various prices.

Law of Demand : A rise in the price of commodities or goods and services is followed by a decrease or reduction of demand and a fall in the price is followed by an increase in the demand in ceteris peribus conditions *i.e.* other things remain same.

Demand Schedule : If prices and their corresponding quantity demand is presented in a tabular form then it is called as demand schedule.

Price (P)	*Demand of Quantity (Q)*
5	90
7	35
9	30
11	26

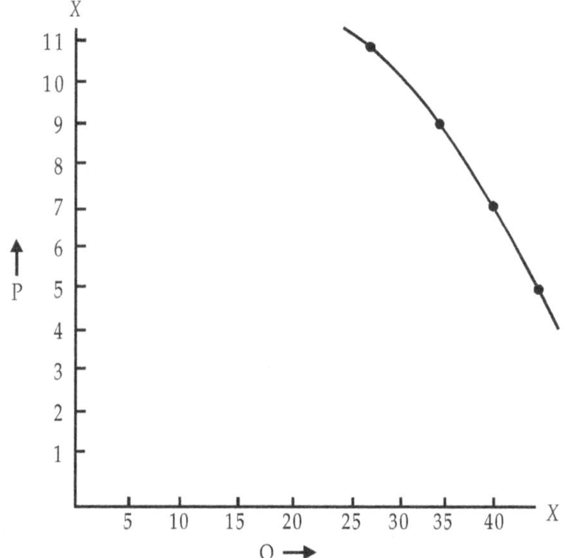

Graphical representation of demand schedule is called demand curve, *i.e.* demand curve is the locus of different points indicating the price of quantity and their corresponding demand.

Factors Affecting Demand : Individual demand factors :

1. **Price of Commodity and Service :** It price increases demand decreases and it price decreases demand increases.
2. **Income :** If income increases purchasing power increases demand also increases.
3. **Price of the Substituent :** *i.e.* it price of alternative commodity or service is low then more people will opt for it.
4. **Time :** *e.g.* in winter woolen cloths sell at a very fast rate. In rainy season umbrella sell at a high rate as compared to other seasons.
5. Individual tastes and preferences
6. Future expectations about price change.

Market Demand Factors :

7. **Population :** Larger the population, larger the demand.
8. **Distribution of Income :** Poor people are only able to buy basic needed items and can not afford luxury items.

Price Elasticity of Demand : It is the responsiveness of the demand to the price change or in otherwords elasticity is the ratio of the relative change in the quantity demanded to the relative change in the price of the commodity.

Price elasticity of demand is denoted by *'ep'*

$$\therefore \quad ep = \frac{\text{relative change in the quantity of commodity}}{\text{relative change in the price of the commodity}}$$

$$\therefore \quad ep = \frac{\frac{\Delta Q}{Q}}{\frac{\Delta P}{P}} = \frac{\Delta Q}{\Delta P} \times \frac{P}{Q}$$

Problem : Suppose it the price of a commodity X has changed from Rs. 5 to Rs. 8 per unit, resulting in the decrease of demand from 60 units to 40 unit per time. Calculate price elasticity of commodity X.

$P = P_1 = 5 \quad \Delta P = P_2 - P_1 = 8 - 5 = 3$

$P_2 = 8 \qquad \Delta Q = Q_2 - Q_1 = 40 - 60 = -20$

$Q = Q_1 = 60$

$Q_2 = 40 \qquad ep = \frac{\Delta Q}{\Delta P} \times \frac{P}{Q} = \frac{-20}{3} \times \frac{5}{60} = \frac{-5}{9}$

Value of $ep = \frac{5}{9}$ minus sign of value indicates the nature of *ep* only.

Types of Price Elasticity of Demand

1. Perfectly inelastic demand.
2. Perfectly elastic demand.
3. Inelastic demand.
4. Elastic demand.
5. Unit elastic.

1. **Perfectly Inelastic Demand** : Whatever be the change in the price, commodity demand remains the same.

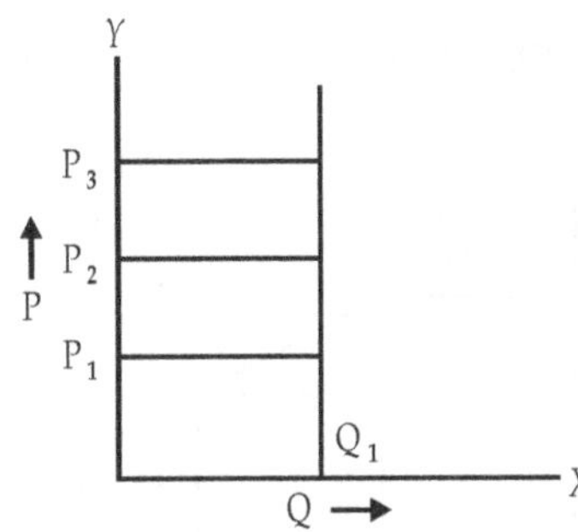

$$\therefore \quad ep = \frac{\Delta Q}{\Delta P} \times \frac{P}{Q}$$

But $\quad \Delta Q = O \;\therefore\; ep = O$

2. **Perfectly Elastic Demand** : *e.g.* salt, with a very small change in price, quantity demand will change very much almost infinity.

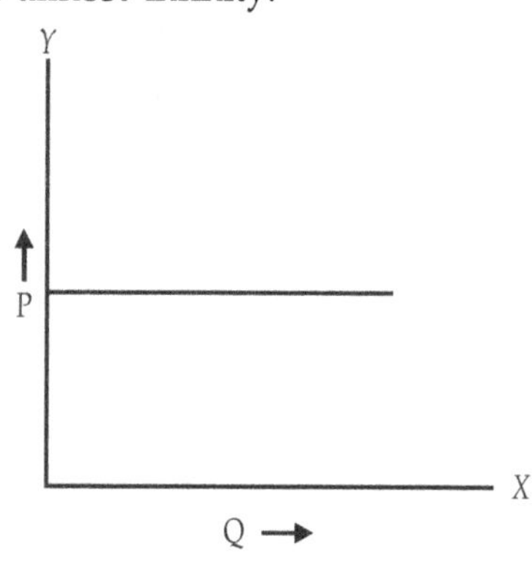

$$\therefore \qquad ep = \frac{\Delta Q}{\Delta P} \times \frac{P}{Q}.$$

But $\Delta P = O$

$$\therefore \qquad ep = \frac{1}{O} = \infty$$ *e.g.* luxury itmes.

3. **In Elastic Demand :** Change in price is greater than change in quantity demand.

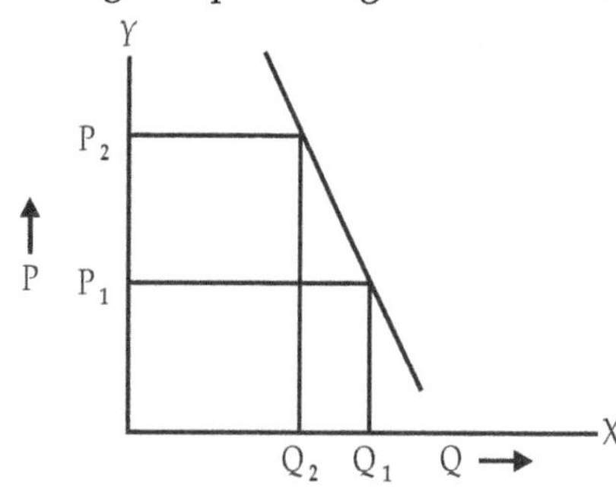

i.e. large change in price will have a small change in the quantity demand.

$\therefore$ Nr. < Dr. $\therefore$ $ep < 1$ *e.g.* edible oils.

4. **Elastic Demand :** Small change in price will cause a large change in quantity demand.

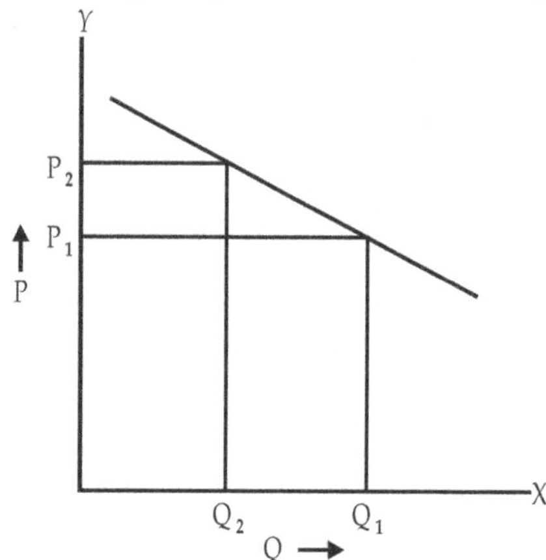

∴ $ep > 1$ ∴ $\Delta Q > \Delta P$

∴ Nr > Dr

5. **Unit Elasticity :** $ep = 1$ change in price will cause a equal change in quantity demand.

   $\Delta Q = \Delta P$

   ∴ $ep = 1$

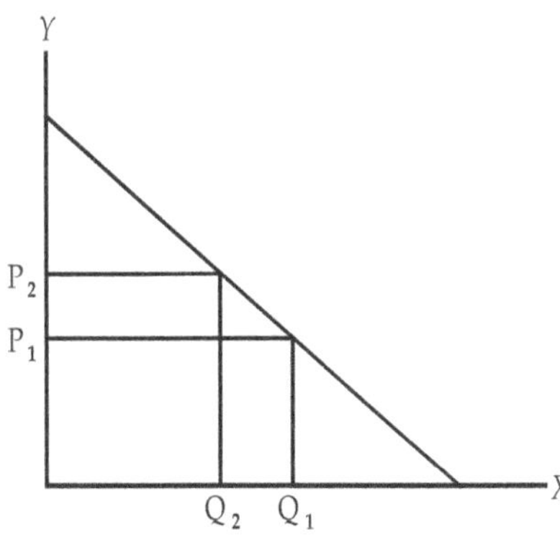

Utility approach of consumer behaviour : This approach is given by 'Marshal'.

Basic Concepts

1. **Utility :** Want satisfying power of any commodity or service.
2. **Marginal Utility :** Satisfaction or utility derived from each successive unit consumed. It decreases as the no. of unit consumed increase. *e.g.* if we eat 2 pieces of bread. Ist will be more satisfying then the 2nd or by the last unit which is considered just worth while.
3. **Total Utility :** Total satisfaction derived by a certain given amount of certain commodity.

   Law of diminishing marginal utility (Marshal). It refers to the belief that addition to the total satisfaction must after a minor interval atleast decrease or decline.

e.g. No of units consumed Marginal utility Total utility

S.No.	*No. of units consumed*	*(Utils) M.U.*	*(Utils) T.O.*
1.	1	20	20
2.	2	18	38
3.	3	14	52
4.	4	9	61
5.	5	5	66
6.	6	1	67
7.	7	0	67
8.	8	–2	65

∴ Total satisfaction after 8 units = 65

Graphical Representation

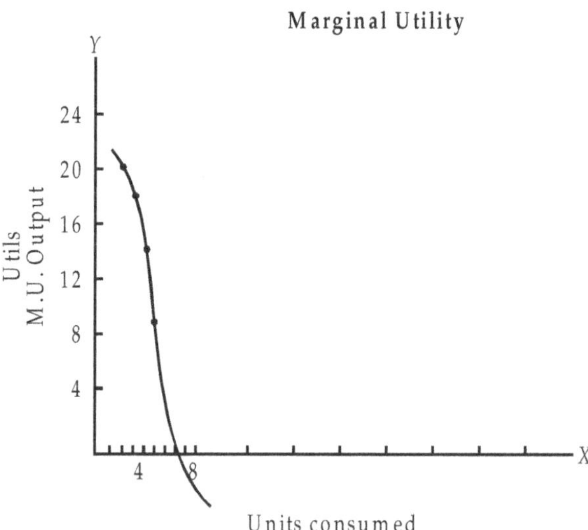

Some times in M.U. the second piece may given more satisfaction then the first. But normally not.

- Marshal took the approximation that satisfaction can be measured. Utility can be measured, unit is 'utils.'

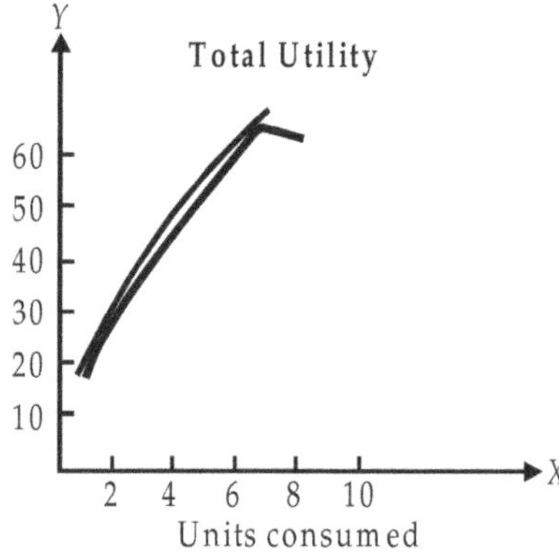

For '*n*' diff. commodities

M.U.a = P_a, (Marginal Ut_a = Price of *a*) satisfaction derived from obtaining any object must at least be equal to dissatisfaction obtained by parting with money (*i.e.* price)

Purchasing pattern *i.e.* how much should one spend on one commodity.

$$\frac{M.U.a}{P_a} = \frac{M.U.b}{P_b} = \frac{M.U.c}{P_c} = \ldots\ldots = M.U.e \text{ (i.e. M.U. of expenditure)}$$

The ratio of *i.e.* in each an every case *M.U.* to prices should be equal. This is called law of equi marginal utility.

When a consumer purchases *n* different commodities then consumer wants to arrange his consumption and purchasing pattern until this following condition is satisfied.

It price increases than $\frac{M.U.}{P}$ ratio is disturbed. Then to decrease *M.U.* less unit will have to be consumed.

*M.U.*e– Marginal utility of expenditure or

Money – should be equal in each case.

$$\frac{M.U.a}{P_a} = \frac{M.U.b}{P_b} = \frac{M.U.c}{P_c} = \ldots\ldots = M.U.e$$

Production : Creation of utility *i.e.* creation of economic utility or value it can be created or increased in the following ways :

1. Form utility.
2. Place utility.
3. Time utility.
4. Possession utility.
5. Service utility.

1. **Form Utility :** Increasing the U by changing the born. *e.g.* wheat when converted to flour and thus to bread, increases the utility. *e.g.* Leather – shoes
2. **Place Utility :** Just changing the place of availability of a good increases utility. *e.g.* sand, increased by taking from sea shore to city. Fish from ponds to market place in a city, increases utility.

   Fish
Wheat
Rice
Sugar Cane
Tea (Assam)
Coffee
Tobacco } To find state of maximum production.

   i.e. goods are transported from place of bulk production *i.e.* place of less demand to a place where demand is more.
3. **Time Utility :** Making it available at a required time. *e.g.* at time of harvesting of rice its price goes down people will store it for leap period when prices are high.
4. **Possession Utility (or Ownership) :** Exchange of commodities from one person, not needing it to another person who is in need of it. Example banks generally work on this principle. Example they take it from people who have a surplus of money and lends to needy people.
5. **Service Utility :** Not physical transformation. But *e.g.* service of doctor, lawyer and teacher. Example people who get paid for their responsibility services.

Factors of Production

1. Land
2. Labour
3. Capital
4. Entrepreneur (organisation).

1. **Land :**

   (*i*) (Marshy) land means the material and the forces which nature gives freely for human aid in land and water, in air, light and heat.

(*ii*) Land stands for all natural resources which yield an income or which have exchange value. It represents those natural resources which are useful and scare, actually or potentially.

Characteristics of Land

1. It's natures free gift.
2. Fixed in quantity.
3. Lacks mobility in geographical sense.
4. It provides infinite variation of degree of fertility, such that no two pieces of land are exactly alike.
2. **Labour :** Any excertion of mind or body under gone partly or wholly with a view to some good other than pleasure, derived directly from the work, is called labour.

Characteristics of Labour

1. Labour is inseparable from labourer.
2. It is perishable in nature. (Labour can't be preserved we can't do the work of a year or week in a day).
3. **Capital :** Capital refers to that part of men's wealth which is used in producing further wealth or which yields an income.

   Capital is not a primary factor of production but a produced mean of production.

   The term capital is used for capital goods for example : plants, machinery, tools and accessories, stalk of raw material and fuel and money.
4. **Organisation or Enterprise or Management :** Supplied by entreprenuer. Entrepreneur has two main functions :
   1. Organising the other three factors of production *i.e.* mixing them in optimum proportion.
   2. Bearing or taking the risk. Risk of loosing or incurring losses.

FISHERIES ECONOMICS

It will be appreciated that no cut-and-dried formula type criteria for investment can be laid down. Instead, a whole host of considerations will have to be borne in mind. Nor can the above criteria be ignored when deciding upon the pattern of investment. For the rest, the best we can do is to offer a few general remarks bearing on the allocation of resources among various investments.

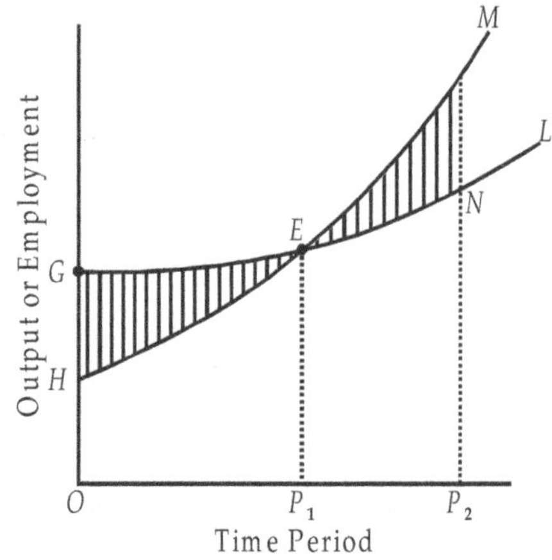

Choice of Technique and Time Series Technique — Amartya Sen : Amartya Sen in his work "Choice of techniques has given the time series criterion. He has compared the labour intensive and capital-intensive technique in relation to total output and employment generation with relation to a time-bound concept. In Amartya Sen's opinion if the society is prepared to wait for certain period of time (say about thirty years) regarding its social welfare function more capital intensive techniques are better otherwise they can opt for labour intensive technique.

On x–axis period time is taken, on y–axis total output or employment generation is taken. '*GL*' curve shows the growth of output and employment by means of labour-intensive techniques where as '*HK*' curve is by capital-intensive technique. At 'OP_1' period of time both the techniques give the same rate of output and employment before '*OP*' period it is labour-intensive technique that is giving greater output where as after '*OP*' period of time it is the capital-intensive technique which giving more output and employment than labour-intensive technique. After OP, period of time it is the capital-intensive technique which is giving more output and employment than labour-intensive technique. At 'OP_2' period of time whatever initial losses of output and employment generation due to capital-intensive technique can be covered between the time period 'P_1P_2'.

That means

$$\Delta HGE = \Delta EMN$$

Initially $OH < OG$

OG continues to be greater till point '*E*' where '$(ON)_L = (ON)_K$

Where ON = output employment generation, '*L*' Labour-intensive technique, '*k*' is capital-intensive technique.

After 'OP_1' time period

$$EM > EN$$

That is $(ON)_K > (ON)_L$

At point 'P_2' "$HG = MN$"

Beyond 'OP_2' period of time capital intensive technique is more fruitful (It will give more output and employment generation).

Prof. Amartya Sen has taken into account output and employment generation as two variables, this can be equated with the income as well. Keynes took that $\Delta O = \Delta Y = \Delta N$ is determined by effective demand.

ΔO = aggregate output

ΔY = aggregate income

ΔN = aggregate employment

The time concept of 'OP_1' or 'P_1P_2' or say 'P_1P_2' as thirty years is an assumption. This is to be taken into account not only the economic variables but also all those depending on the culture, social, political and the desire to opt for better technology, etc.

Same General Guidelines

External Economies : It has generally come to be accepted that the basic consideration in selecting industries for development in an under-developed economy is the prospect of external economies. Allyn Young drew attention to this important consideration in 1928, and Rosentein-Rodan made out in 1943 a strong case of developing those industries which would create condition favourable to the growth of other industries. For example, the development of transport or of sources of fuel and power influences both the costs and the market possibilities of diverse manufacturing industries.

Market : On the demand side, when considering particular industries, one cannot assume that supply will create its own demand. There must be markets for the commodities produced. Where are the potential markets in the poor countries? Investment should be made in those industries which produce commodities having a readily-available demand. The demand for building and construction is likely to be high, since poor countries are deficient in roads, railways, houses and public utilities. Investment in export industries, for which there is foreign demand, is another attractive area, and import competing industries provide still another potential choice of investment. In the initial stages of economic development, it is highly useful to concentrate on certain focal points to have the promise of more rapid growth. From these local areas, a chain reaction usually starts that gradually spreads chain to the remaining areas of the economy. Thus, even an unbalanced process of initial economic growth has every possibility of ultimately merging into the broader requirement of balanced growth.

Balance of Payments Criterion

Investment should also satisfy what we may call the "**balance of payments**" criterion. Alternative types of investment expenditures will have different effects on the country's export capacity and import requirements. One investment project may be more export-creating than another and one project may be more import-requiring than another. Knowing that under-developed countries are particularly prone to balance of payments difficulties, investments should be directed, as far as practicable, to those projects that will reduce imports or increase exports, other things being equal.

Quick-Yielding Investments

Some industries have a long gestation period, while for some others there is a short time-lag between incurring the investment expenditure and the reaping of fruits. While comparing the benefits of the extra output with the costs, the cost of time-lag must never be forgotten. Unless eventual benefits are outstandingly great, priority should be given to those industries or production techniques which have a relatively short time-lag. Mr. Hicks has called such industries **"quick-investment type"**. In a developing economy, with a high inflationary potential and a need for a rapid rise in the living standards, industries which have a high **"fruition co-efficient"** (*i.e.*, a high ratio between output and investment) and also a short "fruition-lag" should generally be preferred. This point is particularly important when one is comparing the two ways of producing the same result. Thus, extra agricultural output may be secured by major irrigation schemes, because of the time-lags and high capital costs of the former unless the eventual benefits are outstandingly great. If extra inputs like good diet, aeration filteration, temperature regulations are given to fish farming the production will increased.

Labour-Intensive vs Capital-Intensive Techniques

Further, there is a problem of choosing between labour-intensive industries or labour-intensive methods and capital-intensive industries or capital-intensive techniques. Since in poor under-developed countries, there is a chronic unemployment and the price of labour is low compared with the price of capital, a relatively high ratio of labour to capital should, as a rule be favoured. In general, where market opportunities exist, and technological restrains are not a problem, the most efficient use of resources in the less developed countries will tend to favour labour-intensive methods. With respect to innovations, it would also follow that capital-saving and labour-using innovations should be favoured as against labour-saving and capital-using innovations.

But it is possible that, as between a technique involving less capital but large labour-employment and another involving a large capital and relatively small labour-employment, there may be such a large differential in productivity in relation to costs that it will be profitable to adopt the capital-intensive technique, despite the high cost of borrowing and amortisation. Again, to strengthen its balance of payments, the country may have to direct some of the new investments into export production. If the export industries are capital-intensive, such as mining and mineral refining, then, even though there is a surplus of labour, investment may have to be directed to these capital-intensive industries for the sake of earning the necessary foreign exchange. *(Dawitt et. al., 2003)*

Local Community Assets

In a country like India, where the problem of disguised unemployment in the agricultural sector is very acute and of wide proportions, there is another important consideration affecting the choice of investment. The building of local community assets, should be a particularly suitable type of investment, since such local assets will absorb the otherwise unemployed and under-employed labour force in the rural areas and add to productive capacity. Minor irrigation works, contour-bunding, land reclamation, village approach roads, bunds, against floods, buildings for schools and health centres are some of the instances of local community works the nature of which amply speaks of their fitness for being undertaken in planned development.

As for the durability of the assets created by an investment, it affects current costs via the rates of depreciation. Less durable project is subjected to a higher rate of depreciation, and vice-versa and hence a larger deduction must be made from its gross output to arrive at the net addition per annum of the project. But the society calculates the rates of depreciation in a different manner from private accounting. The community values capital equipment on the basis of what it can produce relatively to the use of labour involved. Hence if the same equipment can be produced at less cost owing to improvement in labour productivity, the value of the equipment installed earlier will depreciate in terms of its output.

Estimation of the Social Product : Here we repeat that, in the under-developed countries, there is likely to be considerable divergence between the private and social product, especially in the case of building up the necessary infrastructure or the social and economic overheads. This divergence is due ultimately to external economies which in practical life are not easy to define and calculate.

When completed, an investment helps to increase productivity in existing units by either increasing the supply of inputs or making, possible new and more economical combinations of factors. Thus, there is expansion of output, as a result of an investment. Sometimes this expansion needs further investment.

The output of the supplementary investment can be treated as the social product of initial investment if it creates by itself the additional capital required for the supplementary investments. Since one investment leads to another how far can we go on pursuing the effects of an investment? It is better, therefore, to avoid this pursuit and confine ourselves to a definite time period and region and lump together the initial and the likely induced investments and relate this total to the total of expected increases in output resulting therefrom.

Relevant Constraints : These constraints are physical, legal, distributional constraints and budgetary constraints. The most common physical constraint is the production function which relates the physical inputs and outputs of the project. This directly enters into the calculation of costs and benefits. One of the inputs or some inputs may be in totally inelastic supply. Then, the investment must conform to the legal framework. The legal constrainarise, for instance, from regulated pricing. Administrative constraints arise from what can be administratively handled. The distributional constraints arise from the fact that no section should be unfavourably affected in the matter of income distribution. It is not always possible to make the gainers compensate the losers. There are budgetary constraints, since projects have to be executed within the budget allotment.

Where no projects are inter-dependent or mutually exclusive, and where there are no constraints, the projects which maximise the present value of total benefits less total costs can be indicated as under.

1. Select all projects where the present value of benefits exceeds the present value of costs;
2. Select all projects where the ratio of the present value of benefits to the present value of costs exceeds unity;
3. Select all projects where the constant annuity with the same present value of benefits exceeds the constant annuity (of the same duration) with the same present value as costs;
4. Select all projects where the rate of return exceeds the chosen rate of discount.

Capital-output Ratio Criterion

An investment criterion that has often been advocated by various economists is that of capital-output ratio. That is, in choosing among investment projects and in determining priorities, capital-output ratios of different investment projects be compared. Those investment projects (or their technical forms) should be selected that minimize the capital-output ratio. If capital-output ratio of investment A (3 : 1) is less than the capital-output ratio of investment B (5 : 1), then, in development planning, investment. A must get priority over investment B.

The classic case of substitutability is provided by the problem of choosing between alternative techniques to produce the same commodity. Various examples can be given of it. Additional foodgrains production can be had either from constructing major irrigation works or by building small irrigation works or by producing and using more fertilizers. Electricity can be produced either by thermal projects or by hydel projects. Further, more cloth can be produced either in the handloom (khadi-cloth) sector or in the hill sector (mill-cloth). We select a project with a lower capital output ratio. Fish can be produce by culture or capture.

But the criterion of capital-output ratio has been subjected to severe criticism. It is maintained that the economic world is not an abode of perfect or very high substitutability. For example, the allocation of investment between agriculture and industry or between consumption goods and investment goods cannot be adjudged on the basis of capital-output ratio, since the degree of substitutability between these products is very limited. Agricultural products and industrial products are complementary rather than substitutes.

Again, what should be compared in choosing among investment projects is not their capital output ratios, but their contribution to income during a crucial period. The goal of development policy is not the maximum output at a point of time but a maximum rate of growth over time.

In a developing country like India, where fuller employment and better distribution of income and wealth are also the cherished aims of the Five-Year Plans, these other considerations of any investment projects are of paramount importance.

Marginal Social Productivity Criterion

A more general criterion of investment proposed is that of social marginal productivity. According to this criterion, those investment should be made in which social marginal productivity is the highest. Those who advocate social marginal productivity as the main investment criterion have also deduced several corollaries as practical guides to policy. Some of these are : (1) A given value of investment should be allocated in a manner that maximizes the ratio of current output to investment, *i.e.*, capital-output ratio be minimized; (2) Those investment projects should be selected that will maximize the ratio of labour to investment; and (3) To reduce pressures on the balance of payments, investment should be allocated in a manner that will maximize the ratio of export goods to investment.

The use of these specific principles in specific situations is, however, likely to be difficult. For development is a dynamic process which involves changes in the size and quality of population, tastes and pattern of demand, technological knowledge and social and institutional factors. The criterion of social marginal productivity must, therefore, be interpreted within the total dynamic complex. To do this, one must make value judgments regarding the various social objectives some of which can be conflicting. Suppose different projects are likely to result in different distributions of income. If a project maximizes total output or income but at the same time involves a more unequal distribution of income than would another project, should, it be preferred?

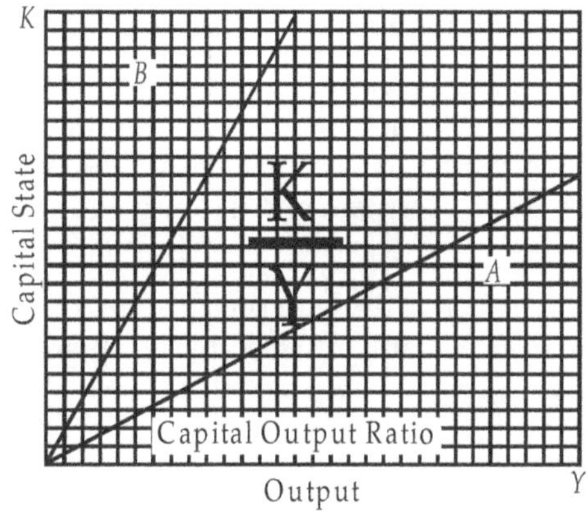

Less the value of $\frac{K}{Y}$ for any project better the prospects

Like the capital-output criterion, the marginal social productivity criterion is also ambiguous as a guide to investment decisions, when the shape of income stream over time is considered. To determine the most productive investment projects, future yields of capital assets must be discounted to their present values, and these discounted values compared with their present costs. Investment decisions will differ according to the future shape of the income stream which is desired.

The other criteria for choosing between techniques of production in an under-developed economy, we may discuss (*a*) The Rate-of-Turnover Criterion, (*b*) The Surplus Rate Criterion, (*c*) Employment Absorption Criterion and (*d*) The Time Series Criterion.

The Rate of Turnover Criterion

According to Prof. J.J. Polak, the investment should be chosen on the basis of the rate of turnover *i.e.*, the ratio of output to capital. We have already discussed above the capital-output ratio. As explained by Prof. Norman S. Buchanan, "If investment funds are limited, the wise policy, in the absence of special considerations, would be to undertake first those investments having a high value of annual product relative to the investment." That is, investment projects with a high rate of capital turnover should be given preference. In other words, capital coefficient is to be minimised in order to maximise output.

But, as a general guide to policy, this criterion suffers from some serious limitations :

(*a*) The high rate of turn-over may entail a high rate of depreciation so that the net output is not necessarily high.

(*b*) Short-function-lag projects' may have a lower capital-output ratio in the short period but not necessarily so in the long period.

(*c*) This criterion ignores the cost of complementary factors like labour used in operating the capital.

(*d*) This criterion also ignores the 'project-complementarity', particularly the vertical and horizontal transmission of external economies. In order to avail of the external economies, it may become necessary to choose an investment with a higher capital-output ratio.

(*e*) In a sector like agriculture, the amount of fixed capital investment is small in proportion to total inputs. Hence, factors other than capital investment may substantially change the fixed capital-output ratio.

The Surplus Rate Criterion

This criterion seeks to maximise the per capita income at some future point of time rather than minimise the national income now. For this purpose, the rate of savings should be maximised so that the rate of reinvestment can be maximised. Hence, for each unit should be chosen "that alternative that will give each worker greater productive power than any other alternative".

The criterion assumes that profits are largely saved and re-invested and that wages are largely spent on consumption. Hence, it is recommended that the capita resources should be so allocated among the alternative uses that the marginal per capital reinvestment quotient is the same in different alternatives. The application of the law of equimarginal return will bring about an optimal utilisation of scarce capital resources.

But the application of this criterion is likely to produce some undesirable social effects : It will accentuate inequalities in income and wealth in the community, because the capitalists will gain at the expense of the wage earners.

Employment Absorption Criterion

It is well-known that in the under-developed but over-populated countries, labour supply is abundant and cheap. There is large-scale unemployment or under-employment especially in the agricultural sector in the form of disguised unemployment. Hence, it is suggested such techniques should be adopted as are labour-intensive. Such techniques will be conducive to a high degree of economic equality by raising the level of income of the working class people and improve growth.

But the defect of such techniques is that they do not necessarily maximise the national output. Labour-intensive techniques are not as productive as the capital-intensive techniques. Low labour productivity may be perpetuated. Besides, the quality or the end products suffers.

The Time Series Criterion

When several techniques are available to choose from, we may estimate real income flows resulting from each technique. For this purpose, we apply the rate of investment with corrections due to the variability of the volume of investments arising from different spending habits and varying import-content of investment. When we have taken two time series of real income flows, we have to apply the relevant rates of time discount. The time discount is necessary because of (*a*) the diminishing marginal social utility of income with the rising income level and (*b*) the uncertainty of the future. In the case of quickly falling marginal social utility, the higher rate of income growth may not mean higher level of social satisfaction. Since future in uncertain, it is necessary to have a valuation of uncertainty discount.

It is not easy to arrive at utility and uncertainty functions. A more practical method seems to be to fix the period of time we want to consider and weigh the loss of immediate output arising from the adoption of more capital-intensive techniques against a gain in increased output later.

Socially Desirable Income Distribution Criterion

Another important investment criterion is the socially desirable income distribution. It means that investment should be so planned as to achieve equitable distribution of benefits. This criterion may be regarded very important because economic development in under-developed countries tends to accentuate disparities of income and wealth distribution in the country. Such disparities cause grave discontent and pose a great threat to political stability. Such desirable investment may be in the form of public utilities, education, public health, improving means of transport and communications, etc. In this type of investment benefits are evenly distributed, may be more in favour of the poor than the rich.

There is no single, simple, precise and objective criterion for planning investments. The best that the planners can do is to strike a balance among the various considerations we have discussed above.

It is clear that investment criteria should not be linked with any one of the objectives. For, promoting economic growth, output, profits, savings and employment all must be increased.

Savings

The total investible resources available at any time in a country at made up of domestic savings and external resources which are obtained from abroad in the form of foreign capital. To take savings first. The aggregate savings of an economy consist of government savings, saving by the business sector and savings by the households. Government savings are the tax revenues minus public expenditure; the business savings are the gross income of trade and industry minus the dividends and the taxes paid and the savings of the households are the disposable income minus consumption expenditure. In India, in 1958–59, government savings accounted for 10.6 per cent, corporate savings 3.5 per cent and the savings of the household sector 85.9 per cent.

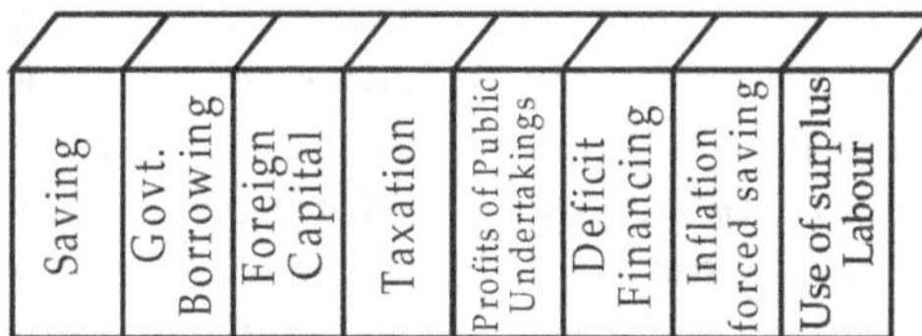

Sources of Finance

Finance is needed both for private and public sectors. So far as the private sector is concerned, it primarily depends on the voluntary savings of the people. Profit of private undertakings can also be ploughed back into investment. Institutions like Finance Corporations set up by the Government can also provide the needed development finance to the private sector.

To finance capital formation and other development activities in the public sector is the responsibility of the Government. Owing to the shortage of voluntary savings, the governments are often compelled to resort to the device of forced saving.

Taxation

There is considerable unanimity among economists about the usefulness, may necessity, of taxation and fiscal policy for mobilising resources for economic development in the under-developed countries. When development has proceeded to achieve a certain rate of growth, the level of savings by households and businesses rises sufficiently to meet the requirements of development. But in the early stages, some measure of compulsion is necessary to compel the people to save by means of suitable taxation measures, because the rate of domestic savings is low and propensity to consume is very high.

Thus, taxation is an important method of increasing the volume of savings by restricting domestic consumption. Both direct and indirect taxes can play a part in augmenting the resources of the governments to be spent on developmental activities. For achieving best results, taxes should be imposed on non-entrepreneurial incomes and luxury consumption. But the taxation of non-functional surplus may not yield a substantial volume of development finance, because most of the income of the vast majority of the people in an under-developed country is devoted to the consumption of necessaries. Thus, the need to raise an adequate volume of development finance makes it inevitable for the government to extend the coverage of indirect taxes to include the staple commodities of mass consumption. Moreover, the taxation of agriculture has to play an important part in the mobilization of resources for the public sector in a developing economy.

Thus, taxation is the most important means available to the State for mobilising nation's resources for economic development. Its yield can be more accurately estimated and its economic effects can be better foreseen. It can be used to finance development with minimum adverse effects on economic stability. Hence, it is necessary to intensify the tax effort, especially because the savings are meager and the rate of capital formation low, whereas the development requirements are very large.

But taxation as a method of development finance has some difficulties. While involuntary savings are increased, voluntary savings may be diminished, since individuals may reduce their voluntary savings in order to maintain their former consumption levels. This may reduce the resources going to the private sector. Another major drawback of taxation is its negative effects which it may have on incentives. If taxes on wage-earners diminish their incentive to work harder, if taxes on profits of the higher income group reduce their incentive to save and to make investments in new enterprises, and if taxes on the output or income of the farmers diminish the incentives to improve agricultural techniques, then the forced savings extracted through these taxes will not be an unmixed gain.

Tax policy of the government can exert a powerful influence both on savings and investment—the two crucial factors determining economic growth. The primary objective of the tax policy in the under-developed countries is to transfer from the community to the State as large a volume of resources as possible with minimum of adverse effects on incentives for production and investment. It is generally agreed that there is a considerable scope for broadening and deepening the tax system by improving the tax structure and by toning up the tax machinery. A sound tax policy can provide incentives for private enterprise. Taxation can be used as means for controlling economic fluctuations, for containing inflationary pressures and to achieve social justice by reducing inequalities in income and wealth. These are principal objectives of tax policy in under-developed countries.

MONEY & BANKING

Meaning and Organs of Money Market

Money market refers to that mechanism whereby borrowers managed to obtain short term loanable funds on the one hand and lenders succeed in getting credit worthy borrowers for their money on the other. In this way any institution or person who is willing to provide short-period monetary debt becomes a part of the money market.

Under Indian money market, the Reserve Bank of India occupies the central position because it regulates and controls the credit supply of the country. Ordinarily, the Indian money market is divided into two parts :

1. The organised sector
2. The unorganised sector.

The organised sector includes the State Bank of India and its 7 associated banks, 19 nationalised banks. Regional Rural Banks, Co-operative Banks, Non-governmental sectors and other banks, whereas the unorganised sector includes the money lenders and indigenous bankers.

Different Stages of the Development of Indian Banking

In order to make the Reserve Bank of India more powerful, the Indian Government nationalised it on January 1, 1949. With a view to have the co-ordinated regulation of Indian banking, the Indian Banking Act was passed in March 1949. According to this Act, the Reserve Bank of India was granted extended powers for the inspection of non-scheduled banks. For the development of the banking facilities in the rural areas the Imperial Bank of India was partially nationalised on 1 July, 1955 and it was named as the State Bank of India. Alongwith it other 8 (at present 7) banks were converted as its associate banks which form what is named as the State Bank Group. They are as follows :

1. The State Bank of Bikaner and Jaipur (In the beginning the State Bank of Bikaner and the State Bank of Jaipur were separate. But they were merged and named as the State Bank of Bikaner and Jaipur).
2. The State Bank of Hyderabad.
3. The State Bank of Indore.
4. The State Bank of Mysore.
5. The State Bank of Saurashtra.
6. The State Bank of Patiala.
7. The State Bank of Travancore.

In order to have more control over the banks, 14 large commercial banks whose reserves were more than Rs. 50 crore each were nationalised on 19th July, 1969. The nationalised banks are as follows :

1. The Central Bank of India
2. Bank of India
3. Punjab National Bank
4. Canara Bank
5. United Commercial Bank
6. Syndicate Bank
7. Bank of Baroda
8. United Bank of India
9. Union Bank of India
10. Dena Bank
11. Allahabad Bank
12. Indian Bank
13. Indian Overseas Bank
14. Bank of Maharashtra

After one decade, on April 15, 1980, those 6 private sector banks whose reserves were more than Rs. 200 crore each were nationalised.

These banks are as :

1. Andhra Bank
2. Punjab and Sindh Bank
3. New Bank of India
4. Vijaya Bank
5. Corporation Bank
6. Oriental Bank of Commerce.

On 4th September, 1993 the Government merged the New Bank of India with Punjab National Bank and as a result of this the total number of nationalised bank got reduced from 20 to 19.

As certain rigidities and weaknesses were found to have developed in the banking system during the late eighties, the Government felt that these had to be addressed to enable the financial system to play its role in ushering in a more efficient and competitive economy. Accordingly, a high-level Committee under the Chairmanship of Shri M. Narasimham on the Financial System (CFS), was set up on 14 August, 1991 to examine all aspects relating to the structure, organisation, functions and procedures of the financial systems. Based on the recommendations of the Committee a comprehensive reform of the banking system was introduced in 1992-93.

A high-level Committee, under the Chairmanship of Shri M. Narasimham was constituted by the Government of India in December 1997 to review the record of implementation of financial system reforms recommended by the CFS in 1991 and chart the reforms necessary in the years ahead. The Committee submitted its report to the Government in April 1998.

Reserve Bank of India

It is the Central Bank of the country. The Reserve Bank of India was established in 1935 with a capital of Rs. 5 crore. This capital of Rs. 5 crore was divided into 5 lakh equity shares of Rs. 100 each. In the beginning the ownership of almost all the share capital was with the non-government shareholders. In order to prevent the centralisation of the equity shares in the hands of a few people, the Reserve Bank of India was nationalised on January 1, 1949.

The general administration and direction of RBI is managed by a Central Board of Directors consisting of 20 members which includes 1 Governor, 4 Deputy Governors, 1 Government official appointed by the Union Government of India to give representation to important stratas in economic life of the country. Besides, 4 directors are nominated by the Union Government to represent local boards. Apart from the central board there are 4 local boards also and their head offices are situated in Mumbai, Chennai, Kolkata and New Delhi. 5 members of local board are appointed by the Union Government for a period of 4 years. The local boards work according to the instructions and orders given by the Central Board of Directors, and from time-to-time they also tender useful advice on important matter.

Functions of Reserve Bank

1. **Issue of Notes :** The Reserve Bank has the monopoly of note issue in the country. It has the sole right to issues currency notes of various denominations except one rupee note. The Reserve Bank acts as the only source of legal tender money because the one rupee note issued by Ministry of Finance are also circulated through it. The Reserve Bank has adopted the **Minimum Reserve System** for the note issue. Since 1957, it maintains gold and foreign exchange reserves of Rs. 200 crore, of which atleast Rs. 115 crore should be in gold.
2. **Banker to the Government :** The second important function of Reserve Bank is to act as the Banker, Agent and Adviser to the Government. It performs all the banking functions of the State and Central Government and it also tenders useful advice to the Government on matters related to economic and monetary policy. It also manages the public debt for the Government.
3. **Banker's Bank :** The Reserve Bank performs the same function for the other banks as the other banks ordinarily perform for their customers. It is not only a banker to the commercial banks, but it is the tender of the last resort.
4. **Controller of Credit :** The Reserve Bank undertakes the responsibility of controlling Credit created by the commercial banks. To achieve this objective it makes extensive

use of quantitative and qualitative techniques to control and regulate the credit effectively in the country.

5. **Custodian of Foreign Reserves :** For the purpose of keeping the foreign exchange rates stable the Reserve Bank buys and sells the foreign currencies and also protects the country's foreign exchange funds.
6. **Other Functions :** The bank performs a number of other developmental works. These works include the function of clearing house arranging credit for agriculture, (which has been transferred to NABARD) collecting and publishing the economic data, buying and selling of Government securities and trade bills, giving loans to the Government buying and selling of valuable commodities etc. It also acts as the representative of Government in I.M.F. and represents the membership of India.

Composition of Banking System in India

Commercial Banking System in India consisted of 286 scheduled commercial banks (including foreign banks) as on 31 March, 2004. Of the scheduled commercial banks, 223 are in public sector of which 196 are regional rural banks (RRBs) and these account for about 77.5 per cent of the deposits of all scheduled commercial banks. The regional rural banks were specially set up to increase the flow of credit to small borrowers in the rural areas. The remaining 27 banks in the public sector (*i.e.,* nationalised banks and SBI Group) are commercial banks and transact all types of commercial banking business.

Amongst the public sector banks, as on March 2004, the nationalised banks group is the biggest unit with 33,090 offices, deposits aggregating Rs. 7,52,558 crore and advances of Rs. 4,10,376 crore. The State Bank of India group (SBI and its seven Associates) with 13,593 offices, deposits aggregating Rs. 3,67,057 crore and advances of Rs. 2,12,420 crore is the second largest unit. The nationalised banks group accounts for around (65.9 per cent of aggregate banking business (aggregate of deposits and advances) conducted by the public sector banks (excluding RRBs) and 46 per cent of the aggregate business of all scheduled commercial banks. The SBI and its Associates as a group accounts for around 34.1 per cent of aggregate banking business conducted by the public sector banks (excluding RRBs) and 23.8 per cent of the aggregate business of all scheduled commercial banks.

Quantitative and Qualitative Credit Control

By credit control we mean control over the quantity and value of credit in the country. Among the functions of Central Bank, one main function is to control and regulate the credit in the country. In India, this function is performed by the Reserve Bank. The measures of credit control can be divided into two types :

(*i*) Quantitative credit control

(*ii*) Qualitative credit control

The main objective of quantitative credit control is to establish control over the total quantity of credit in the country. For quantitative credit control, the Central Bank takes the help of bank rate, open market operations, statutory liquidity ratio and cash reserve ratio, whereas publicity, rationing of credit, regulation of consumer credit, moral suasion, variation in margin requirements are the qualitative credit control methods.

Selective Credit Control

Since 1956, the Reserve Bank is employing the selective credit control. It is used to control the amount of bank advances against the commodities having limited supply. Under it, three measures are taken :

1. Fixation of margin requirements.
2. Fixation of maximum limit of credit given for a special purpose.
3. Discriminatory interest rates on some special types of advances.

While applying selective credit control it is kept in mind that it should not have any adverse effect on the production, and supply position, of the commodities. The main objective of selective controls is to control the credit given to the traders for inventories. To variate the per centage of advances against foodgrains is an examples of such selective credit control.

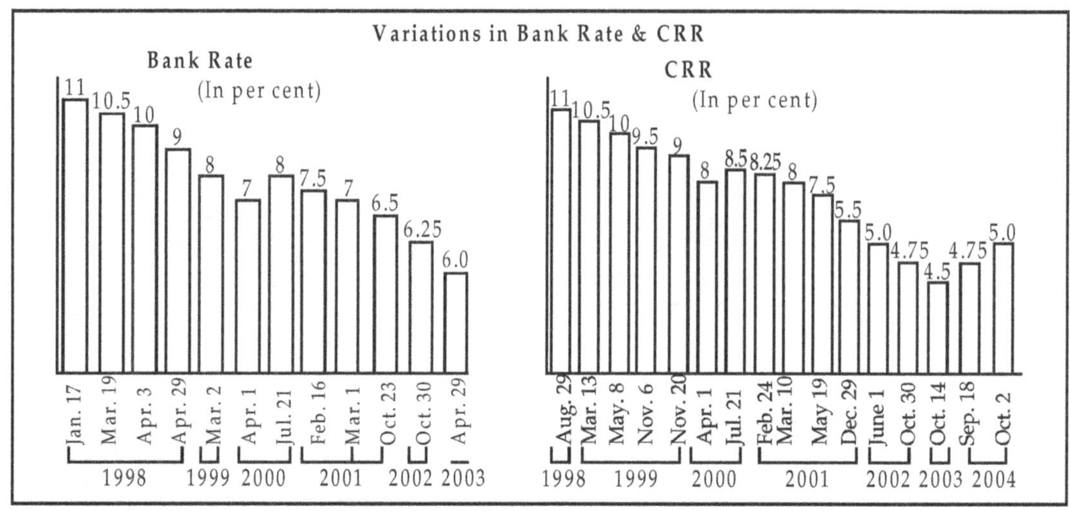

Extension of Priority Sector for Providing Credit

The Reserve Bank of India has recently extended the coverage of priority sector for enlarging credit services of commercial banks. According to the notification of RBI credits given to non-government agencies for making houses for slum-dwellers will be included within priority sector credits but maximum upper ceiling of such credits will be only Rs. 2 lakh per house. Similarly credits given to farmers for purchasing field or goods carrier vehicle will also be included in the category of priority sector credits.

Retailers (who do not have fair price shops and are not trading essential goods) will be given credit upto the limit of Rs. 2 lakh and this credit will also be a part of priority sector lending. Credits given to commercial institutions providing various services will also be included within the category of priority sector lending.

Advances to public sector banks (PSBs) to priority sector increased by 21.0% in 2003-04 compared to an increase of 18.6% in 2002-03. The share of priority sector advances to net bank credit (NBC) of PSB's increased from 42.5% in 2002-03 to 44% in 2003-04. The level of advances is higher than the target of 40% stipulated for domestic banks. For foreign banks the target fixed for priority sector lending is 32% of net bank credit. Total outstanding advances by foreign banks to priority sector increased from 33.9 per cent in 2002-03 to 34.8 per cent of their NBC. The share of priority sector lending by private banks improved from 44.4 per cent in 2002-03 to 47.4 per cent in 2003-04. Thus all banks groups fulfilled the minimum target set for priority sector lending in 2003-04.

While the overall target fixed for priority sector lending had been met by all bank groups, shortfalls were noticed in meeting the sub-targets prescribed under priority sector lending. Within the overall target of 40 per cent of NBC for domestic banks, the sub-target fixed for agriculture is 18.0 per cent and that for weaker sections is 10 per cent. Advances by PSBs to agriculture at 15.4 per cent of NBC in 2003-04 fell short in the sub-target by 2.6 per centage points is 2003-04. PSBs also fell short of the sub-target of 10 per cent set for weaker sections by 2.6 per centage points. Advances to agriculture by PSBs remained within a narrow range of 15.4 per cent and 15.7 per cent of NBC between 2000-01 and 2003-04. Advances by private sector banks to agriculture and weaker sections at 15.8 per cent and 1.3 per cent of NBC, respectively, also fell short of the sub-targets in 2003-04.

Differential Rate of Interest Scheme

In April 1972, the Government implemented differential interest rates scheme in 162 districts of the country. Under this scheme, public sector banks were directed to grant at least 1% of their total deposits of previous year to weaker sections of society at a concessional interest rate of 4%. The poor people eligible for this facility were those whose income was not more than Rs. 6400 p.a. in rural areas and Rs. 7200 p.a. in urban areas. This facility was also provided to those persons who possess landless than 2.5 acres non-irrigated land or 1 acre irrigated land. Such persons are provided the loan of Rs. 6500 as term loan and working capital loan for productive ventures. The public sector banks had an outstanding of DRI credit amounting to Rs. 302.25 crore as at the end of September 2003.

New Strategy for Rural Lending : Service Area Approach

During Nov-Dec. 1987 public sector banks made an empirical study of bank's advances given in rural areas of various districts. On the basis of the results a new strategy for rural lending named **'Service Area Approach'** was adopted. This new approach was implemented under the purview of Lead Bank Scheme since April 1, 1989. Under this new scheme, branches of commercial banks were allotted certain specific semi-urban and rural areas. These banks were made responsible for over-all development in these allotted area. RBI also issued essential directions to commercial banks in Nov. 1992 to eliminate various drawbacks of Service Area Approach.

Credit Flow to Agriculture

In terms of the guidelines issued by Reserve Bank of India in October 1993, both direct and indirect advances for agriculture are taken together for assessing the target of 18 per cent, with the condition that for the purpose of computing their performance in lending to agriculture, lending for indirect agriculture should not exceed one-fourth of the total agriculture lending target of 18 per cent of net bank credit so as to ensure that the focus of Banks on direct lending to agriculture is not diluted. As at the end of September 2003 public sector banks had extended Rs. 76,700 crore constituting 15.85% of the net bank credit to the agriculture sector. Private sector bank extended Rs. 11,873 crore to agriculture sector at the end of March 2003 constituting 10.8% of net bank credit.

Monetary and Credit Policy : Annual Policy Statement 2004-05

Ever since the Reserve Bank of India designated as the Central Bank, Monetary and Credit Policy prepared and announced by it had been crucial part of the management of economy. Upto 2003-04, monetary and credit policy was announced two times in a year, the first one in April, that is for loan season and the second one in Sept.–Oct. that is for busy season. Now the Monetary and Credit policy has been renamed as **"Annual Statement of RBI on Monetary and Credit Policy"**. The first such statement announced by RBI Governor Dr. Y.V. Reddy on May 18, 2004.

The policy documents of the R.B.I. provide a framework for the monetary and other relevant measures that are taken from time-to-time and capture the rationale or the underlying factors at work that affect the macro economic assessments. The documents also set out the logic, intensions and actions related to structural and prudential aspects of the financial sector.

This statement broadly follows the pattern already set in previous years. It delineates and elaborates on various areas in which RBI has been taking measures from time-to-time and provides a focus on our broad policies that are intended to be pursued in due course of the time.

The annual statement's policy measures are based on the following developments during 2003-04 and the outlook for 2004-05.

- In terms of the GDP, India may continue to be among the top performers globally.
- The price situation is unlikely to cause concern to macro stability.
- Credit delivery, in particular to agriculture, small and medium enterprises and infrastructure is critical to sustain growth.
- The financial sector exhibits growing strength, efficiency and stability.
- Current status of as well as the outlook for the external sector accords comfort to the conduct of public policy.
- The real GDP growth rate for 2004-05 will remain between 6.5%–7.0%.
- With a normal monsoon, as forecasted by Meteorological Department, the growth in agriculture can be assumed to be at the trend growth rate of about 3%.
- The inflation rate likely to around 5% on a point-to-point basis.
- Money supply (M_3) will grow at 14.0% during 2004-05.
- Aggregate deposits of scheduled commercial banks is set at Rs. 2,18,000 crore which is higher by 14.5% over its level in previous year.
- Non-food credit is expected to increase by 16.0%–16.5%.

The **overall stance** of the monetary policy for 2004-05 will be :

- Provision of adequate liquidity to meet credit growth and support investment and export demand in the economy while keeping a very close watch on the movements in the price level.
- Consistent with the above, while continuing with the status quo, to pursue an interest rate environment that is conducive to maintaining the momentum of growth and macro-economic and price stability.

The main components and thrust points of Annual Policy Statement (May 18, 2004) are:

- Bank rate kept unchanged at 6.0 per cent.
- Repo rate kept unchanged at 4.5 per cent.
- Revised LAF scheme operationalised.
- The entire amount of export credit refinance to banks and liquidity support to primary dealers made available at the reverse reporate.
- Almost all commercial banks have adopted the new system of BPLR and the rates are lower in the range of 20–200 basis points from their earlier PLRs.
- Banks advised to put in place comprehensive and rigorous risk assessment to relate pricing of credit to risk more appropriately.
- Recommendations of the interim Report of Vyas Committee accepted for implementation in respect of : (*a*) loans for storage facilities under priority sector lending, (*b*) securitised agricultural loans as priority sector lending, (*c*) waiving margin/security requirement for certain agricultural loans up to a limit, (*d*) NPA for crop loans aligned to crop seasons.
- Micro-finance institutions not to be permitted to accept public deposits unless they comply with the extent regulatory framework of the Reserve Bank.

- Development of mechanism for debt restructuring for medium enterprises on the lines of corporate debt restructuring.
- Definition of infrastructure lending broadened to include :
  - (*i*) construction relating to projects involving agro-processing and supply of inputs to agriculture.
  - (*ii*) construction for preservation and storage of processed agro-products, perishable goods such as fruits, vegetables and flowers including testing facilities for quality and
  - (*iii*) construction of educational institutions and hospitals.
- Working group constituted for Credit Enhancement by State Governments for financing infrastructure. Gold Card Scheme for exporters drawn-up
- To rationalise structure of regional rural banks various operations under consideration of the Government and other stakeholders. Vyas Committee is also looking into these aspects.
- Limit on the lending of non-bank participants in the call/notice money market reduced to 45 per cent effective June 26, 2004.
- Automated value free transfer of securities between market participants and the CCIL facilitated. The ECB limit enhanced to US $ 500 million under automatic route with minimum average maturity of 5 years. End use for ECBs enlarged to include overseas direct investment in Joint Ventures/Wholly Owned Subsidiaries to enable them to become global players.
- Resident individuals already permitted to remain freely upto US $ 25,000 per calendar year for any current or capital account transaction.
- Indian corporates and partnership firms allowed to invest overseas upto 100 per cent of their net worth.
- Banks allowed to float long-term bonds to finance infrastructure.
- The extent limit on unsecured exposures for banks withdrawn.
- With effect from April 1, 2005 exposures on all Public Financial Institutions (PFIs) to attract a risk weight of 100 per cent.
- Banks required to maintain capital charge for market risk in respect of securities held for trading by March 31, 2005.
- Banks would be required to maintain capital charge for market risk in respect of the securities included under available for sale category by March 31, 2006.
- Banks to draw a road map for moving towards Basel II by December 31, 2004.
- Banks to make higher provisioning for NPAs in "doubtful more than three years" category.
- Risk based supervision extended to more banks.
- RBI expects most commercial banks to join the RTGS system by June 2004.
- Single window services for all transactions in RBI cash department.
- Operationalisation of On-line Tax Accounting System by June 2004.

Annual Policy Statement of RBI on Monetary and Credit Policy
(Mid-Term Review on Oct. 26, 2004)

- Bank rate kept unchanged at 6 per cent.
- Switching over to the international usage of the term repo and reverse repo from Oct. 29, 2004.

- Fixed reverse repo rate under LAF increased by 25 basis points to 4.75 per cent from Oct. 27, 2004. The spread between the repo and reverse repo rate reduced by 25 basis points to 125 basis points. Accordingly, the repo rate to remain at 6.0 per cent.
- LAF to be operated with overnight fixed rate repo an reverse repo. Accordingly, auctions of 7-day and 14-day reverse repo discontinued.
- With a view to aligning interest rates with international rates, interest rates on Non-Resident (External) Rupee (NRE) deposits raised to US Dollar LIBOR/SWAP rates of corresponding maturities plus 50 basis points.
- Fixation of ceiling on interest rates on FCNR (B) deposits to be shifted from weekly basis to a monthly basis.
- Banks given freedom to reduce the minimum tenor of retail domestic term deposits (uner Rs. 15 lakh) from 15 days to 7 days.
- To improve credit delivery, restrictive provisions of service area approach for banks dispensed with, except for Government sponsored programmes.
- The limit on advances under priority sector for dealers in agricultural machinery raised from Rs. 20 lakh to Rs. 30 lakh and for distribution of inputs for allied activities from Rs. 25 lakh to Rs. 40 lakh.
- Banks advised to make efforts to increase their disbursements to small and marginal farmers to 40 per cent of their direct advances under special agricultural credit plans (SACP) by March 2007.
- The mechanism of SACP extended to private sector banks. Private sector banks urged to formulate SACPs from the year 2005-06, targeting an annual growth of at least 20–25 per cent of credit disbursement to agriculture.
- Composite loan limit for SSI entrepreneurs enhanced from Rs. 50 lakh to Rs. 1 crore.
- The minimum maturity period of commercial paper reduced from 15 to 7 days.
- To promote investment activity, authorised dealer of foreign exchange permitted to issue guarantees/letters of comfort up to US $20 million per transaction for a period up to one year, for import of all non-capital goods permissible under the foreign trade policy and up to three years for import of capital goods, subject to prudential guidelines.
- 100 per cent Export oriented units and units set up under Electronics Hardware Technology Parks (EHTPs), Software Technology Parks (STPs) and Bio-Technology Parks (BTPs) schemes permitted to repatriate the full value of export proceeds within a period of 12 months.
- The limit of outstanding forward contracts booked by importers/exporters increased from 50 per cent to 100 per cent of their eligible limit.
- As a temporary counter-cyclical measure, the risk weight on housing loans to individuals and investments in Mortage Backed Securities of Housing Finance Corporations (HFCs), supervised by National Housing Bank (NHB) increased from 50 per cent to 75 per cent and on consumer credit from 100 per cent to 125 per cent.
- Change in the norms for classification of doubtful assets of financial institutions (FIs). With effect from March 31, 2005, an asset to be classified as doubtful, if it has remained in the substandard category for 12 months. FIs permitted to phase out the consequent additional provisioning over a four year period.
- Asset reconstruction companies (ARCs) required to have owned funds of not less than 15 per cent of the assets acquired or Rs. 100 crore, whichever is less for commencement of business.

- Banks advised to advance loans to distressed urban poor to prepay their debt to non-institutional lenders, against appropriate collateral or group securities.
- Investments by banks in securitised assets representing direct lending to the SSI sector to be treated as their direct lending to SSI sector under priority sector, provided the pooled assets represent loans to SSI sector and originated by banks/financial institutions.

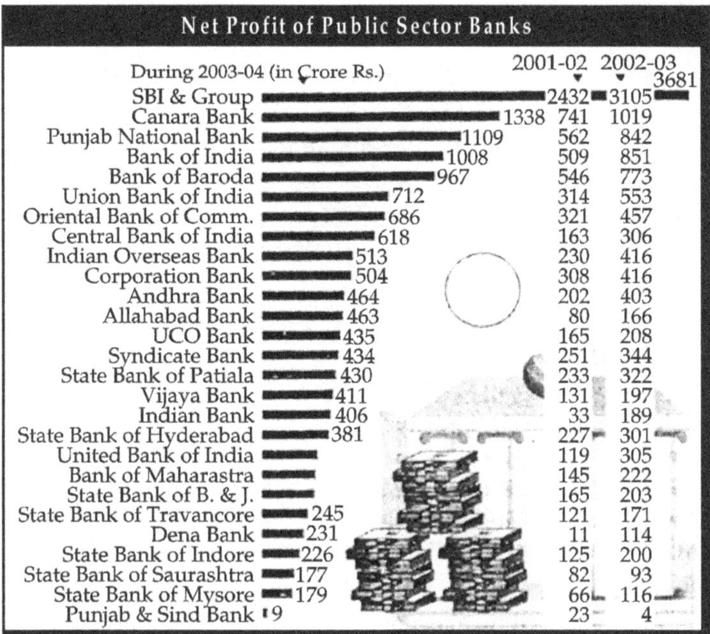

Commercial Banks

At present 27 commercial banks in public sector are working in the country. Out of these 27 banks, 19 banks are nationalised banks (Earlier this number was 20 but New Bank of India was merged with PNB leaving this number to 19).

Commercial Banking system in India consisted of 286 scheduled banks (including foreign banks) and one non-scheduled banks at the end of March 2004. Of the scheduled banks, 223 are in public sector and these account for about 77.5% of the deposits of all scheduled banks. There are 196 RRBs specially set up to increase the flow of credit to small borrowers in the rural areas. These banks are also categorised as scheduled commercial banks. At present the number of nationalised commercial banks is 19.

At the time of bank nationalisation (*i.e.,* July 1969) there were 8262 branches of various commercial banks (1860 in rural areas and remaining 6402 branches in urban areas). In other words, in 1969 only 23% of total bank branches were working in rural areas. But on June 30, 2004, total number of bank branches was increased to 67283. Presently, 47.8% of total branches are working in rural areas. There is one bank branch working for 15,000 population while there was one branch for 64000 population in 1969.

State Bank of India is the largest public sector bank in the country. On June 30, 2004, 13533 branches of SBI & Associates were working in the country. SBI Group also includes 7 associate banks. The share capital of these associate banks has been reserved with SBI. SBI Group (*i.e.,* SBI and its 7 Associate Banks) has about 26.6% share in total banking business in the country. During 2003-04, State Bank of India earned the net profit of Rs. 3681 crore which is 18.6% more than that of 2002-03.

During the year 2003-04, the banking sector witnessed strong growth in deposits and advances. Aggregate deposits of scheduled commercial banks (SCBs) grew by 17.5 per cent compared to 13.4 per cent in 2002-03. Credit and investments by SCBs increased by 15.3 per cent and 25.1 per cent, respectively in 2003-04 compared to 16.1 per cent and 23.3 per cent respectively in 2002-03. These developments coupled with a decline in gross NPAs enabled SCBs to improve their financial performance, despite a financial performance, despite a lower income growth consequent upon low interest rates. Ratio of net profits to total assets of SCBs improved marginally from 1.0 per cent to 1.1 per cent. Ratio of operating profit to total assets improved from 2.4 per cent in 2002-03 to 2.7 per cent in 2003-04.

Total income of SCBs increased by 6.6 per cent to Rs. 1,83.767 crore in 2003-04 as compared to an increase of 14.0 per cent in 2003-03. The lower income growth was on account of lower growth of 2.4 per cent in interest income in 2003-04. Income growth in 2003-04 came mainly from other income, which increased by 25.5 per cent. Total expenditure of SCBs grew at a lower rate of 4.0 per cent in 2003-04 compared to 11.3 per cent in 2002-03.

The ratio of net profits to total assets was the highest for foreign banks (1.7 per cent) followed by old private sector banks (1.2 per cent), PSBs (1.1 per cent) and new private sector banks (0.8 per cent). The ratio of operating profits to total assets followed more or less a similar pattern with foreign banks performing the best (3.7 per cent), followed by PSBs (2.7 per cent), old private sector bank (2.6 per cent) and new private sector banks (2.0 per cent). There has been an overall improvement in the efficiency of SCBs as evident in the declining ratios for operating expenses to net total income (total income–interest expenditure). For SCBs as a whole the ratio declined from 53.0 per cent in 2001-02 to 48.3 per cent in 2002-03 and further to 45.2 per cent in 2003-04.

The profitability of SCBs, though still high, showed a declining trend in the first half of 2004-05 on account of a fall in treasury income. The ratio of net profit to total assets (annualised) was lower at 1.1 per cent in the second quarter of 2004-05 compared to 1.3 per cent in the corresponding previous quarter.

Classification of Commercial Banks

The commercial banking institutions of the country can be divided into two groups—

(*A*) **Scheduled Banks :** Those banks are scheduled banks which have been included in the Schedule (Second) of Reserve Bank Act, 1934. The banks included in this scheduled list should fulfil two conditions :

1. The paid up capital and collected funds of banks should not be less than Rs. 5 lakh.
2. Any activity of the bank will not adversely affect the interest of depositors.

Every scheduled bank enjoys following facilities :

1. Such bank becomes eligible for obtaining debts/loans on bank rate from RBI.
2. Such bank automatically acquires the membership of clearing house.
3. Such banks also get the facility of rediscount of first class exchange bills from RBI. This facility is provided by RBI only if the scheduled bank deposits an average daily cash fund with RBI which is decided by RBI itself and presents the recurring statements under the provisions of RBI Act, 1934 and Banking Regulation Act, 1949.

(*B*) **Non-scheduled Banks :** The banks which are not included in the list of scheduled banks are called non-scheduled banks. The number of non-scheduled banks is continuously declining. Such non-scheduled banks also have to follow CRR conditions. But such banks can have these funds with themselves as no compulsion has been made on non-scheduled banks to deposit CRR funds with RBI. These non-scheduled banks are not eligible for having loans (from RBI for meeting their day-to-day general activities but under emergency conditions these banks can be granted loans by RBI.

Foreign Commercial Banks

The Reserve Bank of India has adopted selective policy for allowing entry to foreign banks in the country. At the end of June 2002, 41 foreign banks and 202 branch offices of foreign banks were working in India. The Mumbai branch of Bank of Credit and Commerce International had been taken over by State Bank of India in March 1993. This Mumbai branch of BCCI works as subsidiary of SBI. RBI releases directions to all foreign banks working in the country after having enquiry at every 2 years interval. During 1995-96, RBI granted permission to 4 new foreign banks for having banking business in the country. On April 4, 1996 (*i.e.*, financial year 1996-97) China Trust Bank—A lead bank of Taiwan established its branch in New Delhi. This branch is the ever first branch of China Trust Bank, established in any foreign country.

Under new provisions every foreign bank has to bring a foreign capital of $ 10 million while opening its first branch in India for business purpose. The capital requirements for opening second and third branch are $ 10 million and $ 5 million respectively. ANZ Grindlleys Bank is a foreign bank having highest number of branches (56) in India.

New Banks in Private Sector

Narasimham Committee had recommended to grant permission for new private banks. On Nov. 19, 2000 the Government decided to reduce govt. holding in nationalised banks from 51% to 33%. For this necessary legislation is to be passed. The RBI released new directions to establish new banks in the private sector on Jan. 3, 2001.

Initially 10 new private banks of such type had started their working in the country. The details of these 10 new private banks have been shown in following table :

	Name of Bank	*Regd. Office of Bank at*	*Date of Issuing Licences*
1.	UTI Bank Ltd.	Ahmedabad	Feb. 28, 1994
2.	Indus Ind Bank Ltd.	Pune	Apri. 02, 1994
3.	ICICI Bank Ltd.	Vadodara	May 17, 1994
4.	Global Trust Ltd.	Secundrabad	Sept. 06, 1994
5.	HDFC Bank Ltd.	Mumbai	Jan. 05, 1995
6.	Centurian Bank Ltd.	Panji (Goa)	Jan. 13, 1995
7.	Bank of Punjab Limited	Chandigarh	April 05, 1995
8.	Times Bank Limited	Faridabad	April 26, 1995
9.	IDBI Bank Ltd.	Indore	Sept. 28, 1995
10.	Development Credit Bank Ltd.	Mumbai	May 31, 1995

Indian Banks Abroad

Indian banks had their presence in 42 countries with a network of 93 branches (including six offshore units), five joint ventures, 16 subsidiaries and 18 representative offices. As on 30 June, 2004, ten Indian banks—eight from the public sector and two from the private sector—had operations overseas. Besides, another three private sector banks had four representative offices abroad. Bank of Baroda had highest concentration, with 38 branches, six subsidiaries and one joined venture in 17 countries, followed by State Bank of India with 21 Branches, five subsidiaries, two joint ventures and eight representative offices in 28 countries and Bank of India with 18 branches, two subsidiaries, two joint ventures and three representative offices in 16 countries.

Among the countries, UK registers the largest presence with 19 branches, Fiji a distant second with nine branches. While the off-shore banking units are located in important of-shores financial centres such as Bahamas. Cayman Islands, Channel Islands and Mauritius, the subsidiaries owned mainly by the bigger banks (*viz.*, SBI, BOI and BOB) are located in countries such as Guyana, Hong-Kong, Uganda, Kenya, Botswana, Canada, Mauritius and Nigeria.

Establishing Local Area Banks in Private Sector Allowed

The Government has granted permission to establish **Local Area Banks** to meet the local credit requirements by exploiting local resources itself. Such local area bank can be established by private promoters under Company Act. 1956. The minimum paid up capital limit of such bank will be Rs. 5 crore out of which private promoter will bear the share of at least Rs. 2 crore. These banks have also been allowed to open their branches but only within the limits of adjoining 3 districts. These banks will be regulated under RBI Act 1934, Banking Regulation Act 1949, and Regional Rural Bank Act 1976. These banks also have to maintain 8% capital adequacy standard form their inception.

Till now RBI has granted its approval, in principle, for establishing 3 such local area banks– Andhra Pradesh, Maharashtra and Karnataka (one each).

Co-operative Banks

Co-operative banks in India also perform fundamental banking activities but they are different from commercial banks. Commercial banks have been constituted by an Act passed by parliament while co-operative banks have been constituted by different States under various Acts related to co-operative societies as various states. **Co-operative Bank organisation in India has three tier** set **up.** State Co-operative Bank is the apex co-operative institution in the state. Central or District Co-operative Bank works as district level. At the lowest level co-operative setup is Primary Credit Agency which works at village level.

Commercial banks have been constituted under unitary basis and every commercial bank has been given authority to seek refinance facility from RBI while only State Co-operative Bank has been provided this facility under co-operative banking structure.

Commercial bank can establish its branches in any district/state of the country while, contrary to it. Co-operative bank can operate its activities only within limited area. For example, District Co-operative Bank can perform banking activities with the boundaries of the concerned district. Similarly. Primary Credit Societies can perform banking services within concerned villages. Co-operative banks cannot open their branches in foreign countries while commercial banks can do that.

Banking Regulation Act 1949 is fully applicable to all commercial banks while it is partially applicable to co-operative banks. In other words, **RBI has partial control on Co-operative banks.**

Co-operative banks work on principles of co-operation while commercial banks adopt pure commercial principal in their operation. That is the reason why Co-operative banks succeed in getting financial assistance from RBI on concessional rate.

Urban Co-operative Banks : There were 2104 Primary (urban) Co-operative Banks in the end of Dec. 2003 working in the country. Primary Co-operative Banks (PCBs) have to disburse 60% of their total advances to primary sector and at least 25% to weaker section of the society. Total deposits and advances of UCBs were Rs. 103478 crore and Rs. 61930 crore, respectively on Dec. 31, 2003. Out of 2104 UCBs, 176 were under liquidation and 636 had turned weak/sick.

Primary Credit Societies : These societies short term credit facilities to agriculture sector. Minimum 10 persons of a village (or area can form a primary credit society. These PCSs are also called Primary Agriculture Credit Societies. These societies grant short-term loans (generally one year period) for productive activities but his period can be extended upto 3 years under special circumstances.

The various reconstruction and revival programmes for PCSs adopted by Indian Government and RBI have considerably reduced the number of primary credit societies over past two decades. There were 161000 PCSs in 1970-71 which got reduced to 88000 in 1994-95, but again it increased to 93000 in 1995-96. As on March 31, 2001 about 1 lakh primary agricultural credit societies with membership approximately 10 crore, have outstanding deposits and loans of Rs. 13481 crore and Rs. 34522 crore respectively. However, a large number of PACs, face severe financial problems primarily due to significant erosion of own funds-deposits and low recovery rates.

Central (or District) Co-operative Bank

The working area of these banks is limited to one district only. Centery Co-operative Bank can be divided into two parts:

1. Co-operative Banking Union
2. Mixed Central Co-operative Bank

The membership of Co-operative Banking Union is given to Co-operative Societies only, while the membership of mixed Central Co-operative Bank can be granted to both Co-operative Societies and individuals. Generally all states in India are having Central Co-operative Banks with mixed membership and are providing sufficient financial assistance to both PCSs and individuals.

Central Co-operative Banks get loans from State Co-operative Bank and give loans to Primary Credit Societies. The duration of such loans vary from one year to three years. In this way Central Co-operative Bank plays a bridge role between State Co-operative Banks and Primary Credit Societies. At the end of March 2004 there were 366 Central Co-operative Banks working in the country with deposits of Rs. 73000 crore.

State Co-operative Bank (SCB) : It is the apex Co-operative Bank of the state. It grants loans to Central Co-operative Banks and regulates their activities. State Co-operative Bank gets loans from RBI. Hence, SCB acts as a link between RBI and Central Co-operative Banks.

State Co-operative Bank raises its current capital by shares and loans. RBI generally provides loans to SCB on interest rate, one or two per cent lower bank rate.

At the end of March 2001, 30 State Co-operative Banks are working in the country with deposits of Rs. 39000 crore.

LOAN

Regional Rural Banks

Regional Rural Banks (RRBs) were established since 1975 under the provisions of the RRB Act 1976 with a view to developing the rural economy as well as to creating an alternative channel to 'Co-operative Credit Structure' in order to ensure sufficient institutional credit for rural and agricultural sector. In other words, Regional Rural Banks (RRBs) were established to take the banking services to the door steps of rural masses, especially in remote rural areas with no access to banking services. There banks provide institutional credit to the weaker sections of the society at concessional rate of interest. These banks were also intended to mobilise rural saving and channelise for supporting the productive activities in the rural area. On October 2, 1975, initially 5 RRBs were established at Moradabasd and Gorakhpur (UP), Bhiwani (Haryana), Jaipur (Rajasthan) and Malda (West Bengal). Later on RRBs were extended to other districts of the country.

Though RRBs were initially intended to support productive activities in the rural areas, with effect from March 22, 1997, the RRBs were allowed to lend outside the target group by classifying their advances into 'priority sector' and 'others'. Similarly the interest rates on term deposits offered and interest rates on loans charged by RRBs have also been freed.

At the end of June 2004, there were 196 RRBs covering 516 districts with a network of 14507 branches. 82.6% of total branches have been opened in rural areas. The rural branch network of RRBs at 12003 constitutes 36.9% of the total rural credit outlets of the scheduled commercial banks in the country. The deposits of RRBs aggregated to Rs. 48900 crores in around 5 crore deposits as on 31 March, 2003 registering growth rate of around 16%. Out of total deposit, term deposits constitute 56%, saving deposits 40% and current deposits 4%. **As at the end of March 2003, RRBs' advances** and investments **stood at Rs 20700** crore and **28400** crore respectively. The gross **NPAs of RRBs** stood at Rs. 3200 crore, **14.4% of** their total loans.

RRBs are working in all states of the country except in Sikkim and Goa. At present, 196 RRBs are working in the country. **Since April 1987, no new RRB has been opened keeping in view the recommendations of Kelkar Committee.**

The working group on RRB under chairmanship of Mr. Kelkar recommended to increase paid up share capital of all banks to Rs. 100 lakh. The Government accepted this recommendation and increased the paid-up share capital from Rs. 25 lakh to Rs. 50 lakh is first phase and from Rs. 50 lakh to Rs. 75 lakh in second phase. The Government also extended the share capital of 20 RRBs from Rs. 75 lakh to Rs. 100 lakh in the third phase.

Narasimham committee, constituted for financial reforms, recommended to give freedom for all types of banking activities to RRBs. The RRBs should be made economically viable, but they should continue to grant loans to target group as usual.

In 1994-95 budget proposals, the Government had introduced an action plan to strengthen RRBs. RBI appointed a committee under chairmanship of Mr. M.C. Bhandari, Chief General Manager of NABARD, for seeking suggestions regarding restructue of RRBs. This committee selected only 49 RRBs (**Out** of 50 RRBs referred to the committee) for reconstruction purposes. On the basis or recommendations made by Bhandari Committee, the **RBI** permitted these RRBs to invest non-SLR surplus funds in profitable areas.

In the second phase of reorganising RRBs, the Government appointed another committee in 1995-96 under the chairmanship of Mr. K Basu, Chief General Manager, NABARD. This committee was given responsibility to select **50 RRBs** for re-organisation purpose. Basu Committee Selected 68 RRBs instead of 50. The Government made provisions of Rs 200 crore in 1996-97 budget for reorganising **RRBs.** Again, the Government provided Rs. 200 crore in 1997-98 budget proposals for the same purpose.

Consequent upon the permission of RBI to determine their own lending rate with effect from 26th August, 1996, most of the RRBs have been charging interest rate on loans.

Capital Adequacy Standards for Banks and No-banking Financial Institutions

Capital adequacy means the availability of sufficient capital as a per centage of risk weighted assets RBI on the recommendation of Narasimham Committee directed all Banks and Non-banking Financial Institutions to attain capital adequacy within a certain period.

RBI fixed following mentioned Capital Adequacy ratios for banking institution—

1.	**All Foreign Banks working in India**	8% Capital Adequacy Ratio (till March 1993)
2.	**All Indian Banks**	4% Capital Adequacy Ratio (till March 1993)
3.	**All Indian Banks having branches in foreign countries**	8% Capital Adequacy Ratio (till March 1995)
4.	**All other Indian Banks**	9% Capital Adequacy Ratio (till March 2000)

RBI extended the time limit for 4 banks of public sector to March 31 1998. These banks were—Vijaya Bank, Indian Bank, United Commercial Bank and United Bank of India.

RBI has directed Non-banking Financial Companies to attain capital adequacy as per following rates :

Till March 1995	6%
Till March 1996	8%

These capital adequacy ratios will be applicable on those NBFCs which are registered with RBI and which possess net funds of at least Rs. 50 lakh.

Accepting the recommendation of Narasimham Committee (II), it was decided to raise CRAR from 8% to 10% RBI has decided to raise CRAR from 8% to 9% *w.e.f.* March 31, 2000.

RBI has made capital adequacy norms more tough since September 1994. Under new guidelines, banks have been directed to reconstitute their capital structure having tier-2 capital not more than 50% of tier-1 capital. It means that tier-2 capital should not be more than 33% of total capital of bank. (Tier-1 capital includes paid up capital and reserve while tier-2 capital include shares, debentures and re-evaluated assets). These new provisions have come into force since April 1, 1994. These new provisions will create obstacles for bank to attain capital adequacy norms because now banks will have to maintain at least 67% of their capital in tier-1 capital.

The concept of minimum capital to risk weighted assets ratio (CRAR) has been developed to ensure that banks can absorb a reasonable level of losses. Application of minimum CRAR protects the interest of depositors and promotes stability and efficiency of the financial system. At the end of March 31, 2004, CRAR of PSBs stood at 13.2 per cent, an improvement of 0.6 per centage point from the previous year. There was also an improvement in the CRAR of old private sector banks from 12.8 per cent in 2002-03 to 13.7 per cent in 2003-04. The CRAR of new private sector banks and foreign banks registered a decline in 2003-04. For the SCBs as a whole the CRAR improved from 12.7 per cent in 2002-03 to 12.9 per cent in 2003-04. All the bank groups had CRAR above the minimum 9 per cent stipulated by the RBI.

During the current year, there was further improvement in the CRAR of SCBs. The ratio in the first half or 2004-05 improved to 13.4 per cent as compared to 12.9 per cent at the end of 2003-04. Among the bank groups, a substantial improvement was witnessed in the case of new private sector banks from 10.2 per cent as at the end of 2003-04 to 13.5 per cent in the first half of 2004-05. While PSBs and old private banks maintained the CRAR at almost the same level as in the previous year, the CRSR of foreign banks declined to 14.0 per cent in the first half of 2004-05 as compared to 15.0 per cent as at the end of 2003-04.

Debt Recovery Tribunals

The Government has established Debt Recovery Tribunals in six cities (Calcutta, Delhi, Jaipur, Ahmedabad, Bangalore and Chennai) to speed up recovery of loans disbursed by banks and other financial institutions. An Appellate Tribunal is also working at Mumbai. The first Debt Recovery

Tribunal was established in June 1994 at Calcutta. These tribunals have been established under 'Recovery of Debts due to Banks and Financial Institutions Act 1993.

Ordinance Issued for Recovering Debts of Banks and Financial Institution

"The Securitisation and Reconstructions of Financial Assets and Enforcement of Security Interest Ordinance 2002" was promulgated on August 22, 2002. It is a comprehensive piece of legislation to tackle the problems of non-performing assets (NPAs) or bad debt of banks & financial institution. The ordinance is based on the recommendations of the Narsimham & Andhyarujina Committees of financial sector reforms.

This ordinance allows banks & financial institution to take possession of securities of defaulting companies and sell them to enable them to recover their debts from defaulters quickly.

This ordinance also enables banks & financial institutions to even take over the management of companies that have defaulted continuously leading to accumulation of NPAs and bad loans.

Under this ordinance banks & financial institution are empowered to constitute ARCs (Asset Reconstruction Companies) which will focus on taking over bad debts of banks and financial institutions at a negotiated value.

Computerisation in Banks

Under schemes of modernisations computerisation in banks has been felt essential to dispose of day-to-day work with higher efficiency in banking sector. Computerisation will prove an accelerating force in obtaining better productivity and profitability in banks. A committee on computerisation in banks was constituted in 1989 which suggested an action plan for ensuring bank computerisation between a five year's span 1990-94. The committee strongly recommended cent-per-cent computerisation of those banks which deal in atleast 750 or more vouchers daily.

Agriculture and Rural Debt Relief Scheme (1990)

The Government of India declared Agriculture and Rural Debt Relief Scheme in 1990 to write off loans upto Rs. 10,000 (as on Oct. 2, 1989) taken from commercial banks and RRBs. Farmers, rural artisans and weavers of cottage and rural industries were covered under this scheme. State Government also adopted this concept and introduced the similar plan for co-operative sector. This scheme was introduced on May 15, 1990 which ended of March 31, 1991. (In Assam and J.K. this scheme was ended on June 30, 1991). Under this scheme 316.29 lakh debtors were benefited and they were granted a debt relief of Rs 7744.09 crore. This debt relief definitely put a financial burden on the Government.

Narasimham Committee Recommendations on Financial Reforms

The Government of India constituted a 9-member committee under the chairmanship of Mr. M. Narasimham, Retired RBI Governor, on Aug, 14, 1991 for making recommendations on existing financial system and to give suggestions for improving the existing structure. The committee submitted its report to the Finance Minister in November 1991 which was placed on the table of Parliament on December 17, 1991.

The salient recommendations are :

1. 4-tier banking system should be introduced in the country.

   I tier 3 or 4 International Banks — II tier 8 or 10 National Banks

   III tier Regional Banks — IV tier Rural Banks
2. Branch licensing system for opening new bank branches should be abolished.

3. A liberal view should be adopted for allowing foreign banks in the country. Both foreign and domestic banks should be treated at par.
4. SLR for banks should be curtailed to the level of 25% within next 5 years. CRR should also be curtailed in various phase.
5. Banks should be given more autonomy and the directed credit should be abolished.
6. Primary targets for credit should be redefined and such credit should not be more than 10% of total credit.
7. Computerisation in banks should be promoted.
8. Banks should be authorised to appoint banking official at their own discretion.
9. The dual control of RBI and Finance Ministry on banks should be abolished and RBI should function only as a regulatory authority of banking system in the economy.
10. RBI's representative should not be included in the management boards of banks. Only Government representative should be there.
11. Granting resources to development finance institutions on concessional rates of interest should be abolished in phases within next 3 years. These institutions should be allowed to mobilise resources from open market on competitive rates.
12. Quick and effective liberal attitude should be adopted in the policy related to capital market. System of getting prior permission by the companies for their new share issues should also be abolished.

Recommendations of Goiporia Committee on Consumer Service Improvement in Banks

RBI constituted a committee under the chairmanship of Sri M.N. Goiporia, the then President of SBI in September 1990 on making recommendations for consumer service improvements in banks. The committee submitted its report on December 5, 1991. The main recommendations of the committee were as given ahead.

1. Extension of banking hours for all works excluding cash payment.
2. Re-adjustment of bank opening time for staff so as to ensure start of work at bank counters well in time.
3. Spot deposit of outstation cheques of Rs. 5000/- (instead of exiting 2500/-) in bank account.
4. Increase in bank interest rates on saving accounts.
5. Providing tax benefit on bank deposit amounts.
6. To ensure optimum use of powers available with bank staff.

The recommendations of Goiporia Committee are still under consideration of the Government and the RBI.

Non-Performing Assets (NPAs) of Scheduled Commercial Banks

There was a significant decline in the non-performing assets (NPAs) of SCBs in 2003-04, despite adoption of 90 days delinquency norm from March 31, 2004. The gross NPAs of SCBs declined form 4.0 per cent of total assets in 2002-03 to 3.3 per cent in 2003-04. The corresponding decline in net NPAs, was from 1.9 per cent to 1.2 per cent. Both gross NPAs and net NPAs declined in absolute terms. While the gross NPAs declined from Rs. 68.717 crore in 2002-03 to Rs. 64.787 crore in 2003-04, net NPAs declined from Rs 32,670 crore to Rs. 24,617 crore in the same period. There was also a significant decline in the proportion of net NPAs to net advances from 4.4 per cent in 2002-03 to 2.9 per cent in 2003-04. The significant decline in the net NPAs by 24.7 per cent in 2003-04 as compared to 8.1 per cent in 2002-03 was mainly on account of higher provisions (up to 40.0 per cent) for NPAs made by SCBs.

The decline in NPAs in 2003-04 was witnessed across all bank groups. The decline in net NPAs as a proportion of total assets was quite significant in the case of new private sector banks, followed by PSBs. The ratio of net NPAs to net advances of SCBs declined from 4.4 per cent in 2002-03 to 2.9 per cent in 2003-04. Among the bank groups, old private sector banks had the highest ratio of net NPAs to net advances at 3.8 per cent followed by PSBs (3.0 per cent) new private sector banks (2.4 per cent) and foreign banks (1.5 per cent).

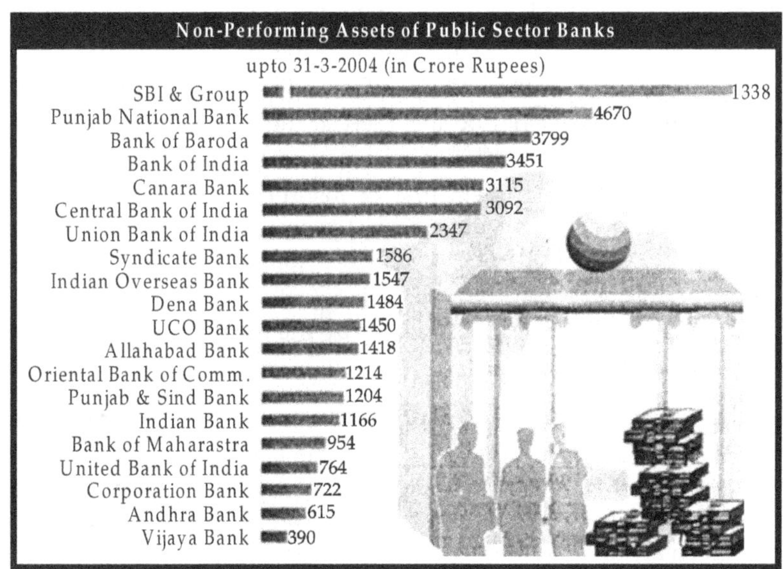

An analysis of NPAs by sectors reveals that in 2003-04, advances to non-priority sectors accounted for bulk of the outstanding NPAs in the case of PSBs (51.24 per cent of total) and for private sector banks (75.30 per cent of total). While the share of NPAs in agriculture sector and SSIs of PSBs declined in 2003-04, the share of other priority sectors increased. The share of loans to other priority sectors lending also increased. Measures taken to reduce NPAs include reschedulement, restructuring at the bank level, and recovery through Lok Adalats. Civil Courts and debt recovery tribunals and compromise settlements.

Non-performing Assets of Scheduled Commercial Banks

Items	*Gross NPAs (Rs. crore)*			*Per centage to gross advances*			*Per centage to total assets*		
	2001-02	*2002-03*	*2003-04*	*2001-02*	*2002-03*	*2003-04*	*2001-02*	*2002-03*	*2003-04*
Bank Group									
1. Public sector	56473	54090	51538	11.1	9.4	7.8	4.9	4.2	3.5
2. Private sector	11662	11782	10355	9.6	8.1	5.8	4.4	4.0	2.8
3. Foreign	2726	2845	2894	5.4	5.3	4.6	2.4	2.4	2.1
4. SCBs (1+2+3)	70861	68717	64787	10.4	8.8	7.2	4.6	4.0	3.3

	Items	*Net NPAs (Rs. crore)*			*Per centage to net advances*			*Per centage to total assets*		
1.	Public sector	27958	24867	18860	5.8	4.5	3.0	2.4	1.9	1.3
2.	Private sector	6676	6882	4857	5.7	5.0	2.8	2.5	2.3	1.3
3.	Foreign	920	921	900	1.9	1.8	1.5	0.8	0.8	0.1
4.	SCBs (1+2+3)	35554	32670	24617	5.5	4.4	2.9	2.3	1.9	1.2

Shares Scam in India

14 banks in 1969 and 6 banks in 1980, *i.e.*, total 20 commercial banks were nationalised to promote economic development in the country with a view to provide better assistance to weaker sections of the society without under direct control of the Government and efforts were made to channelise resources to more productive sectors of the economy.

A big shares scan was noticed in April 1992 which developed a feeling of destruct among investors. This shares scam created a big question mark on the working of banking system in the country. This scam conveyed a message to the bank customers that their savings with the bank are not safe. Share brokers of Mumbai withdraw millions of rupees from banks by illegal practices with the co-operation of corrupt high bank officials and directors. The committee set up under chairmanship of Mr. R. Janakiraman revealed on April 30, 1992 in its report a scam of Rs. 4024.45 crore (CBI estimated this amount to be Rs. 8383 crore).

From the stock exchange crash in 1988-89, there was an upward swing in the stock market, spearheaded by excellent performance by market leaders and industry-friendly policies by successive government such as liberalisation of industrial licensing, liberal fiscal measures, liberal and positive export-import policy, etc. RBI's all India index number of ordinary share prices (1981-82=100) rose from 189 at the end of March 1988 to 528 at the end of March 1991 and crossed 1000 at the beginning of February 1992. The sensex-the index of 30 of the key and most traded scrips in the Mumbai Stock Exchange was nearly 2000 in January 1992, crossed 3350 on March 9 and 4300 on April 20, 1992.

The RBI conducted preliminary investigations into certain aspects of securities trading by Indian and Foreign Banks that led to the exposure of serious irregularities and frauds in the securities transactions of different banks. Unscrupulous brokers in the stock exchange, colluding with some bank officers violated established rules and guidelines and siphoned off bank funds for speculative transactions in the stock market.

Recommendations of Jankiraman Committee

RBI set up a high level enquiry committee on April 30, 1992 under the Chairmanship of Mr. R. Jankiraman. The committee submitted the fifth and final report on May 7, 1993. The committee identified several types of irregularities in securities transactions which were used to siphon off funds out of the banking system :

1. Purchase of securities and other instruments were made by banks and their subsidiaries where the counter party was ostensibly another bank but when in reality the proceeds were directly or indirectly credited to the accounts of brokers.
2. Ready forward (Sale and purchase) transactions were entered into either on their own or on client's accounts by banks with brokers who used these funds for speculative activity.
3. Brokers in the stock exchanges were directly financed by banks by discounting bills not supported by genuine transactions.

4. Banks and other institutions showed large payments as call money to other banks. However, in the books of the receiving banks, there was no record of call money acceptances. Instead, the amounts were credited to the accounts of individual brokers. On the due date, these alleged call loans were repaid by payment out of the broker's accounts in the name of other banks.
5. Banks and other institutions rediscounted bills of exchange held by other banks and institutions but the proceeds and repayments were routed through broker's accounts.
6. Sums received as inter-corporate deposits and under portfolio management schemes (PMS) by merchant banking subsidiaries of public sector and other banks were passed on to brokers through ready forward deals.

There were other types of frauds too. The Jankiraman Committee estimated that the extent of unreconciled amounts would be around Rs. 4,000 crore. These irregularities were committed by public sector banks, private sector banks and foreign banks.

Major Recommendations of the High Level Committee on Agricultural Credit through Commercial Banks :

- Simplification of procedures regarding loan application forms, agreements/documents etc.
- Rationalisation of internal returns of banks.
- Banks may ensure that pre-sanction appraisal of the borrower, his capability for taking up the activities proposed, integrity etc. and technical viability of the proposal.
- Delegation of powers to branch managers to ensure quick disposal; at least 90% of loan applications should be disposed at the branch level.
- Introduction of Composite Cash Credit limits to all agricultural borrowing families.
- Introduction of new loan product with savings component.
- Cash disbursement of loan.
- Dispensation of No Due Certificate as a compulsory requirement. Obtaining No Due Certificate is now left to the discretions of lending banker.
- Discretion to banks on the matters relating to margin/security requirements for agricultural loans above Rs. 10,000.
- The target of agricultural lending should be based on the flow of credit through preparation of Special Agricultural Credit Plans (SACP), the objective of which should be to accelerate the flow as well as to substantially improve the quality of lending.
- Addressing a host of HRD related with regard to bank officials posted at rural branches.

Joint Parliamentary Committee

All the Members of Parliament raised a demand for an equiry into the securities and shares scam. Accepting this demand, a 30 member Joint Parliament Committee, under the chairmanship of Mr. Ram Niwas Mirdha, a senior congress leader was constituted. This 30 member committee included 20 members from Lok Sabha and 10 members from Rajya Sabha. This committee was constituted on August 10, 1992. The committee submitted its report (Total 474 pages) in two parts on December 21, 1993. It was an unanimous report submitted to the Government, but 13 opposition members of the committee demanded a separate enquiry of Harshad Mehta's allegation about handing over Rs. one crore to Mr. P.V. Narasimha Rao, the then Prime Minister of the country.

National Stock Exchange

Stock exchange or share market plays a dominant role in mobilising resources for corporate sector. It constitutes an organised market for exchanging securities and shares. There are 24 Stock exchanges in the country, 20 of them being regional ones with allocated areas. Three stock exchanges—National Stock Exchange (NSE), over the Counter Exchange of India Ltd. (OTCEI) and Inter-connected Stock Exchange of India limited (ISE) have mandate to have nationwide trading network. The Nifty index, which shown the biggest 50 liquid stocks in the country, experienced a sharp growth in market capitalisation form Rs. 285007 crore in 2001 to Rs. 902831 crore in 2004. Strong returns of 71.9% in 2003 were followed by modest returns of 10.7% in 2004.

In 1991 Pherwani Committee made recommendation for establishing National Stock Exchange. In 1992 the Government authorised IDBI for establishing this exchange. IDBI is the main promoter of National Stock Exchange (NSE).

NSE has initial authorised capital of Rs. 25 crore.

NSE includes trading of equity shares, bonds and government securities NSE's head office is situated in South Mumbai at Worli.

Names of Share Price Indices Changed

On 28 July, 1998, main Share price indices have been renamed as follows:

Old Name	*New Name*
NSE—50	*S & PCNX Nifty*
Crisil 500	*S & PCNX—500*

Two New Share Price Indices for Mumbai Share Market

For representing changes in share prices in a better way, two new share price indices have been introduced in Mumbai share market—**BSE 200** and **Dollex.**

BSE 200 includes the shares of 200 selected companies (85 of specified list A and 115 of non-specified list B). Ths index also includes the shares of 21 public sector enterprises.

Dollex is an index dealing in dollar prices. In other words. Dollex is the dollar price index of BSE 200.

BSE 200 and Dollex—both the indices have 1989-90 as the base year.

[It should be kept in mind that share sensex of BSE includes 30 shares and it has 1978-79 as base year. The base year of National Index is 1983-84.]

All over world share market has gone down in December 1998 Jan 2009.

Main Share Price Index in Famous Share Market of the World

Mumbai	DOLEX SENSEX S & PCNX NIFTY FIFTY
New York	DOW JONES
Tokyo	NIKKEI
Frankfurt (Germany)	MID DAX
HongKong	HANG SENG
Singapore	SIMEX STRAITS TIMES

Securities and Exchange Board of India (SEBI)

SEBI (Securities and Exchange Board of India) was initially constituted on 12 April, 1988 as a nonstatutory body through a resolution of the Government for dealing with all matters relating to development and regulation of securities market and investor protection and to advise the Government on all these matters. SEBI was given statutory status and powers through an ordinance promulgated on January, 30, 1992.

The statutory powers and functions of SEI were strengthened through the promulgation of the Securities Laws (Amendment) ordinance on January 25, 1995 which was subsequently replaced by an Act of Parliament. In terms of this Act. SEBI has been vested with regulatory powers over corporates in the issuance of capital, the transfer of securities and other related matters. Besides, SEBI has also been empowered to impose monetary penalties on capital market intermediaries and other participants for a range of violations.

SEBI is managed by six members—one chairman (nominated by Central Government), two members (officers of central ministries), one member (from RBI) and remaining two members are nominated by Central Government. The office of SEBI is situated at Mumbai with its regional offices at Calcutta Delhi and Chennai. In 1998 the initial capital of SEBI was Rs. 7.5 crore which was provided by its promoter (IDBI, ICICI, IFCI). This amount was invested and with its interest amount day-to-day expenses of SEBI are met.

All statutory powers for regulating Indian capital market are vested with SEBI itself.

Functions of SEBI

1. To safeguard the interests of investors and to regulate capital market with suitable measures.
2. To regulate the business of stock exchanges and other securities market.
3. To regulate the working of Stock Brokers, Sub-brokers, Share Transfer Agents, Trustees, Merchant Bankers, Underwriters, Portfolio Managers etc. and also to make their registration.
4. To register and regulate collective investment plans of mutual funds.
5. To encourage self-regulatory organisations.
6. To eliminate malpractices of security markets.
7. To train the persons associated with security markets and also to encourage investors' education.
8. To check insider trading of securities.
9. To supervise the working of various organisations trading in security market and also to ensure systematic dealings.
10. To promote research and investigations for ensuring the attainment of above objectives.

SEBI Regulations on Collective Investment Schemes (CIS)

The salient features of SEBI Regulations notified on October 15, 1999 are the following:

- CIS includes any scheme or arrangement with respect to property of any description, which enables investors to participate in the scheme by way of subscriptions and to receive profits or income or produce arising from the management of such property.
- Schemes structured for investment in shares/bonds and other marketable securities would not be treated as CIS.
- CIS can be floated only by companies registered under the Companies Act, 1956; the company floating CIS has to seek registration with SEBI as Collective Investment Management Company (CIMC).

- CIS shall be constituted as a two-tier structure comprising a Trust and a CIMC. At the time of registration as CIMC, the company should have a minimum net worth of Rs. 3 crore, which has to be raised to Rs. 5 crore.
- The CIS is prohibited from guaranteeing assured returns; indicative returns, if any, should be based on the projections in the appraisal report.
- The duration of the scheme shall be for a minimum period of three years.
- The assets of the scheme would be covered by compulsory insurance.
- Units issued under CIS should be listed on recognised stock exchanges.
- Entities operating CIS on the date of notification of SEBI Regulations would be treated as existing CIS, who should seek registration from SEBI within two months from the date of notification of the Regulations.

Decisions Taken by SEBI for Ensuring a Healthy Capital Market

SEBI has adopted a number of revolutionary steps to reestablish the credit of capital market, which include the following:

1. **Control on Utilizing 'Application Amount' having no Interest by Companies Releasing Public Issues :** At the instance of SEBI, commercial banks introduced Stock-Investment Scheme under which investor has to submit stock-invests, purchased from banks, with their share application. If the investor in allotted share/debentures, the required amount is transferred in concerned company's account by the bank issuing 'Stock invest'. In other case (if share/debenture is not allotted) investor gets a pre-determined interest rate on invested capital. This step of SEBI ensured interest earning to the investor until he got share/debenture allotment. It also ensures the refund of invested amount to the investor in case shares are not allotted.
2. **Share Price and Premium Determination :** According to the latest directions of SEBI, Indian companies are now free to determine their share prices and premium on those shares. But determined price and premium amount will be equally applicable to all without any discrimination.
3. **Underwriters :** The minimum asset limit has been fixed to be Rs. 20 lakh to work as underwriter. Besides, SEBI has warned under-writers that their registration can be cancelled if any irregularity is found in the purchase of unsubscribed part of the share issue.
4. **Control on Share Brokers :** Under new rules every broker and sub-broker had to obtain registration with SEBI and any stock exchange in India.
5. **Insider Trading :** Companies and their employees usually adopt malpractices in Indian capital market to variate share prices. To check this type of insider trading. SEBI introduced SEBI (Insider Trading) Regulation 1992 which will ensure honesty in the capital market and will develop a feeling of faith among investors to promote investments in capital market in the long-run.
6. **SEBI's Control on Mutual Funds :** SEBI introduced SEBI (Mutual Funds) Regulation 1993 to take over direct control of all mutual funds of government and private sector (excluding UTI). Under this new rule, the company floating a mutual fund should possess net assets of Rs. 5 crore which should consist of atleast 40% contribution from promoter's side.
7. **Control on Foreign Institutional Investors :** SEBI has made it compulsory for every foreign institutional investors to get registered with SEBI for participating in Indian capital market. SEBI has issued directives in this regard.

SEBI—More Powerful

Under new provisions SEBI has been given powers for granting recognition to any stock exchange in the country. SEBI has also been given the authority of deciding voting right for any member and also of amending it (Till now this authority was exercised by Central Government). SEBI will now settle the disputes relating to regulation of transactions spot delivery and non-listing of share by the company.

Concept of Depository System

Depository system is that system in which ownership of security is changed by an electronic account entry and physical transaction of securities does not take place. The main functions of depository are as follows :

1. To accept deposits for ensuring safe custody of securities.

FISHING ALONG EAST COAST COMMENCES WITH HIGH HOPE

Normal Monsoon Prediction Brightens Outlook

Fisherfolk along East coast have resumed fishing from 1 June 2006 after observance of a 45-day ban.

The pre-monsoon rainfall and the enforcement of the ban after a long time in Orissa coast are the twin factors which have bolstered the spirit of the fisherfolk of upper east coast. The prediction of a normal monsoon has brightened the hopes of fishers all along the coast towards an encouraging harvest.

The ban was observed along the coasts of Tamil Nadu, Pondicherry, Andhra Pradesh, Orissa and West Bengal from April 15 to May 31 to allow shrimps and other species to grow aided by the ban enforced by the respective governments.

Over 1500 mechanised boats are estimated to have resumed fishing operations from various ports along the east coast from the early hours of first June, as the ban came to an end at the stroke of midnight of 31 May.

According to the reactions of fishermen, they were expecting a good catch of shrimps. As before, the boat owners utilised the ban period for annual maintenance of the boats. It is said that each of the boats need fuel worth Rs. 1 lakh for a seven to 10-day long voyage and about Rs. 50,000 towards ration, spare nets, wire rope and other material. Each of the boats, will carry a crew comprising around eight members. The mechanised boats provide direct and indirect employment to a large number of fishermen and others involved in supplies and services.

Tragic Incident : A fatal incident that took place on 1st June at about 8 pm out at sea off Visakhapatnam marred the enthusiasm, particularly because it took place just 48 hrs after the capsize of a boat at the fishing harbour. On the ill fated day (1st June), the mechanized boat bearing no. 272 which set out for fishing at one a.m. on that day became a victim of a fire accident off Visakhapatnam coast in which two fishermen were burnt alive, stated to be because kerosene leaked from a stove and got ignited by the sparks that came out of the generator. Four others who were on the same boat jumped into the sea and were rescued by men in a passing fiberglass boat.

Keeping in view the multidimensional role of women and full utilization role of women and full utilization of the women's potentiality in the development process, it is necessary to enhance their capacities, which will contribute to efficient human resources deployment. There is sufficient scope to improve the exposure of the developed technologies among the womenfolk engaged in fishing, aquaculture and other activities. This exercise needs to be undertaken in conjunction with identified self-help groups in selected districts. Keeping this in view, as a pilot attempt, this exercise was carried out in detail for three days at CIBA, Chennai, during 21-23 February, 2006 and 16-18 March, 2006.

The SHGs are small informal associations created for enabling members to reap economic benefits out of mutual help, solidarity, and joint responsibility. The benefits include, basically, obtaining savings and credit facilities, and pursuing group enterprise activities among the poor rural women. SHGs are also activity specific.

The benefits of the training workshop are related to mobilising the women SHGs for constructive group action, improving their social and economic conditions bringing changes in socio-cultural attitudes among men and women and creating awareness on selected technologies of CIBA, like shrimp feed development and crab fattening and stimulating government institutions to respond to the objectives and felt needs of the stakeholders.

There exist golden opportunities for Indian seafood packers and exporters for marketing their products in the Republic of Korea [South Korea] and these can be exploited to their great advantage if they adopt the simple business axiom with perseverance-sell the right product to the right customer at the right time. Let us try to briefly understand the Korean seafood industry. With the Yellow Sea on the West and the East Sea/Sea of Japan on the East and the Korea Strait on the South, Korea has 6,228 km of coastline and nearly 3,000 islands scattered around the region. The fishing industry in Korea is the seventh largest in the world with a catch of 2.9 million metric tons per year Korea, has, in recent years, become one of the world's leading traders of fishery products. The country's tight import regulations had generally discouraged the importation of fish for domestic consumption in the past and had encouraged importing for the purpose of re-processing to produce value-added items and re-exporting to earn valuable foreign exchange. Most of Korea's seafood import form the United States were earmarked for re-export to Japan where they competed with US seafood products. Although Korea still imports and re-exports a large quantity of fishery products, a booming domestic economy (with real GNP growth of about 10 per cent) and a higher standard of living has greatly affected this pattern. Per capita disposable income in the country is more than $2,100 per year. This has resulted in a growing population and an increasing consumer demand for fish and processed fishery products. Korea's fishing industry has not been able to keep pace with this demand and as a result, South Koreans have, over a period of time, substantially increased the imports of seafood products into the country to meet the domestic demand. During the last decade, the volume of imported products consumed domestically multiplied form 16 to 42 per cent—and this trend is continuing (R.A.M. Varma 2005).

The strong recovery of the South Korean economy and the complete liberalization of fish imports since 1997 have provided good opportunities for the U.S. seafood products. Currently, tariffs for most fish and seafood products range from 10 to 20 per cent, although for about a dozen products, higher adjustment tariffs are applied. Over-fishing has severely depleted Korea's own fishery resources. As a result, the government has downsized Korea's fishing fleet and plants to reduce it even further over the next several years. The Korean government is trying to purchase additional fish quotas from other countries but in the end, Korea will have to import more just to meet the current demand. In 2006, Imports of fish and seafood products are expected to rise steadily and substantially on account of considerable decline in the domestic production. Moreover, Korean consumers are increasingly turning to seafood as a law-fat, affordable alternative to more traditional sources of protein. Fish and marine products are an important component of the Korean diet as they are in the other regions of Asia. Korea developed a powerful fisheries industry by intensively exploiting its inshore fishery sector and establishing a large offshore fishery capability that actively sought opportunities worldwide.

The production-oriented policy has led to the over-exploitation of costal and offshore fishery resources. The politics of the international fisheries has reduced Korean access to remaining offshore stocks. As a result, Korean fishery production began to decline significantly in the 1990s and since 1995, Korea has been relying heavily on imported fish and other seafood. The total size of the fish and seafood market based on production in Korea in around 2.5 million metric tons. The estimated value of the fish and seafood import market in 2005 was $1.8 billion. All of the above factors make South Korea an excellent market for fish and seafood despite the predominant position held by the USA. Imports from the other countries of world as well as that of the United States are likely to have reached record figures in 2005. However, competition is strong: as many as 70 countries already export fish and seafood products to Korea. China continues to be the largest supplier accounting for more than 40% of the market share mostly yellow corvina, frozen squid and frozen hair tail.

The United States holds the second place and Russia is ranked the third largest supplier. Both China and Russia have the advantage of proximity. Moreover, the prices of some species from these countries are reportedly half those of the U.S. counterparts. However, the quality of the Chinese and Russian products does not compare very well with that of the U.S. products.

The major items that Korea import are :

- yellow corvina, frozen
- Alaska pollock roe, frozen
- hair tail, frozen [ribbonfish]
- Alaska pollock surimi, frozen
- shrimp, frozen
- mackerel, frozen

Historically, Koreans preferred to buy fresh whole fish at wet markets which they cleaned at home. But these days, most people don't have the time or the space for cleaning fish. So they go to super markets, buy fish and have it cleaned there itself. Therefore the present trend shows that consumers will increasingly purchase filleted, processed products. Though the U.S, Russian and Chinese packers/exporters will have ample opportunities to introduce more processed, convenience-oriented seafood items, such as halibut, cod, pollock, monkfish and salmon etc., it is also the right time for Indian processors to enter the Korean market in a big way with valued added products. Korea also exports a large volume of fish products, some of which are imported, processes and re-exported. Korean importers operate about 20 processing plants in China which produce fish fillets and imitation crab meat. Korean exports comprise tuna, oysters, conger eels, squid, imitation crabs flat fish etc. the major importing countries of Korean seafood products [in the order of importance] are Japan, the US, EU countries, China, Thailand and Taiwan. Marine fish represent the main growth sector due to the strong demand for raw fish.

The main species cultured are olive flounder, rockfish, sea bream, sea bass, yellow tail, shrimp, prawn lobster, and sea squirt (turicate). These species are sold live to fish mongers in the open markets as well as to hotels and upper-class Japanese restaurants for making raw dishes (sashimi or sushi). Canadian live lobster suppliers are successful in the Korean market. The market potential for live lobsters and snow crabs is still growing. The leading species for frozen in terms of consumption are : Alaskan pollack, Pacific cod, squid, Japanese pilchard (herring), mackerel, chub mackerel, Spanish mackerel, largehead hairtail, monkfish, globe fish, skate, pomfret, tuna and smoked salmon.

Oyster, clams, cockle, mussels, arkshell and various types of seaweed are widely consumed in South Korea. However, there is an abundant local supply of these products an therefore, there is only very limited business opportunity to import them. Frozen lobsters are used mainly in hotel buffets and in up-scale Japanese restaurants, especially Tepanyaki seafood restaurants. Fresh lobsters and rock lobsters are consumed principally in raw form (sashimi) in the more affluent household. Lobster is an essential ingredient, along with locally cultured shrimp and prawns in the Tepanyaki restaurants. Canadian lobsters have been gaining a high profile in the Korean market due to it low mortality rate and low mercury and lead levels in comparison with Australian and U.S lobsters.

The consumption of fresh salmon is extremely limited due to its high costs. Those who consume salmon products have them fresh, cooked or smoked. Smoked salmon is mainly consumed in tourists hotels, family restaurants and buffet restaurants and department stores/discount outlets. Norway and Chile are the major competitors for frozen and smoked salmon with their salmon supply generally being sourced from farms. Even though Koreans consider salmon a luxury food, Canada should be able to change the general perception of consumers on salmon through targeted promotional efforts. These efforts would help expand.

India was the second in ranking followed by Indonesia and Vietnam. The latter had a significant jump in production with large supplies of catfish and shrimp targeted export markets. China also became the world's leading producer and exporter of farmed shrimp and tilapia in less than five years.

	in Quantity	*in Value*
China	68.3%	51.6%
India	5.2%	4.1%
Indonesia	2.4%	2.8%
Vietnam	2.2%	3.2%

It is interesting to note that among the world's top ten producers of aquaculture products, eight are from Asia namely, China, India, Indonesia, Vietnam, Japan, Bangladesh and Thailand.

Fishery Highlights : 2003-2005

Wild Catches & Aquaculture

1999: 126.65 million MT (33.3 mill. MT)

2000: 130.43 million MT (35.6 mill. MT)

2001: 142.08 million MT (37.8 mill. MT)

2002: 145.94 million MT (39.8 mill. MT)

2003: 146.29 million MT (42.3 mill. MT)

Fishery Highlights : 2003-2005

Fishery Trade

Production costs in the fishery sectors has increased which are hardly compensated by import market particularly for traditional block frozen products.

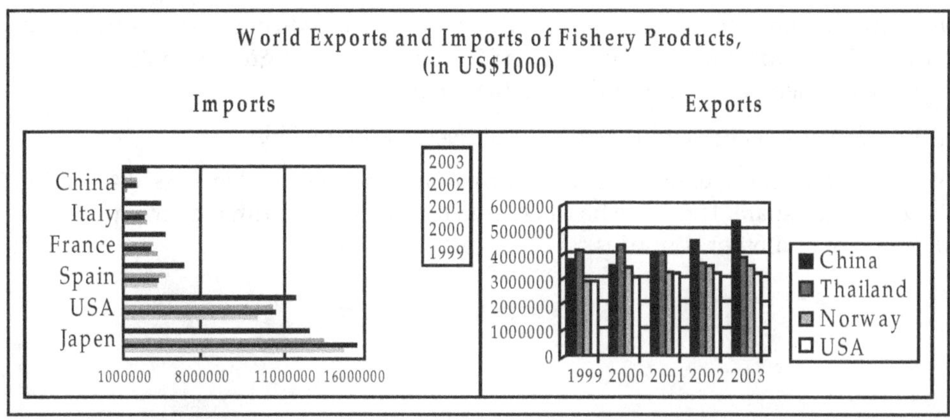

Exports from developing countries are constantly hampered by increased tariffs and technical barriers.

Trade flow increased in domestic and international markets supported by better consumer demand and rising demand for higher value products. The importance of domestic market has increased and there are also Free Trade Agreements (FTAs) and augmented trade among developing countries.

Imports

The international fishery trade increased from US$60 billion in 2000 to US$ 68.26 billion in 2003, report, UN FAO. Japan, the USA, Spain, France, Italy, Germany, the UK and China were the leading importers. There was a negative growth rate in the world's largest markets Japan. Imports increased in the other markets including Australia and developing East Asian countries including China. The import values in these markets in 2003 were:

Japan	:	US$ 12.62 billion
USA	:	US$ 11.75 billion
Spain	:	US$ 4.92 billion
France	:	US$ 3.80 billion
Italy	:	US$ 3.57 billion
Germany	:	US$ 2.65 billion
UK	:	US$ 2.53 billion
China (main land)	:	US$ 2.42 billion

In 2004 the global fishery trade increased again with import recovery in the Japanese market and positive growths in the USA, European Union and in the Asia Pacific region. But the growth in the important market. The USA has been affected due to the imposition of the punitive anti-dumping duties on shrimp from China, Thailand, India, Vietnam, Brazil and Ecuador with a maximum import duty of 113%.

Exports

China ratained the position of world's leading exporter of fishery product is in 2003 the USA and Canada.

China	:	US$ 5.36 billion
Thailand	:	US$ 3.92 billion
Norway	:	US$ 3.66 billion
USA	:	US$ 3.46 billion
Canada	:	US$ 1.45 billion

Following the imposition of high import duties on Chinese shrimp, exports from China to the US declined in 2004. Nonetheless, China had a 27% annual export growth in 2004 and it's fishery export revenue reached nearly US$7 billion in that year.

Export value also increased in Thailand but declined in the USA.

Notably, Asian fishery exports to the international market totalled US$20.48 billion in 2003 with a 31.19% market share. The value has increased further in 2004 with higher exports from China, Thailand, Vietnam and other countries.

Japan – The World Leader in Fishery Imports: 2000-2004

2000 : Q = 3.54 mill. MT; V = US$16.12 bill.
2001 : Q = 3.82 mill. MT; V = US$14.23 bill.
2002 : Q = 3.82 mill. MT; V = US$14.08 bill.
2003 : Q = 3.82 mill. MT; V = US$13.51 bill.
2004 : Q = 3.82 mill. MT; V = US$14.24 bill.

Japan is the third largest fishery producer in the world with a total catch of 6 million tons (live weight) in 2003, accounting for 4.10% if global fishery production. In the same year the market imported 3.32 million tons (product weight) of fishery products positioning itself as the number one importer of fishery products in the world.

The country's annual fishery production has halved since the mid-1980s while fishery imports have increased steadily till mid 1990s. Currently domestic landings meet only 53% of the Japanese market requirement. The balance is imported and the Japanese fish consumption in one of the highest in the world.

The decade long economic recession have had dwindling effect on the Japanese fishery imports. Imports, however, are showing some signs of recovery.

Among the product group, sashimi grade tuna, shrimp, salmon, squid/cuttlefish! octopus and crab are the important ones in Japan's annual imports. However, import market growth for traditional block frozen products has been showing falling trends in the recent past.

Domestic demand and imports of processed and value-added products are increasing. Last year the market imported 413 447 tons of high value prepared fishery products (excluding raw products) at a value of US$2.85 billion compared to 355 271 tons and US$2.35 billion in 2003. A large quantity of these are shrimp, fish and cephalopod based products.

During 2001-2004, imports of Prepared (value added) Fishery Products into Japan increased by 20.5% or 70230 tons in quantity and 26% in value.

JAPAN : January-September 2005 *Import* Trends

		Total Fishery Imports
Quantity	:	2.48 million MT
Value	:	US$ 11.03 billion
AGAINST	:	January-September 2004
Quantity	:	2.51 million MT
Value	:	US$10.79 billion

During the first nine months of the year, Japanese fishery imports declined marginally by 1.2% in quantity but the value increased by 2.70%, although the Yen remained relatively weak compared to last year.

Imports of raw frozen shrimp, air flown fresh sashimi tuna, fish fillets etc. declined during this period. For reference, supplies of raw frozen shrimp fell from Indonesia, Vietnam, India and Bangladesh. Higher exports only came from Thailand which could be attributed to increasing exports of vannamei based nobashi (peeled tail-on shrimp used for processing tempura or breaded products) shrimp.

On the other hand, imports of prepared shrimp, squid/cuttlefish and other fish based products, regarded as value-added items showed increasing trends. This trends translated to higher value in imports during this period. Thailand, China, Vietnam and Indonesia are the leading suppliers of many high value products to the Japanese market.

US Consumers are a Record Amount of Seafood in 2004.

- ❑ Shrimp : 4.20/lb
- ❑ Tuna : 3.30/lb
- ❑ Salmon : 2.15/lb
- ❑ Pollack : 1.28/lb
- ❑ Catfish : 1.09/lb
- ❑ Tilapia : 0.07/lb

Total Seafood : 16.60/lb

US consumers spent more money for fish and seafood last year; seafood consumption was record high at 16.3 lbs per capita, a trend noticed since 2002. Shrimp continued to be the most favorite seafood.

US imports of edible fishery products in 2004 increased marginally to US$ 11.3 billion compared to US$ 11.1 billion the year before. In quantity imports increased by only 20073 tons to 2.245 million tons compared to 2.22 million tons imported in 2003. However, the growth in shrimp market has been hampered by the US anti dumping Law and the punitive import duties on the six major suppliers.

Increased supplies from other sources such as Indonesia, Malaysia, Bangladesh, Mexico etc., compensated the decline in supplies from China and Brazil. As a result, shrimp imports recorded high again at 517617 MT against 504495 MT in 2003.

However, the value fell to US$ 3.7 billion compared to US$ 3.8 billion tons recorded in 2003. Imports of semiprocessed and processed shrimp such as peeled, cooked and peeled and breaded shrimp also increased during this time. Supplies of breaded shrimp, a product group not included in the anti-dumping tariff list, increased significantly last year and also this year particularly from China. Notably imported shrimp has a 90% share in the US market. US domestic landings of shrimp fell by 2% in 2004. Local supplies this year will be much less due to the Hurricanes that affected the southern states.

In response to the increased demand for tilapia, whole and fillet, the market imported 112 939 tons of this species in 2004 at a value of US$ 297.41 million. Nearly 50% of these were fillets. For frozen tilapia, whole and fillet, China had a 52% market share. Per capita consumption of tilapia in the USA increased from 0.35lb in 2001 to 0.70lb, replacing cod as the 6th most popular "seafood".

Good demand for fresh and frozen tuna (for non-canned consumption) also led to higher imports, particularly from Asia.

Canned tuna market remained weak with lower imports which was also reflected on its consumption pattern. Per capita canned tuna consumption fell from 3.5 lbs in 2001 to 3.3 lbs in 2004. Per capita canned seafood consumption also declined. High raw material prices for frozen skipjack throughout 2004 also affected the traditional canned tuna market.

The number of US domestic wholesale and processing facilities declined by nearly 2% compared to 2003. The rate of employment in this sector also fell by 3.2% from 67 472 in 2003 to 65 36 in 2004. High import duties on shrimp raw materials was one of the reasons for these declines.

USA : Higher Imports Continue in 2005

Despite the anti-dumping duties, shrimp imports during January-September 2005, increased by 3.78% against the same period last year. Supplies from Thailand, Vietnam and Ecuador recovered compared to last year but declined from India, China and Brazil.

Due to the high import duties on its raw shrimp exports, Chinese producers diversified product lines to breaded shrimp (not subject to anti-dumping duties) for the US market. US Imports of breaded shrimp totalled 30,000 tons during January-September 2005 compared to an annual import of 9,000 and 17,000 tons in 2003 and 2004. China became the major supplier followed by Thailand.

Imports of tilapia crossed 90,000 tons during this first nine months of the year.

The EU block (15 members) is the largest market area of fishery product with an annual combined import value of US$ 26 billion in 2003. The top importers (with annual value ranging from US$I-5 billion) were Spain, France, Italy, the UK, Germany and Denmark, the Netherlands, and Belgium. Imports from developing countries mainly consist of frozen fish fillets, shrimp, canned tuna and frozen squid/cuttlefish.

There was a 25% growth in imports of fishery products into these countries during 1999-2003. This could be attributed to the steady increases in imports of value-added fishery products, particularly from Asian sources. China has become the major reprocessing centre for coldwater species such as cod, pollack and salmon processed into fillets.

However, there is strong trend in imports of higher value products (ready-to-cook and ready-to-eat) particularly from south East Asian countries. Although, Spain, France and Italy are top leading seafood exporters in Europe, many of these value added products are directed to consumers in the UK, Germany and the Netherlands. Belgium also imports a wide variety of value-added products for the wider EU markets.

Revised GSP in the EU from 2006: The EU will implement the revised GSP scheme from January 2006 which will be valid till 2008. It will also introduce the autonomous and temporary measure, cutting import tariffs from 12 to 4.2% for frozen shrimp and 20 to 7% for processed shrimp from most of the countries including Thailand.

Sri Lanka has received blanket zero duty rate for its fishery products to the EU markets.

EU 10

The 10 new members who joined the EU in May 2004, imported 619 tons of seafood with a total value of US$ 900 million. Consumption and imports of fishery products are still low in these countries. However, seasonal imports of shrimp, fillets and some convenient foods are expected to grow in these markets particularly for the catering and tourist industries.

The 10 new members are Estonia, Latvia, Lithuania, Poland, Czech Republic, Slovakia, Hungary, Slovenia, Malta and Cyprus with a total population of 455 million.

Poland one of the new members of the EU, harvested 215800 tons of fishery products in 2004. In the same year imports into the country totalled nearly 460 000 tons at a value exceeding US$ 500 million. The total consumption of fishery products was 422 900 tons last year. Poland also exported 252 600 tons of fishery products in 2004 valued at US$334 million. The popular Asian food service outlets in metropolitan cities are Chinese restaurants and sushi bars where fish are widely used. According to the US GAIN Report, Polish reatail, Sales have increased by 10% each year. The per capita seafood supply in Poland is 9.6 kg.

Russia : A Growing Non-EU Market

Russian Federation is a major producer of fishery product in Europe with an annual exports of US$1.48 billion. Russian fishery exports, however, dwindled over the last few years-US$ 1.52 billion in 2000 to US$ 1.42 billion in 2002.

On the contrary, Russia's fishery imports increased by 183% during 2000-2003, from US$ 193.8 million to US$547.4 million confirming to a rapid rise in consumption of fishery products in the Federation.

In 2004, Russian GDP growth rate was 7.1%, higher than the production that year. Actual income also increased by 7.8% which has resulted in an increase in food consumption in the country. The trend is specially seen in consumption of fishery products. Last year imports grew again totalling US$643 million, close to a 60% increase over the previous year. In comparison, fishery export value declined by 19% in that year.

Coldwater fish species are traditionally popular in Russia. But there is a striking increase in the consumption of exotic species such as shrimp, squid tropical fish fillets, oysters etc. Demand for surimi-based product in also increasing.

According to the Russian Federal Fishery Agency, per capita fish consumption during January-June 2005 reached 16 kg. Boosted by the grow disposable income the yearly market growth is expected to be high.

Among the other countries in Eastern Europe, the important importers are Croatia (US$ 90.85 mill.) and Romania (US$ 56.5 mill.)

Asian fishery exports increased from US$ 19 billion in 2000 to US$ 20.6 billion in 2003 registering a 8.42% sales growth. Imports into the region were also high at US$ 22 billion in 2003.

Excluding Japan, where fishery imports fluctuated over the years, import of fishery products into the eleven countries and territories increased by 21.13%. during 2000-2003. These markets are China, Hong Kong, Macao, Taiwan, South Korea, Malaysia, Philippine, Singapore, Thailand and Vietnam.

In value total fishery imports in these markets reached US$ 9.02 mill. in 2003 against US$ 7.43 bill. in 2000. China is the major consumer market in the region, whereas Thailand continues to import a large quantity of raw material for reprocessing and exports.

The other growing consumer markets are South Korea, Hong Kong, Malaysia, Taiwan and Singapore. Some of them are also regional leaders in fishery exports.

China: Fishery Exports & Imports

Exports	
	2002 : Q=2.08 mill MT; V=US$ 4.69 bill.
	2003 : Q=2.10 mill MT; V=US$ 5.48 bill
	2004 : Q=2.42 mill MT; V=US$ 6.96 bill
Imports	
	2002 : Q=2.49 mill MT; V=US$ 2.27 bill
	2003 : Q=2.33 mill MT; V=US$ 2.48 bill
	2004 : Q=2.98 mill MT; V=US$ 3.23 bill

Outlook : Opportunities and Threats in Global Fishery Trade

Global demand and supply for foodfish are growing but at a different pace. Opportunities and challenges in growing markets are associated with convenience, healthy, functional natural or organic and Food safety, sustainability, environment friendly etc..

Tariff barriers are also in place such a the US anti dumping law for imported tropical shrimp from Asia and Latin America, catfish/rom Vietnam. salmon from Norway and the US custom bond on imported shrimp.

Shrimp market growth in the USA is over-shadowed by the anti-dumping import duties, causing serious hardship to exporters and importers.

Tariff barriers are in place not only for imported tropical shrimp from Asia and Latin America but also on catfish from Vietnam and salmon from Norway.

The US Customs Bond on shrimp also adds fuel to the fire for imported products. The bond is determined by multiplying imported products value from the previous year by the tariff. If an importer buys US$ 10 million of shrimp, at a 10% tariff, he or she must pay a US$ 1 million bond. The US Customs and Border Protection, the agency responsible for tariffs and bonds collection, is yet to return the first round of bond imposed for 2005 imports. Usually it takes 3-5 years to return the money. Reportedly, the second round bond for imports in 2006 is due early next year.

Anti-dumping Tariff Threats in the US Market

Other shrimp supplying countries are under investigations. Tropical shrimp and Vietnamese catfish are already facing high anti-dumping duties in the US market. US farmed catfish and wild salmon producers may come up with a similar proposition for imported tilapia. Farmed tilapia has gained so much prominence that the Alaska Seafood' Marketing Institute is actively studying the fish, its increasing production, and potential effect on the market for Alaska seafood.

Opportunities

More value-added fishery products are processed for the traditional developed markets namely Japan, the USA and the EU.

USA : Consumption and imports fishery products are already on the rise. Supplies of higher value imported products will continue.

WalMart, the largest retail importer in the USA sources US $1.2 bill. WalMart plans to increase its import value by 30% from India near future.

"Tasty fish dishes" are very popular with customers. However, many people are no longer confident in preparing fish themselves at home. They increasingly buy ready-to-cook products at the retailer's or go to a restaurant if they want to eat fish.

In high-end gastronomy sector consumers are looking for fish. In contrast canteens and restaurants in tourist areas, prefer frozen products that could be used with flexibility. They also alternate between fresh and frozen fish depending on demand and supply.

Opportunities in Europe

In the EU, total shrimp supply exceed 500000 tons in 2004. Tropical shrimp takes 70% of volume sales. Demand for tropical fin fish, particularly fillet in increasing Gourmet fastfood sector in Europe has added fishery products. The 10 new EU members offer opportunities for fishery exports. However, products from certain countries will also be subject to higher import duties in the EU territory.

Russia's Imports are growing. Exports from Vietnam increased by 150% during Jan-July 2005 compared to last year.

Asia Opportunities

In Asia, Japan offers greater opportunities for marketing processed and higher value fishery products.

Fishery trade is also expanding in the other Asia Far East and South Asian markets. Seafood sales are expanding at food service and retail sectors.

Fish consumption and trade (exports & imports) are also growing in West Asian.

Opportunities in China

At present it is the No. 1 for producing seafood for export markets. With higher fishery imports for domestic consumption and re-processing/exports, China remains the main attraction in the international fishery trade!

Because of the consumption revolution in China with 95 million middle class, the country is expected to become less dependent on fishery exports in future.

Some others points to consider...

Asian style snackfood, finger-food, appetizers for western entertainment & retail markets made of broken shrimp, fish paste, skewers etc., should be considered. Seafood is added in the Asian Fast-Food shrimp burger and salmon pizza in Japan, shrimp nuggets in China and shrimp spring roll in South East Asia.

Bluefin tuna prices have declimed due to high supplies of farmed tuna.

Value creation through ready-to-eat and ready-to-cook products will be a key marketing element.

Fish and Seafood exports to the EU are expected to increase with better GSP schemes for 2006-2008. The EU market for frozen fishery products, particularly for processed products are growing... However, more challenges are to come on the food safety issues.

Chapter 8

Economic Reforms and WTO Implications

The massive reduction in foodgrains absorption is the outcome of economic reforms programme which has further intensified the persisting crisis of the agrarian economy, entailing more and more decline in purchasing power of the rural people. The meagre employment provisions have also been curtailed. The cultivation has become uneconomic. The vast majority of the farmers, particularly small and marginal farmers and tenant cultivators have to depend more and more on private money lenders for loan at an excessive rate of interest.

Agriculture has an important place in the Indian economy. At present, It accounts for 22 per cent of the GDP and provides livelihoods to 58 per cent of the country's population. In 1990s, two significant developments took place. These development have had a profound impact on Indian agriculture. The first development relates to liberalization of economic policy as a part of the economic reforms programme initiated in July, 1991. Some of these reforms were directed towards opening up of the Indian economy through numerous economic policies.

The Second development concerns to the new international trade regime following Uruguay Round agreement and formation of World Trade Organization (WTO). The Uruguay Round agreement, signed in Marrakesh in April 1994 and WTO came into existence on January 1, 1995. Agreement on Agriculture (AOA) under WTO regime has fundamentally changed the global trade picture in agricultural sector.

Economic Reforms in Agriculture

Since July 1991, the country has taken a series of measures to structure the economy and improve the balance of payments position. The New Economic Policy 1991 introduced changes in the areas of trade policies, monetary and financial political, fiscal and budgetary policies and pricing and institutional reforms. These policies are known as economic reforms. The salient features of economic reforms were liberalization (internal and external), privatization, redirecting scarce public sector resources to areas, where the private sector is unlikely to enter and globalization of economy.

The economic reforms did not include any specific package for agriculture. Rather, the presumption was that freeding agricultural markets and liberalizing external trade in agricultural commodities would provide price incentives, leading to enhanced investment and output in that sector. However, the pattern of structural adjustment and the government's macro-economic strategy since 1991 have actually been associated with a reduced rate of overall agricultural growth, decline in per capital foodgrain output and inadequate employment generation.

Government curtailed subsidies on seeds, fertilizer, pesticides, electricity etc. These vital input of agriculture were handed over to private companies. Government did not even introduce any system to control the quality of these inputs. By government's opening of the domestic market for the multinational companies (MNCs), which control the international market of seeds, fertilizer and pesticides, they took the opportunity and supplied spurious seeds and pesticides. Consequently, the prices of seeds, fertilizers and pesticides increased and yield per hectare decreased.

The massive reduction in foodgrains absorption is the outcome of economic reforms programme, which has further intensified the persisting crisis of the agrarian economy, entailing more and more decline in purchasing power of the rural people. The meagre employment provisions have also been curtailed. The cultivation has become uneconomic. The vast majority of the farmers, particularly small and marginal farmers and tenant cultivators have to depend more and more on private money lenders for loan at an excessive rate of interest.

Thus, any strategy of economic reform can not succeed without sustained and broad based agricultural development, which is critical for raising living standards, alleviating poverty, assuring food security and making substantial contribution to the national economic growth.

Agreement on Agriculture Under WTO Regime

The Agreement on Agriculture, which forms a part of final act of Uruguay Round of Multilateral Negotiations (1986-93), was signed by member countries including India in April 1994 at Marrakech in Morocco. The AOA contains provisions in three broad areas:

Tariffication

Tariffication means conversion of all non-tariff barriers into tariff. It means that non-tariff barriers such as quantitative restrictions and export and import licensing etc. are to be replaced by tariffs to provide the same level of protection. Tariffs, resulting from this tariffication process together with other tariffs on agricultural products, are to be reduced by a simple average of 36 per cent over 6 years in the case of developed countries and 24 per cent over 10 years in the case of developing countries.

Domestic Support

All domestic support is qualified the mechanism of total Aggregate Measurement of Support (AMS). AMS is a means of quantifying the aggregate value of domestic support or subsidy given to each category of agricultural product. Each WTO member country has made calculations to determine its AMS wherever applicable. Commitment made requires a 20 per cent reduction in total AMS for developed countries over 6 years. For developing countries, this per centage is 13 per cent and no reduction is required for the least developed countries. The base period on which the reductions are calculated is 1986-88.

Export Subsidies

The agreement requires developed countries to cut the value of export subsidies by 36 per cent and to reduce the volume of subsidized exports by 21 per cent over six year. The base period is 1986-90. Countries may keep their existing export support programmes, but may not introduces new programmes. Developing countries must reduce the value of subsidies by 24 per cent and volume of subsidized exports by 14 per cent over ten years. Because of developing countries, higher marketing costs, subsidies to reduce the costs of marketing and transporting exports both domestically and internationally are excluded.

Implications of Economic Reforms and WTO

Agricultural sector is the mainstay of the rural Indian economy around which socio-economic privileges and deprivations depends and any change in its structure is likely to have a corresponding impact on the existing pattern of social equity.

During 2002-03, the gross capital formation (GCF) in agriculture and allied sector at 1993-94 prices was Rs. 20066 crore. The per centage shares of public sector in gross capital formation were 6.0 per cent and 5.4, respectively. Investment in agriculture and allied activities as a per centage of total GCF declined from 9.0 per cent to 5.0 per cent during the period 1990-94 to 2002-03, respectively as shown in Table 1.

Table 1 : Gross Capital Formation Agriculture and Allied Sector (At 1993-94 Prices)

(Rs. Crore)

Year	*GCF in Agriculture and Allied Sector*			*Share of Agriculture and Allied Sector in Total GCF (%)*		
	Public Sector	*Private Sector*	*Total*	*Public Sector*	*Private Sector*	*Total*
1990-91	4992	11424	16416	7.1	11.9	9
1991-92	4376	10589	14965	6.6	9.9	8
1992-93	4539	11602	16141	6.7	10.5	9
1993-94	4918	10331	15249	6.9	9.4	8
1994-95	5369	11416	16785	6.7	7.7	7
1995-96	5322	12367	17689	7.1	5.9	6
1996-97	5150	13176	18326	7.0	7.5	7
1997-98	4503	13791	18294	6.2	7.5	7
1998-99	4444	13026	17470	5.7	7.8	7
1999-00	4756	15268	20024	5.4	8.5	7
2000-01	4435	15374	19809	5.4	8.5	7
2001-02	4658	14821	19479	6.2	5.8	5
2002-03	4950	15116	20066	6.0	5.4	5

Source: GOI (2004), *Agricultural Statistics at a Glance 2004,* Department of Agriculture and Cooperation, Ministry of Agriculture.

The reason given is that the agriculture is basically in a private sector. Public investment has a critical role to play in creating the infrastructure in terms of irrigation, markets, storage facilities, electrification and technology development, besides education and health.

India has removed quantitative restrictions (QRs) for 1429 items *i.e.* 715 items by March 2000 and 714 items by March 2001. Of these, 208 items belongs to agriculture sector. India has earlier successfully revised the binding on 15 tariff lines, which included skimmed milk powder, spelt wheat, etc. Following the provisions of WTO, the government of India opened up the domestic market of agricultural and agro-based products to world markets. It removed quantitative restrictions on imports of those products.

In India, product specific support is less as compared to non-product specific support. Subsidies on agricultural inputs such as power, irrigation, fertilizers, etc. is well below the permissible level of 10 per cent of the value of agricultural output. India is, therefore, not under obligation to reduce domestic of developed countries spent huge amount on agricultural subsidies. These resulted in fall in prices of agricultural commodities in the world market.

The policies of the government have compelled the farmers of the country to compete with cheaper foreign commodities when they have to spend more for ever-increasing cost of agriculture inputs like seeds, fertilizers, pesticides, electricity etc. Increasing cost of inputs, decline in growth rates and lower prices of outputs have most adversely affected the farmers. Due to above policies, exploitation farmers has intensified causing indebtedness, desperation, destitution and starvation to vast majority of rural people, particularly the small and marginal peasants and tenant cultivators.

Two major factors are responsible for the present downfall of Indian agriculture. First, in its eagerness to reduce fiscal deficit. The government has substantially reduced the development expenditure in attributed in a big way to reduction in prices of agricultural products. Having failed in getting remunerative prices for their products, many farmers have curtailed their farm operations, which in turn have increased unemployment among the agricultural workers. Thus, import liberalization is a major cause of the existing plight of the peasantry.

Indian farmers are worried as they do not precisely know about the WTO, GATT, Dunkel and Genetically Modified Crops. They are trying to understand what it means to live under the WTO regime. They have not yet achieved clarity, on future trends in Indian and international agriculture. Should they continue to grow existing varieties of crops, which have given them adequate returns in the past? Or, should they abandon them and sow new varieties of crops which are being successfully grown in more developed countries? It so, should these be the Genetically Modified ones or conventional seeds? will the Government permit foreigners to grab the Indian market by virtue of superior quality and competitive prices? Who exactly will help them to reach international standards of quality and cost? These and several related questions trouble their minds all the time.

Two major factors are responsible for the present downfall of Indian agriculture. First, in its eagerness to reduce fiscal deficit, the government has substantially reduced the development expenditure in agriculture sector. Secondly, import liberalization has contributed in a big way to reduction in prices of agricultural products. Having failed in getting remunerative prices for their products, many farmers have curtailed their farm operations, which in turn have increased unemployment among the agricultural workers.

Keeping in view the current agricultural situation and the long-term interests of farmers the following steps should be taken :

- Reduce import duties on essential agricultural inputs such as fertilizers and agricultural machinery, except under special cases of dumping. At the same time rationalize the subsidies on fertilizers so that it would benefit only farmers.
- Give further incentives to food processing industries, so that demand for agricultural produces would improve with better prices.
- Initiate measure to ensure supply of quality seeds and other inputs.
- Gross capital formation in the public sector in agriculture both in absolute and relative terms should be increased.
- Raise import duties on farm products, protect the domestic agricultural goods.
- Encourage upgradation of agricultural technology such as bio-technology to enhance farm productivity. Special attention may be given towards enhancing production of pulses and oilseed crops.

Conclusion

The overall impact of economic reforms and WTO on agriculture is positive as well as negative. There are positive impacts in some areas like floriculture, horticulture, diary etc. due to opening up of the agricultural trade. But the reforms measures have not yielded the desired results in agricultural exports and improving the production and productivity. Further, the WTO regime brought new problems like insecurity in food and seeds, loss of bio-diversity ecological imbalance and bio-piracy.

The poor farmers may be benefited in the era of globalization. But, the paradigm shift is needed in domestic food and agricultural policy. This would go a long way to assuring that small and marginal farmers and other rural poor people have a stake in export opportunities. These opportunities should be given by assuring them access to infrastructure, inputs, credit that can facilitate their participation in markets. For that success of this shift, developing countries like India must compel the developed countries, to reduce their barriers on developing country's exports and other trade distorting policies under WTO regime, because the root cause of distortion of international trade in agriculture has been the massive domestic subsidies given by the developed countries to their agricultural sector over many years.

Women Food Security

Women's knowledge of local plants and animals as sources of food and medicine often ensure household food security in communities dependent on forest resources and subsistence agriculture. Yet, rural women's traditional knowledge of the local biodiversity is often ignored as the biotechnological innovations take hold of agriculture development.

To date the impact of the Asian economic crisis on rural women is presented in anecdotal evidence, but no systematic study has been undertaken. Anecdotal evidence suggests that as the informal economy shrank, the rural household economy managed by women provided a household safety net. Agriculture sustained the poor, marginally skilled and unskilled from the urban sector whose livelihood was threatened. Hence the region's national economic policies are turning towards revitalizing sustainable agriculture, developing rural poverty alleviation strategies and instituting safety nets. With increasing migration from rural to urban centers by men and educated young women, the success of sustainable agriculture and rural poverty alleviation strategies will increasingly depend on women who have opted to remain in rural areas.

Economic prosperity and vulnerability seem to be two sides of globalization-driven economic development. Rural women who have neither local opportunity nor educational preparedness to shift skills therefore be adversely affected, and women who provide food to a global market may face economic vulnerability in the local economies. Globalization also creates links with the household economy through migration of workers and their overseas remittances. Such income could diversity household livelihood strategies so that investment in farming may become one among several components in rural livelihood strategies.

It is relevant to rural women's productivity that human righted to access to property and education are stressed. Among others freedom from hunger and starvation is also relevant human rights associated with human resource development. Thus the role of rural women in contributing to food security is linked to rights to education and property. A rights-based approach to development and integration of women in development in particular will demand education of society and changing attitudes that undervalue girl children and women. The link between women's right to education, and gender equity in human resource development will require consistent national commitment and considerable investment of resources.

At the subsistence level technological interventions should take into consideration the drudgery of women's work. The modernization of rural economies should balance technological support both on the home front and on the farm front. In the modern agriculture sector disparity caused by lower wages paid to women should also be addressed with a focus on provisions for equity in agricultural labour.

Innovative and gender-equitable approaches that recognize gender roles in rural livelihood strategies, and related demand for productive resources will be needed to avert the depletion and degradation of natural resources. This demands guaranteed rights for women to access and/or own land and forest resources in order to responsibility to manage these with the support of relevant technologies and information. The undervaluing of women's indigenous and their agricultural knowledge in relation to biodiversity management should also be documented to inform national policies and strategies for biodiversity conservation. The role of rural women in local biodiversity conservation should therefore be strengthened with appropriate recognition and incentives.

Globalization will generate demand for advanced techniques and semi-skilled expertise in agricultural production to meet global market standards. This will occur in the prevailing context of high illiteracy rates among rural women in the Asia-Pacific region. The agricultural extension systems in the region have to adopt strategies to improve rural women's knowledge and skills in a more competitive global market. It is also imperative that the skills and knowledge of the extension system professional should be upgraded to take on tasks of improving capacity among rural women in a local economy linked to world markets. Information and communication technologies should therefore be harnessed to achieve the two-fold purpose of improving the knowledge base of extension professionals as well as upgrading the technical expertise of rural women.

Rural education systems should be improved and opportunities for rural girl children's education should be expanded. This demands interventions to alter current biases that impede women's education and the rural girl child's right to learn. The cost-effectiveness of communication technologies and the infiltration of popular media into rural areas could thus support a public education strategy that creates awareness of the biases against education for rural girl children and women.

Among the countries is the region an explicit, comprehensive policy framework to address issues related to the advancement of rural women is yet to evolve. Though mainstreaming approaches are prohibited these have not made a substantial impact on the advancement of rural women in most developing countries in the region. The situation is made more complex with the devolution of power to local governance. Commitments made at global fora and in national capital have not translated into concrete action with allocation of found at the local level. A consistent institutional approach to develop stronger and more effective linkages between gender mainstreaming, activism in national capitals/urban centers and priorities set in local governance agency programs will be needed, with provisional and district level governance bodies serving rural clients directly.

There is seldom any interaction between national machineries for women and technical ministries that have the mandates to improve agricultural productivity and food security. The former are responsible for the National Plans of Action for women and the latter for allocating resources for development in agriculture and related sectors. Although they are on parallel courses, they seldom come together to confront the issues related to rural women in agriculture and food security. This communication-cooperation gap has to be explicitly addressed to formulate strategies to improve that programs directed to rural women's role in order to achieve food security both locally and globally.

Enhancing Rural Women's Knowledge of Nutrition and Food Safety

Food and nutrition are the basis for life. Food is essential is essential for humans, starting from fertilization and continuing throughout the whole of life. This means that food and nutrition are important through the total life cycle. As a matter of fact these two terms "food" and "nutrition" may not be separated. Food is what we eat and drink, while nutrition is concerned with the utilization of nutrients and non-nutrient chemicals of the food in out body. In other words, nutrition is the link between food and human health. We eat food in order to get nutrients needed for growth and development during the growing periods of life, and for all function of the body to work properly throughout one's life span.

The daily consumption of various foods should supply all nutrients required by humans for the purposes mentioned above. The recommended daily intake (RDI) for various countries therefore is designed appropriately for the national population and should be followed. In case the RDI is not met, nutritional problems of individuals or of the entire population may arise.

It is commonly said that "you are what you eat". This implies that diseases related to improper food intake and nutrition can develop. The phrase may be elaborated or described in terms of three aspects :

1. *Insufficient or Not Enough Consumption*

This will result in nutritional deficiencies. These nutritional deficiencies will cause so-called **undernutrition**, leading to abnormal growth and various deficiency diseases. The main deficiency diseases include:

(*a*) *Protein Energy Malnutrition (PEM)* : The growth and development, in case of children, will be hampered and be lower than the standards established for the population. For adult, the body mass index (BMI) may be lower than the standard which indicates too much leanness of individuals.

(*b*) *Nutritional Iron Deficiency Anemia (IDA)* : Iron is a most essential element for the production of heme which is important for the function of red blood. With a lack of iron the red blood cell will be abnormal leading to signs of anaemia. The effect of iron the red blood cell will be abnormal leading to signs of anaemia. The effect of this on growth and development can be observed, and furthermore the cognitive function is impaired. We can obtain iron from foods such as meats, chicken or pork blood, egg and some dark green leafy vegetables.

(*c*) *Vitamin A Deficiency Disease (VAD)* : This vitamin is very important for children's growth, normal function of eyes and efficiency of immune function. There will be night blindness in both children adults with moderate deficiency. Children may experience stunted growth and permanent blindness with severe and prolonged deficiency. We can obtain vitamin A from such food as liver, egg, milk and fish liver oil. Vitamin A can be produced in the body from precursors such as beta-carotene which can be obtained from yellow and yellowish orange-coloured fruits and vegetables (e.g., ripe mango, papaya, pumpkin and carrot). It also can be found in yellowish and green leafy vegetables as ivy gourd and others.

(*d*) *Iodine Deficiency Disorder (IDD)* : This deficiency will lead to the formation of goitre. If the deficiency occurs during the period of pregnancy and in early life, brain development is affected resulting in permanent brain damage, an intelligence quotient (IQ) deficit and also growth retardation. Iodine can be obtained from seafood and importantly, from iodized salt. A rich amount can also be found in seaweed.

(*e*) Some other common deficiencies such as vitamin B_1 and vitamin B_2 deficiencies can result in various diseases and malfunctions of the body. Beriberi is a result of vitamin B_1 deficiency which causes weakness of muscles and abnormality in carbohydrate metabolism which can lead to cardiac disease. Vitamin B_1 can be obtained from pork and unpolished rice. Vitamin B_2 deficiency will result in very painful angular stomatitis.

If the dietary intake is insufficient in both in quantity and diversity, many other deficiencies will occur and lead to the malfunction of various systems in humans :

2. *Over-consumption*

This will result in overnutrition, to overweight at the beginning and lead to obesity later on. There are many other complications resulting from obesity, such as hyperlipidemia including hyperchoresterolemia which may lead to arteriosclerosis and infarction of the heart. Brain problems can occur from brain vascular abnormality. Blood vessels may harden with less elasticity, becoming very fragile and restricted. Insufficient blood supply will cause many brain and neuromuscular problems.

The most important practice to avoid overweight is to always watch one's body weight. The simplest index normally used is BMI which can prove useful in controlling weight. The BMI can be calculated easily from weight and height by the following formula: MBI = weight (in kg)/height (in m).

3. Food Safety

There are many instances where food can become contaminated with microorganisms and toxic or hazardous substances. The consumption of contaminated food will cause some forms of toxic disease. It can be either acute or chronic toxicity depending on the severity of contamination and the presence of toxic agents. Diarrhoeal disease is often encountered from eating heavily bacterial-contaminated food. Some bacteria may be vital such as *vibrio cholera, chlostridium botulinum, salmonella* and some strains of *E. coli.*

Chronic diseases may result from some types of chemical, and that can be as serious as cancer if foods contaminated with carcinogens are continuously consumed. Chemical contamination can occur during agricultural production, storage and transportation, processing, cooking and in various methods of consumption. The problem of aflatoxin contamination from mouldy food is a good example.

It is practical for the population of a country to have recommended national food-based dietary guidelines. In Thailand there are nine recommendations for daily food consumption and dietary practices called FBDGs. For a healthy life for the people these are as follows :

(1) Eat a variety of foods from each of the five food groups, and maintain proper body weight;

(2) Eat adequate amount of rice of alternative carbohydrate sources;

(3) Eat plenty of vegetables and fruits regularly;

(4) Eat fish, lean meats, eggs, legumes and pulses regularly;

(5) Drink milk in appropriate quality and quantity for one's age;

(6) Eat a diet containing appropriate amounts of fat;

(7) Avoid sweet and salty foods;

(8) Eat clean and safe food; and

(9) Avoid or reduce the consumption of alcoholic beverages.

With these guidelines a nutrition flag has been developed for use as a nutrition education tool for the general population. The nutrition flag demonstrates pictorially the relative proportion of various foods to be consumed daily from the five food groups. Adoption of educational tools such as this is helpful in ensuring that the population maintains good health with proper nutritional intake. In this context, rural women have an important role in promoting the proper consumption of safe, quality foods.

Gender-Disaggregated Data for Women's Empowerment and Participation in Food Security Strategies

The scarcity of data on rural women in most countries of the region is exacerbated by the often misleading information on women's role in the agriculture sector. The United Nations System of National Accounts (UNSNA) and definitions of economic activity and *"value"* decreed by the International Labour Organisation (ILO) and the International Standards for Industrial Classification (ISIC) have conspired to make the majority of women food producers invisible. The flow-on form this ensures their contributions are ignored in policy and planning.

Data bases, generally thought of in a somewhat narrow way as statistics or facts, play a major in part in developing policy and plans that determine how national resources are used to meet national needs defined by those same makers and planners. These are overwhelmingly economic, providing the information that is used to report of goals represented in the UNSNA as GDP. Scant attention is paid to social political objectives the ensure quality of life beyond basic human needs, and both foreign and national agencies concerned with development base their decisions about who will get help and support, and what kind of help will be offered and when, on such data. In agriculture and therefore in food security that data is both inadequate and misleading because it overlooks most of the work done by half the world's farmers.

Sex-disaggregated data is essential to sensitize agricultural policy makers and planners to gender disparities so the they can be addressed with a view to achieving gender equity. The imperative however may be more compelling when the new data shows how critical women farmers are to food security. To this end, research methods for the collection of data on the role and contribution of women farmers needs radiation; revision. Researchers and agricultural planners who refuse to see, or continue to ignore what is emerging as an enormous challenge to traditional research will continue to plan for only half the farmers who produce half the food in the world.

In addition to the routine disaggregation of data sex and by rural-urban, traditional definitions and methods of counting dictated by the UNSNA and the ILO exclude most of rural women's contributions to food security, and must be changed. Surveys and the resulting data must value women farmers' work by quantifying their production and documenting constraints resulting from their differential access to and use of land, water, technologies, information and extension services. Data is them more likely to reflect actual contributions to GDP and to national and hoursehold food security, pointing the way to policy changes and new approaches to planning. Researchers need to consult with women farmers in the analysis of sex-disaggregated data to define barriers around their access to factors of agricultural production, and to articulate these for policy makers and planners. Agricultural policy makers and planners will then have no reason to ignore women's critical contributions to food security through subsistence production and gathering.

Gender-sensitive questionnaires can be designed for much of the necessary data collection, but both men and women in the household must be interviewed separately about their own activities. Accuracy increases for example, when a female farmer is interviewed by a trained female enumerator, and a male farmer by a trained male enumerator. Proper selection and training of those responsible for data gathering pays dividends in the accuracy and depth of information collected. Participatory analysis of raw data with men and women farmers further increases accuracy, and provides additional insights.

Qualitative data collection and analysis using analytical frameworks and matrixes throw light on many gender issues, and time use studies are an effective tool for providing answers to the critical questions raised in the analysis of both quantitative and qualitative gender data. Data on farmers thus should be routinely disaggregated be sex, as should *all* data on people if development planning is to be gender responsive. Analysis be a competent, gender-sensitive and experienced analyst is, of course, essential to proper identification of the issues implicit in sex-disaggregated data, and the ways in which these are best addressed.

The most accurate data on time use by women and men in rural areas in studies conducted in many countries was obtained using non-participant observation methods. Other studies however can be enriched by adding limited, complementary time use studies to a traditional socio-economic household survey. Using the non-participant observation methods, a person observes and records all activities by a worker observed over one normal working day, without the observer participating in any of the activities her/himself. As long as the observer is of the same sex, and of similar socio-economic and ethnic background as the observed person and is properly trained, the non-participant observation method has consistently proved better than other methods at capturing the reality of rural women's household, farm community activities. The method causes minimal distortion in normal patterns of work, and records both productive and reproductive contributions to the economy.

The greatest danger to the provision of good data on gender roles and issues in agriculture is often in poor analysis by inadequately trained and inexperienced "*gender specialists*"– sometimes so–called simply because they are not a male.

The empowerment of women farmers may not be a priority for those who favour the rural *status quo,* but to anyone with a primary concern for food security the empowerment agenda is not a threat, but a tool for the elimination of hunger and the achievement of food security. The disaggregation of good data on the role, contributions and constraints to both male and female farmers' production and productivity is a first step towards addressing inequity. More than that however, it provides the best basis for policy makers and agricultural planners to unleash both halves of the human potential for solving world food insecurity.

Prospects of Rural Women's Contribution Sustainable Food Security (Ms Puengpit Dulay-apach)

Food security has already been defined by the FAO as the economic and physical capacity of a household to continually acquire and provide family members with sufficient food for individual bodily needs, with no threat of shortage in any period. In addition, the family can exploit its food availability to the advantage of the individual nutritional needs of the individuals.

The significance of food security however depends largely on how nourishing the food is from a family's point of view. In fact, food is vital all throughout one's life to maintain regular living and good health in order to be able to carry on efficient activity. The three major factors in making food security possible at the household level are:

(1) The capacity to produce sufficient food;
(2) Economic capacity to always acquire and provide adequate food; and
(3) Ability to utilize to fullest extent the food based on the individual nutritional needs.

Taking into account the above three factors, the people who are in the best position to ensure food security of the family are women—rural women in particular.

An FAO survey reveals that 42 per cent of women in 82 developing countries are engaged in farming. Although it is customary for men to be in charge of producing major foods, women actually have to work hard on crop cultivation, orchard operation and taking care of small livestock for home consumption. In addition, any surplus production requires that women preserve or prepare those foods for local village trade. As men migrate to seek jobs elsewhere, the burden of farm production upon women is expected to double.

In Thailand, farm women account for 43 per cent of the total population. It is therefore evident that they play a crucial role in farm production processes. Usually 65 per cent of them take part in paddy farming. After land preparation by men, women sow the seeds, prepare seedlings, transplant, apply fertilizers, and undertake harvesting, threshing, storing, milling and selling. Furthermore 90-100 per cent of farm women are engaged in gardening, livestock raising, aquaculture, food storage, processing and preserving, and other activities for supple mentary income. An additional and important role of women in food and agriculture in Thailand is assisting in the decision-making processes, especially regarding seed and livestock selection, the choice of agricultural chemicals and the pricing of the farm products.

The chain from farmer to consumer is long and often complicated, and women's work along this food chain does not end with cultivating crops, gathering, fishing and tending animals. In developing regions once the harvest is in or the catch landed, it is almost exclusively women who process, preserve and store the food as well as sell it at the market. All of these activities are essential for ensuring that food will be safe, nutritions and available throughout the year, while also serving to generate substantial income for the family. Across the developing world, most food processing activities are performed by women. Processing improves the convenience and marketability of the food, which together with other post-harvest activities strengthens household food security just as effectively as increasing production in the fields. Women all over the world are also actively involved in the trade of food, primary agricultural products and processed food products, which translate directly into improved household food security.

In addition, women also take responsibility for food preparation for family members. They select the daily menu and provide the nutritional food balance in meals. They are also directly involved in food safety, clean water supply, collecting fire wood, soil conservation and crop cultivation. Rural women are often the custodians of knowledge regarding crop varieties and their use as food, in their applications in medicine, and their use in crafts and cultural practices.

Farm women's working house in the field amount to 2,250 hours per year while work at home amounts to 1,664 hours per year. That means a total of 3,894 hours per year compared to 2,250 hours per year for men in the fields alone. Yet in seeking farm credit and information, women farmers have far less opportunity.

Farm women's labour's unpaid labour, having a direct impact on economic activities at all levels, especially on national food production. However, farm women's role in terms of socio-economic development and environmental management is often neglected by national policy makers. In order to empower farm women in the economy, national policies should be developed with farm women included as specific target groups. Harnessing the development potentials of farm women requires an emphasis on farm women's role in the process of planning and decision-making involving, among others, the use of agricultural technology, marketing, production and post-production activities and investment. While farm women play a crucial role in national food production and national socioeconomic and environmental development, sustainable food security requires both men and women to exchange views, ideas and experiences. Including and integrating farm women's experiences and skills in decision-making, planning and implementation will enhance sustainable food security.

Agriculture continues to be an important sector of the economy in practically all countries of Asia and the Pacific. Even in those countries where its share in GDP has significantly declined, agriculture has remained a major policy concern, particularly in light of the recent Asian financial crisis, ongoing liberalization of agricultural trade and the renewed call for sustainable food security. The former in fact has prompted some countries in the region to take a more serious look at the role of agriculture, with a view to according it a higher priority in national development. Agriculture of course contributes not only food and fiber for the population, but also provides employment and livelihood options, and vital foreign exchanges for socio-economic development.

In many agricultural operations undertaken on the farm, rural women play a major role. Rural women in fact comprise the majority of food producers in the region. Virtually all of them are also responsible for food preparation for their families. These important contributions unfortunately go largely unrecognized due to a variety of factors including policies and programs, cultural traditions and legal provisions that discriminate against rural women. Rural women therefore have remained largely an invisible in the development process.

Women's responsibility for food preparation has also highlighted the importance of ensuring the safety and quality of foods consumed by members of the household. The increased application of chemicals in food production for instance, as well as their use in food processing and preservation, has raised concerns about food contamination. More recently the expanded production of genetically modified food items has further added to public concern about food safety. Given these developments the need to enhance rural women's knowledge about nutrition and food safety will become increasingly important for ensuring sustainable food security.

Rural women are involved in a variety of tasks both off the farm, and in the household. Aside from doing household chores and attending to the needs of family members, particularly the children, they are also engaged in various farm and non-farm economic activities. To a varying extent these include not only primary production and economic activities. To a varying extent these include not only primary production and processing, but marketing and the distribution of farm products. These multiple roles have been defined largely by customs and traditions characteristic of Asian culture which places on women the responsibility of rearing children and performing household chores while at the same time helping with the farm work. It is therefore not uncommon for women to work more than 12 hours a day. Much of the work however represents unpaid labour, and even when paid women generally receive much less pay compared to men for the same work.

Many developing countries of the region face a number of constraints to increasing the productivity of rural women. The first and foremost is poverty. Poverty has severely limited women's capacity as productive members of the rural community. It is also the main cause of low social and economic status of women. Their low literacy levels and limited educational attainment is to some extent another outcome of poverty. These combine to ensure the much more limited access of women to employment and income opportunities compared to their male counterparts.

Most countries have constitutional or legal provision requiring the state to provide free and/ or compulsory education to their youth up to secondary or high school level. Participation of rural females in education and training programs however has been relatively low due to, among others, poverty, parental preference for boys to attend school, and the expectation in some societies that women do not need education as they are not required to be income earners.

Various measures have been taken in a number of countries to enhance the school attendance of young girls in rural areas. As a result significant changes in women's education over the past three decades or so, have occurred. Enrolment of women in schools for example has increased, and governments have started to spend more on basic education. Despite the improvement however, high illiteracy rates in developing countries, especially in rural areas. Thus programs that are aimed at increasing women's participation in economic activities, particularly in food production, but also in their control over economic and neutral resources, need to continue to address the more basic problems of illiteracy, and society's traditional and conservative view of women's roles.

The lack of gender-disaggregated data on rural women is a constraint to recognition of the importance of women's contribution to food security. This in turn means women's concerns are ignored in policy and in planning, and they are often bypassed in extension programs. A gender-responsive data base is needed to target women as critical food producers in order to ensure and sustain security. Other constraints to increasing rural women's productivity include"

(1) Gender-insensitive policies and programs that limit their access to credit, education and training, and other government services;
(2) Social values and traditions that inhibity their participation in the economy;
(3) Lack of gender-based technology for women's work in agriculture; and
(4) Weak institutional mechanisms for implementing gender-focused policies.

It is clear that for rural women to become better food producers their economic empowerment is necessary. This would enable them to participate more effectively in the planning and decision-making processes that are important for improving their productivity. Such empowerment requires that the more general associated with gender discrimination and inquality are addressed adequately. Improving literacy and the educational attainment of rural women is a critical factor in their empowerment. Education and training specifically would need to focus on developing occupational skills as well as leadership and communication, interpersonal and organizational skills. From a macro perspective gender mainstreaming needs therefore to be further promoted in all aspects and at all level of economic and social life.

For the future the high population growth of some countries in the regions will remain a major food security issue. Other issues include :

(1) The impact of globalization and trade liberalization on domestic food production;

(2) The aging of the farm population as more and more of the younger generation migrate to urban areas or foreign countries;

(3) Emerging technologies such as biotechnology that are increasing people's concern about food safety; and

(4) Growing pressure for conservation of the environment.

Some technologies however including information technology could have a significant positive role in improving rural women's productivity and thus in promoting sustainable food security.

Chapter 9

Economic Empowerment of Women Through SHGs

Though the Government has continued to allocate resources and formulated policies for empowerment of women, it has become strikingly clear that political and social forces, that resist women's rights in the name of religious, cultural or ethnic traditions, have contributed to the process of marginalization and oppression of women. The basic issue that prevents women from playing full participatory role in nation building is the lack of economic independence. Planners and policy makers have been eagerly searching for certain alternatives. The participatory approach to development has emerged as a vital issue in developmental policies and programmes for women.

Under the trickle town theory in the planning process, it was expected that women will equally benefit along with men. This has been belied by actual developments. The ninth plan document recognizes that in spite of development measures and the Constitutional legal guarantees – women have lagged behind in almost all sectors. In the past decades, there have been various forces and pressures which are more dominant than those which have tried to push women towards growth and development.

Though the Government has continued to allocate resources and formulated policies for the empowerment of women, it has become strikingly clear that political and social forces, that resist women's rights in the name of religious, cultural or ethic traditions, have contributed to the process of marginalization and oppression of women. The basic issue that prevents women from playing full participatory role in nation building is the lack of economic independence. Planners and policy makers have been eagerly searching for certain alternatives. The participatory approach to development has emerged as a vital issue in development policies and programmes for women.

Self Help Groups (SHGs) are considered as one of the most significant tools to adopt participatory approach of the economic empowerment of women. It is an important institution for improving the life of women on various social components. The basic objective of an SHG is that it acts as the forum for members to provide space and support to each other. SHGs comprise of very poor people who do not have access to formal financial institutions. It enables its members to learn to co-operate and work in a group environment.

An SHG is a group of people that meets regularly to discuss issues of interest to them and to look at solutions of commonly experienced problems. The group may or may not be promoted by Government or non-government institutions.

The SHGs may have completed two cycles of one-year duration each successfully where they were receiving inputs from the implementing agency in the form of credit and marketing help. After one more cycle, when these SHGs will have generated sufficient money and developed required expertise, they will be working independently. It is the normal tendency.

Benefits

Successful working of these three SHGs has given enormous benefits. Organized working of the women through these SHGs has increased the income of the families involved. Most of them are now able to repay their old debts and started asset building. The existing enterprises of beneficiaries are better managed now. Success of these SHGs not only improved the economic status of the women concerned but there is also a drastic change in their social status.

Now these women have better say in their family matters. Success of these SHGs has given an amazing confidence in the women concerned. Many of them are now coming forward to help other women of that area. The overall changes in the Janata Colony due to these SHGs again substantiate the saying *Educating a women means educating a family.*

Future

After one more cycle, these SHGs will have generated sufficient money and development required expertise that they will be working independently. Two SHGs involved in the stitching work (Shakti and Pragati) are trying to get contract of stitching uniforms for hospitals and various other institutions. Prerna SHG has sold its Murabba and Pickles successfully till now. They are having talks with the authorities to allot them a permanent space. They have now started Tiffin system and trying to expand their enterprise in larger area. Implementing agency has adopted Dhanas village for furthering the scheme of micro credit through SHGs.

SHG

Empowerment is a multidimensional process, which should enable the individuals or a group of individuals to realize their full identity and powers in all spheres of life. It consists of greater access to knowledge and resources, greater autonomy in decision making to enable them to have greater ability to plan their lives, or have greater control over the circumstances that influence their lives and free them from the shackles imposed on them by custom, belief and practice. Empowerment of woman may also mean equal status to the woman, opportunity and freedom to develop her. Empowering women socio-economically through increased awareness of their rights and duties as well as access to resources is a decisive step towards greater security for them.

'The status of women is a barometer of the democratism of any state, an indicator of how human rights are respected in it", according to Mikhail Gorbachev. The root cause of women's oppression in India is patriarchy, which has snatched legitimate powers off, leaving them completely defenseless and weak. The unrealistic way in which women are depicted in literary works and firms by male chauvinists and misinterpretation of women in epics and scriptures contributed much to the poor self-image, suffering nature, defeatist attitude and lack of assertiveness on the part of women. Of late, lot of awakening is found among them due the entry of emancipated women-writers in the literary field.

Empowerment of women is aimed at striving towards acquisition of the following :

- Higher literacy level and education,
- Better health care for her and her children,
- Equal ownership of productive resources,
- Increased participation in economic and commercial sectors.
- Awareness of their rights,
- Improved standard of living,
- Achieve self-reliance, self-confidence and self-respect amongst women.

Empowerment of women would mean equipping women to be economically independent and personally self-reliant, with a positive self-esteem to enable them to face any difficult situation. Moreover they should be able to contribute to the developmental activities of the country. The empowered women should be able to participate in the process of decision-making. Women empowerment is a dynamic process that consists of an awareness-attainment-actualization cycle. Again, it is a growth process that involves intellectual enlightenment, economic enrichment and social emancipation on the part of women.

Education is one factor that plays the most crucial role in empowering women. Schools, colleges and other professional bodies are persistently trying to educate, motivate and train the women in their chosen areas of career through curriculum, training, field-exposure and other practical methods. Research and publication in the areas of women's problems, social evils and their eradication and women empowerment are the hot topics of the present. Media-coverage aiming at attracting the attention of the policy makers and authorities is at its highest level now.

Women are in for a new deal today as they are the focus of economic development. All possible steps are being taken to strengthen them to achieve their economic, social, cultural and political growth and welfare. Projects such as Rashtriya Mahila Kosh (March 1993) and Indira Mahila Yojana (August 1995) are specially designed for empowerment of the economically backward women through micro-credit in order to promote self-employment.

A National Plan of Action for the Empowerment of Women with measurable goals to be achieved in a time frame of the next 10 years is being formulated in consultation with the State Governments and various Ministries and Departments of Government of India.

The Government of India had declared the year 2001 as the Year of Women's Empowerment. The year was formally launched by the Prime Minister in a function held at Vigyan Bhavan on 4th January, 2001 when he also awarded the first "Stree Shakti Puraskars" to five distinguished women from the grassroots who had made outstanding services for the social, educational and economic empowerment of women in remote and difficult areas.

The purpose of declaring the year 2001 as the Women's Empowerment Year was as follows:

1. To create and raise large scale awareness of women's issues with active participation and involvement of all women and men;
2. To initiate and accelerate action to improve access to and control of resources by women;
3. To create an enabling environment to enhance self-confidence and autonomy of women.

The Government approved, for the first time, a National Policy on Empowerment of Women in order to mainstream gender into all activities of the Government and other agencies. The main objectives of the Policy are :

1. Creating an environment through positive economic and social policies for full development of women to enable them to realize their full potential;
2. The de-jure and de-facto enjoyment of all human rights and fundamental freedom by women on equal basis with men in all spheres-political, economic, social, cultural;
3. Equal access to participation and decision making of women in social, political and economic life of the nation;
4. Equal access to women to health-care, quality education at all levels, career and vocational guidance, employment, equal remuneration, occupational health and safety, social security and public office etc.;
5. Strengthening legal systems, building and strengthening partnership with civil society, aimed at elimination of all forms of discrimination against women;

6. Changing societal attitudes and community practices by active participation and involvement of both men and women;
7. Mainstreaming a gender perspective in the development process;
8. Elimination of discrimination and all forms of violence against women and the girl child.

Measures Suggested for the Empowerment of Women

In order to bring women to the center-stage of development and thereby ensuring better participation in the developmental efforts of the nation and enabling them to take a lead role in the social and economic system, the following suggestions may be conceded.

1. Compulsory Education

Eradication of illiteracy is the first step towards empowerment of women. Knowledge is power. When a woman is educated, it is in effect the whole family is educated. Researches show that in families where the women are educated, social evils such as illiteracy of girl-children, child labour, female infanticide and other superstitious practices are much less. It is education that kindles the urge for independence, hard work, achievement and self-actualization. It may be mentioned that education that inculcates human and spiritual values is of great significance for the empowerment of women. These are needed not only in educational institutions but also in every walk of life. The entire population is to be involved to create a sense of awareness about values and the need to empower women through quality education. Education upto a minimum of 10th standard must be made available and compulsory for every child.

2. Gainful Employment

Women should find appropriate employment/occupation to support themselves and lead a life contributing to the economic status of her family as well as the nation. Under the present condition, self-employment is the only feasible answer that warrants economic power to the millions of women in the unorganized sector of our economy.

Money is strength. Though education is the primary ingredient for empowerment, it is the economic power that acts towards it instantly. All-out efforts are required to introduce the womenfolk towards various kinds of business that can provide gainful occupation with less risk. EDPs, awareness programs, conferences, workshops etc., can help them start their own industrial/business units. Having an occupation of their own would provide them with ample opportunities to prove their mettle, resulting in moving towards higher levels of achievement.

3. Formation of Self-Help Groups

Women should unit themselves into social groups called Self-Help Groups for their own progress as well as that of the community. These SHGs have a common perception of need and the advantage of collective action. It is easy for the Government, banks and other development agencies to have access to the grass-root level in order to spearhead development process. Being a member of a group engaged in collective effort towards social and economic progress, the women can enjoy security and, be guaranteed of their emotional, intellectual and financial well-being to a great extent. Formation of SHGs is an easy way to enroll women into the organized sector.

4. Credit Facilities

Liberal supply of credit along with other financial and nonfinancial incentives will go a long way in promoting self-employment among women through micro-enterprises and SSI units. Loans (micro-credits) must be sanctioned to them on their own capacity and security, without must of hurdles. NGOs, SHGs and other development agencies could initiate/recommend/monitor the disbursement of credit as per the project-needs of these borrowers to the satisfaction of the lending agencies. The appropriate authorities should provide managerial and marketing facilities wherever necessary. Training and technical consultancy services must be made available close to their station. Adequate supply of information is also essential for their success.

5. *Mental Revolution*

"The greatest discovery of any generation is that a human being can alter his life by altering his attitude"–William James. There should be a revolutionary change in the perception and attitude of both men and women towards women. Women are in no way inferior to men and they have already imprinted their mark in almost all walks of life. Their capacity to endure and persevere is an accepted fact. Being the better half of the total population, it is upto them to grab the opportunity now abundantly available to realize their goals, working shoulder to shoulder with men or independently, towards empowerment. They should cherish emotional, intellectual and economic freedom. Growth and development must be their primary motto. Emotional maturity and progressive thinking will take them to a life more rewarding and self-satisfying.

Perception about the Modern information Communication Technology (MICT)	0.068^{NS}
Innovativeness	0.139^{NS}
Social participation status	0.023^{NS}
Value orientation	0.106^{NS}
Self confidence	0.215^{NS}

NS : Non-significant, ** : Significant at 0.01 per cent level, * : Significant at 0.05 per cent level, @ : found to have missing relationship and not amendable for statistical analysis.

From the above results, it could be inferred that rubber being the traditional crop in the study area, most of the subjects were born and brought up in the predominantly rubber belt and hence, their prime age and experience in rubber cultivation might have increased their knowledge gain. Likewise, the respondents who were having more area under rubber cultivation might have had high information seeking behaviour to know more about their plantation. The similar trend of increased possession of modern electronic gadgets and familiarity in using computer would have increased the knowledge gain related to pests and diseases of rubber crop.

The influence of the 15 independent variables towards the knowledge gain was studied and the results presented in Table 1.

It could be observed from the Table that all the 15 independent variables together explained 25.80 per cent of variation towards knowledge gain which was significant at one per cent level.

The partial regression co-efficient value was found to be positive and significant for the variable experience in rubber cultivation at 0.01 level of significance. Majority of the subject possessed 11-12 years of experience in rubber cultivation. The practical experience put forth by them would have influenced their knowledge gain.

The findings however were in contradiction to the findings of Anandaraja (2002) who reported that farming experience had no significant influence on knowledge gain.

Table 1 : Influence of Independent Variables Towards Knowledge Gain (Y1)

Variables	*Partial regression coefficient*	*Standard error*	*'t' value*
Age	0.04878	0.038	1.281
Eduction status	–0.04825	0.348	–0.139
Occupational Status	0.16700	0.453	0.369
Area under rubber cultivation	0.30500	0.178	1.715
Experience in rubber cultivation	0.07649	0.033	2.322
Annual income	–0.22900	0.290	–0.788
Communication status	–0.08621	–.092	–0.933
Information seeking behaviour	0.05367	0.034	1.573
Possession of modern electronic gadgets	0.03229	0.118	0.273
Training undergone on computer	@	@	@
Familiarity in using computer	0.29800	0.378	0.790
Perception about the Modern Information Communication Technology (MICT)	0.24500	0.321	0.764
Innovativeness	0.59500	0.443	1.343
Social participation status	0.15700	0.165	0.950
Value orientation	–0.00190	0.073	–0.026
Self confidence	0.22400	0.277	0.809

R2 = 0.258, F = 2.405**, NS = Non-significant, ** = Significant at 0.01 per cent level, * = Significant at 0.05 per cent level @ found to have missing relationship and not amenable for statistical analysis.

Broadly farm women perform economic, home making, caring and nurturing roles. Although equality of rights for men and women is guaranteed in our Constitution, in reality it is far off. The role differentiation between the two sexes continues with men being regarded as the providers and women treated essentially as homemakers.

Despite multiple role of women in agricultural operations and household chores, their work is generally underestimated and undervalued. The status of women in general is much lower than that of their male counter parts, largely because of customary male dominance in society, inherent shyness of farm women and lack of opportunities for education and training. Instead of all efforts there is continued inequality and vulnerability of women in all sectors *viz.* economic, social, political, education, health care, nutrition and legal. As women are oppressed in all walks of life, they need to be empowered. There is need to enhance the participation and leadership role of women in different walks of life and in the development process.

Women empowerment is synonymous with the achievement of equality and equal mindedness in society. It is an active process, which enables them to realize their identity and power in all aspects of life. It enables them to have more access to knowledge and resources, greater autonomy in decision making, greater ability to plan their times, free them from the clutches of irrelevant custom builds and practices. All over the world efforts are being made to empower women through literacy, education and training, health support and entrepreneurship development for economic freedom.

Empowerment of women could be in any sphere of life; legal, social, political and economic. The contribution of women in rural areas is multifold, therefore, economic empowerment directly affects all other areas of empowerment.

As women play an active role in the economy of their families, they are wise enough to invest money and lead better life. There is a linkage between woman's access to independent income and her position in the family. It is believed that when women's are provided credit and they take up income generating activities, their income is expected to increase. When they earn money, their say in the decision making in the house improves.

To give rural women visibility, they must get organized into self help groups. Group approach is a viable set up to empower women economically, socially and technologically for improved of life. Role of Self Help Groups (SHFs) is emerging as promising tool in this context. The Self Help Groups are created to enable the members to reap economic benefits of mutual help, solidarity and joint responsibility towards self and sustainable development.

Concept of Self Help Group (SHG)

Self-help Group is a voluntary and self-managed group of women who come together to promote savings among themselves as well as pool savings for activities benefiting either individuals or communities economically Members support each other and are accountable to one another through the sharing of information and resources and assist in decision-making in individual, family and community matters. Formation, promotion and nurturing of Self Help Groups is a time intensive process, for which a number of steps are to be followed :

- Mobilization of farm women through informal meetings and mass communication.
- A group of 10-20 farm women who are basically homogeneous in nature is the pre-requisite of forming a Self Help Group.
- Members are encouraged to save on a regular basis. The amount of savings is within the range of Rs. 20 to Rs. 100. They rotate this common pooled resource within the members with a very small rate of interest.
- Each roup has a leader who is called as the President and an executive committee. They usually maintain records of transaction on daily basis in written format and take initiative for developmental activities.
- Members are encouraged to shoulder responsibility equally right from the beginning. Most of the operating rules and norms are established in the initial stage by common consensus. The members of group should have sense of realism, strong ownership, cohesiveness and intensive mutual interaction.
- The external intervener should be guide and facilitator without interfering in the autonomy of the group.
- The function of self help group should be to pool small savings and lend as per need and open account in a bank.
- The groups are encouraged to have regular meetings once in 15 days or monthly on day and time suitable to all the members. Decisions and actions regarding financial transactions are taken in these meetings. These meetings play a very crucial role in developing strong ties among members and sharing of problems and their solutions.
- Linking groups to a bank and getting bulk credit for the group is very important for grow of the group.
- As groups grow stronger, other developmental issues are taken up. These may be related to technology use, entrepreneurship development, social problems of women, issues related to health and nutrition, child-care education of girls, etc.
- Programmes for farm women have to be tailored to their needs and their variable time schedules must be kept in mind. Vocational training and education for rural women should seek to enhance their natural skills an aptitude so that it is meaningful and relevant to their life situations.

Role of Self Help Groups in Empowering Farmwomen

The self help groups empower women and train them to take active part in the socio-economic progress of the nation and make them sensitized, self made and self-disciplined. The SHGs have inculcated great confidence in the minds of rural women to succeed in their day-to-day life. SHGs enhance the quality of status of women as participants, decision makers and beneficiaries in the democratic, economic, social and cultural spheres of life. The SHGs bring out the capacity of women in molding the community in right perspective and explore the initiative of women in taking the entrepreneurial ventures. SHGs also organize women to cope with immediate purposes depending on the situation and need.

Participation of women in SHGs makes a significant impact on the empowerment in social aspect also Participation helps women come out in open and discuss their problem. It also helps to bring about awareness among rural women about savings, education, health, environment, cleanliness, family welfare, social forestry, etc. Researches also reveal that increased participation of women in decision making at all levels will help to adjust the goals pursued through development.

Empowerment should be externally induced so that women can exercise a level of autonomy. There should also be **'self empowerment'** so that women can look at their own lives. The process of **'learning by doing and earning'** would certainly empower rural women. More and more rural women need to be involved in self employment. Self employment in agriculture, village and small industries and retail trade and services should be expanded. Self employment is also conducive to the development of individual initiative and entrepreneurial talent and offers greater personal freedom. The added advantage is that the institution of family remains undisturbed. The emergence of self help groups in this context is a welcome development. The groups would provide a permanent forum for articulating their needs and contributing their perspectives to development.

Self help group should be developed as an institution for financial intermediation as well as people's network rather than a vehicle for credit disbursal only.

Experiences of a project on 'Empowerment of Women in Agriculture'

A mission mode National Agriculture Technology Project entitled 'Empowerment of Women in Agriculture' initiated in September 2001 was completed successfully in March 2005. The main objectives of the project were to identity and promote need – based drudgery reducing tested technologies in agriculture animal husbandry and fisheries to promote entrepreneurial activities for economic empowerment of women. Technology envisaged for technological and economic empowerment of women through education, training by fishery participatory approach involving women self help groups. The major thrust was on drudgery reduction in agriculture, animal husbandry and entrepreneurship development for economic empowerment.

Need based enterprises were selected and established at different centers. Massive training programmes should be organized for capacity building of farm women in the areas of – drudgery reduction, entrepreneurship development and fruit and vegetable preservation. As a result of concerned efforts through training, intervention and field demonstrations drudgery reduction with use of improved technologies ranged from 40 to 75 per cent can be done. The most accepted implements were improved sickle, tubular maize sheller, weeders, fertilizer broadcaster, groundnut decorticator, cleaner graders, manual rice transplanter, pedal operated paddy thresher and rake other technological tools. Savings of majority of SHGs were between Rs. 10-20 thousands and 72 per cent of groups provide loan to the members. Farm women can earn a net profit of Rs. 200-600 per month from various enterprises like vermi-composting.

Participation of women in SHGs makes a significant impact on the empowerment in social aspect also. Participation helps women come out in open. They can discuss and put forward their problem. It also helps to bring about awareness among rural women about savings, education, health, environment, cleanliness, family welfare, social forestry, fishery etc. If women play important role in decision making at all levels will help to adjust the goals pursued through development.

Empowerment is an externally induced force so that women can exercise a level of autonomy. There should also be **'self empowerment'** so that women can look at their own lives. The process of **'learning by doing and earning'** would certainly empower rural women. More and more rural women need to be involved in self employment. Self employment in agriculture, village and small industries and retail trade and services should be expanded. Self employment is also conducive to the development of individual initiative and entrepreneurial talent and offers greater personal freedom and family remains undisturbed. The emergence of self help groups is a welcome development. The groups would provide a permanent forum for articulating their needs and contributing their perspectives to development.

Self help group should be developed as an institution for financial intermediation as well as people's network rather than a vehicle for credit disbursal only nursery raising, masala, *papad, badi, dalia, suji* potato chips making, tailoring, animal feed making, leaf cup plate making, bakery, goat rearing, diary cooperative, pisciculture, floriculture. Apart from this, 100 trainings covering 3205 beneficiaries were conducted in nutrition education and fruit and vegetable preservation for awareness generation about nutrition. The pre-post knowledge and skill analysis of these trainings elicited that 45 per cent of women started preservation of different foods which improved health status of children and family members. The project had marked impact on socio-psychological aspects and social empowerment of women farmers. SHG approach has developed leadership qualities, group confidence, strength and enabled farm women to initiate any new venture in a group. Majority of the women have become member of SHGs for the first time and expressed satisfaction to be a member of the group. The system of mandatory contribution strengthened the habit of saving/thrift leading to capital augmentation. The small earnings gained were being utilized purposefully for betterment of their home and living conditions. The project has resulted into a successful model of women empowerment which can bring miracles in improving the quality of life of rural women.

The SHGs are a viable alternative to achieve the objectives of rural development and to get community participation in all rural development programmes. The possible outcome of women's empowerment through SHGs at household level are self-employment (assured wage employment through the year), sustainable livelihoods, improved health and education, enhanced social dignity and better status of women. The empowerment of women through SHGs would lead to benefits not only to the individual women and women's groups but also for the family and community as a whole through collective action for development. They assume the role of decision makers in major and deciding aspects of the family and village. Organization of farm women into self help groups can go a long way towards bringing women in the mainstream of development. "Empowerment is not just for meeting their economic needs but also for more holistic social development". (AER)

Conclusion

The findings reveal that effective use of electronic media such as the Expert System in this information age would serve as a handy tool the farmers and agricultural advisors who are in need of immediate answers. In RUBEXS–04 the delivery of information through text, pictures and audio were tailored to suit the pace of learning of the subject. Hence, RUBEXS–04 with discussion has contributed to the maximum knowledge gain among all the other treatments used.

Further important variables which were found to have significant relationship with the knowledge gain such as age, area under rubber cultivation, experience in rubber cultivation, information seeking behaviour, possession of modern electronic gadgets and familiarity in using the computer, have to be given in-depth focus while planning and implementing training programmer for rubber growers in future, (AER).

REFERENCE

A Successful Approach for Women Empowerment. Agriculture Extension Review.

Chapter 10

Rural Jobs, Agribusiness Centres

Agriculture farms are the backbone of the Indian economy and despite planned industrialization in the last five decades, Agriculture occupies a significant place, agriculture being the largest industry it provides employment to around 65 per cent of the total workforce in the country. The current Agriculture Extension worker and farm holding ratio in the country is about 1:833, which is quite inadequate to disseminate the first changing Agriculture technology to the rural people including farmers. Under the new economic policy of GATT and WTO, the Agribusiness has come under the strong and direct influence of international markets. Indian farmers have to produce the quality of goods to compete with international standards of the markets. The present system of help and assistance to farmers are not sufficient to meet the required standard of international markets. Existing Government Machinery is not properly equipped, nor adequately qualified and trained for rendering the specialized advice to the farmers.

At present there is a need to supplement the efforts of Government. Extension system by the private partners to accelerate the process of science and technology transfer in Agriculture field. Specialized Agri-services are needed for the Agri processing, Milk processing, artificial Insemination, Liquid Nitrogen Plant, Agriculture Insurance, Soil and Input Testing Maintenance and Repairs. Seed and Fertilizers and Pest Harvest Management etc. It is felt that this gap can be filled by qualified agricultural graduates. If they are provided right kind of support system to start their own business or job or enterprise on various agriculture and allied activities in the country. These opportunities are also for fisheries graduates.

Agri-Education and Employment Scenario

The turnout of graduates in Agriculture and allied subjects are around 1,19,000 per year. The intake capacity in post-graduate programmes in the country's universities are around 5500, leaving about 6400 graduates of which hardly sectors. Thus a reservoir of 9900 graduates every year is available for supporting agriculture production process, which is important for quality. It viable business opportunities are to be provided to them. To provide gainful self-employment opportunities to Agriculture and allied graduates. For this Ministry of Agriculture, Government of India has launched unique scheme in Association with NABARD, called Agribusiness and Agriclinic Centre. In this paper I want to highlight the scope for Agri graduates to in their own business or enterprise and their contribution to create the self help employment for the younger generation. This programme aims to tap the expertise available in the large pool of agriculture Graduates irrespective of whether they are fresh graduate or not, or they are currently employed or not, they can set-up their own Agriclinic or Agribusiness Center and offer the professional services to the farmers. The government is also providing start up training to graduates in agriculture or any subject allied to agriculture activities like; Horticulture, Sericulture, Veterinary Sciences, Forestry, Dairy Poultry Farming and Fisheries etc.

Concept of Agribusiness, Agriclinic and Self-Help Groups

Agribusiness centers are envisaged to provide input supply, farm equipment on hire and other services. In order to enhance viability of the venture, Agriculture graduates may also take up on Agriculture and allied areas along with Agriclinic and Agribusiness centres.

Agriclinics are envisaged to provide expert services and advice to farmers on cropping practices, technology dissemination, crop protection from pests and disease, market trends and prices of various crops in technical markets and also clinical services for animal health etc. which would enhance productivity of crops and animals. The Self-Help Group (SHG) concept helps members to develop both economic and social strengths. The collateral substitute positively influences the resources of the SHG, whereby members meet their contingent obligations without going to moneylenders or private sources. The banks financing them are convinced about the collective wisdom of the groups, helps in developing trust and confidence. Such-confidence building process involves a marginal period of 6-8 months.

Objectives

The objectives of Agribusiness centres are as follows :

- To provide gainful employment to agriculture graduates is new emerging areas in agricultural sector.
- To supplement the efforts of Government extension system, and
- To make supplementary sources of input supply and services to needy farmers by agri-professionals.

Eligibility and Activities

This Project is opened to Agriculture graduates in their professional subjects, allied to Agriculture activities like; Horticulture, Animal Husbandry, Forestry, Dairy Veterinary, Poultry Farming, Pisciculture and other allied activities are most eligible for this project.

An illustrative list is given below (including all the activities):

- Provision of Extension Consultancy Services.
- Soil and Water Quality cum input testing laboratories (with Atomic Absorption spectrophotometers.
- Hatcheries and production of fish finger-lings for aquaculture.
- Pest Surveillance, diagnostic and Control Services.
- Provision to livestock health cover, setting up veterinary fish dispensaries and services including frozen seems banks and liquid Nitrogen supply.
- Maintenance repairs and customs hiring of Agricultural implements and machinery including Micro irrigation system (Sprinkler and Drip).
- Setting up of information technology kiosks in rural areas for access to various Agricultural related portals.
- Agri Service Centres including three activities mentioned above (Group activity).
- Feed processing and testing units.
- Seed processing units.
- Value addition centres.
- Micro propagation through plant tissue culture labs hardening units.
- Setting up of vermiculture units, production of bio-pesticides, bio-control agents.

- Retail marketing outlets for processed Agri-products.
- Setting up of Apiaries (bee-keeping) Honey wax and Bee product processing units.
- Rural Marketing dealership of farm inputs and outputs, supply system, availability of customers.

Even, any combination of two or more of the above viable activities along with any other economically viable activity can be selected by the Agri graduates, which is most acceptable to the Bank.

Cost and Approach

The project can be taken up by Agrigraduates or allied professionals either individually or on cooperative basis. The outer ceiling for the cost of a project by individual would be Rs. 10 lakhs and for the Group would be Rs. 50 lakhs. The Group may normally consist of 5 members, of which one could be Management graduate with qualification or experience in business development and management. The rate of interest to be charged by the financing bank to the ultimate beneficiary depends upon the amounts of loan, types of bank as per the guidelines circulated to these Banks by NABARD and RBI from time to time. The period of loan will very between 5 years to 10 years depending on the activity. The schedule of repayment period may include a grace period, to be decided by the sponsoring Bank as per the individual project of a maximum of 2 years. The selection of borrower and location of the projects may be done by the Banks in consultation with Agricultural Colleges, Agrimanagement institutions, universities and Agriculture departments of the state etc. in their area of operations is necessary. All other terms and conditions, Banking procedure and leading norms which are normally applicable to project lending and systematic refinancing will also be applicable to financing of agribusiness centres.

Application Procedure

All the applications are invited by SFAC (Small Farmer's Agribusiness Consortium) by open advertisement in print and electronic media, giving all the project outlines as per the standards and norms decided by the Government and Banks.

Even Model project outlines of about 20 identified projects are made available by SFCA at the office, at district level Agricultural offices NABARD, LBO etc. Model project outlines helo facilitate the Agricultural graduates, who seek Assistance under this scheme, to identify the area in which they want to develop their venture. It also enables them to prepare their own Techno Economic Feasibility (TEF) reports for projecting to the Banks for availing finance.

Conclusion

In the main findings and suggestions, it may be suggested that the Farmers awareness must be created by way of imparting farmer's education in the rural areas and this responsibility can be given to the Agricultural graduates. At present, there is a need to supplement the efforts of Government, extension system by private partners to accelerate the process of technology transfer in Agriculture field, Specialised Agri Services (e.g. Agro processing, Soil and input testing, Seed, Artificial Insemination, Milk Processing Liquid Nitrogen Plant, Agricultural Insurance, Information Technology, Post Harvest Management etc.) also need supplementing as the present infrastructure for providing these services is infrastructure for providing these services is inadequate in the country. One of the important conclusions of this paper is that rural farmers purchase required goods from retail shops in the villages. Whenever, purchasing is done from weekly Bazaars, and Taluka places, rural farmers are subject cheating by way of similar packing, colour and size. In this regard, suggestion can be made that farmer's protection movement should be activated. The distribution should be made effective and efficient through Agribusiness Centres in rural areas so that none is deficient of essential goods. By doing all these things in a systematic way, the employment opportunities will increases and unemployed Agrigraduates will be benefited by operating these centres.

The emerging changes in the values and attitudes of the members of the SHGs are a clear manifestation of socio-economic empowerment interventions yielding relatively quicker results. The socio economic programmes reinforce each other and promote all-round development of the children, the women, the households and the communities. It is a process which ultimately leads to self-fulfillment of each member of the society. It is in this direction that SHGs are moving towards fulfilling their objectives with a meaningful strategic direction, the authors opine.

Andhra Pradesh has been in the forefront of the Self-Help Movement in India. Alongside women's own savings a major initiative in providing SHG members with bank credits was introduced in 1992. Apart from banks, some NGOs are also striving to empower the unorganized women to organize them financially so that the "investment-deprivation" may not be an impediment to conquer the "opportunity powerty". Powerty continues to be a serious problem in both rural and urban India and its intensity varies widely across the States to our country. One of the reasons for the perpetuation of poverty is that some sections of the society are excluded from growth and not necessarily due to lack of growth. Hence, the gender streamlining strategy aims at provision of benefits of growth proves to women by socio-economic background in terms of labour, income, credit and investment etc. 'Self Help Group' strategy also adopts the same philosophy and aims at bringing the excluded and neglected women into mainstream of economic development through "savings-investment-employment and income generation" strategy.

The effective organization of Self Help Groups (SHGs) is a significant instrument in the process of empowerment. The emerging changes in the values and attitudes of the members of the SHGs are a clear manifestation of socio-economic empowerment interventions yielding relatively quicker results. The socio-economic programmes reinforce each other and promote all round development of the children, the women, the households and the communities. It is a process which ultimately leads to self-fulfillment of each member of the society. It is in this direction that SHGs are moving towards fulfilling their objectives with a meaningful strategic direction.

A SHG is an informal association of 10-15 women, who have voluntarily come together for the business of saving and credit and to enhance the member's financial security as primary focus and other common interests of members such as area development, awareness, motivation, leadership, training and associating in other social intermediation programmes for the entire community.

Objectives

- To include the savings and banking habits among members.
- To secure them from financial, technical and moral strengths.
- To enable availing of loan for productive purposes.
- To gain economic prosperity through loan/credit and
- To gain from collective wisdom in organizing and managing their own finance and distributing the benefits among themselves.

It is seen from Table 1 that an amount of Rs. 4,24,76.405 was mobilized through savings and Rs. 4,52,82,930 is taken as loan amount from the PASS as micro finance through its 6 branches to take up various income generating activities.

Case Study

Provision for credit and generation of savings have long been recognized as an essential element in any developing country. Credit plays a crucial role in the modernization of agriculture, but in its role in the fight against rural poverty, the SHG strategy avoids credit system for rural unemployed women to ensure the best satisfaction in a credit programme. The membership in a group activity gives a member a feeling of cooperation and protection. SHGs as a strategy holds power and provides strength and acts as an antidote to the helpless poor. The group savings of SHGs serve a wide range of objectives other than immediate investment.

Keeping in view the role of SHGs in the development of women in rural areas, a micro-level fields study in Renigunta Mandal of Chitoor District was conducted during the month of December, 2005 by a personal visit to SHGs managed by RASS. For assessing the impact of SHGs on the status of women, Karkambadi village was selected for the fields study, which is at a distance of 2 kms from Renigunta mandal. Women in this village have successfully demonstrated how to mobilize and manage thrift, appraise credit needs, maintain linkage with the SHG and enforce financial discipline. It is observed that there are 169 SHG groups in Karakambadi village and 2020 women were enrolled as members with a savings of Rs. 57,64,830 with a core fund of Rs. 23,98,416. The total loans disbursed account for Rs. 2,68,97,116 from the inception of SHGs.

Methodology

Total 202 (01 per cent of the total women group members) group members from 17 SHG groups were randomly selected for this micro-level study. Care was taken to attach importance to their activity and literacy status. A well structured schedule was canvassed to them and interview method was adopted to record their opinions on the impact of SHGs on income and employment generation and appraise the performance of SHGs on alleviating rural poverty. The information collected refers to a period of 12 months, *i.e.*, form January 2005 to December, 2005.

Demographic and Social Features

It is customary to present the demographic and social background of the women group members as they exert tremendous influence of availing credit and its utilization.

Age Profile: In general, the unemployed women belong to the age group of 20 to 50.

The details of age-profile show that 67.3% of the selected women members belonged to the age group 26-40 years and 11.9% of them to the age group upto 25 years. This distribution reveals that 79% of the beneficiaries belong to less than 40 years of age, who represent the economically active segment of population.

Social Status

The distribution of women-beneficiaries according to their social groups reveals that about 53.4% of the members belonged to weaker sections, i.e., SCs, STs and Backward Caste in rural areas.

Literacy Status

Generally, the literacy rate among rural women is very low, which influences the use of credit for economic betterment of women.

It seems that the SHGs are giving equal importance to illiterate and literate women. The data shows that 35.6% are illiterate members and 64.4% are literate women members.

Family Size

Employment of credit in any productive activity and its beneficial effects depend upon the family size and dependency load.

The average size of the women member's family is 5 members consisting of 2 children and 3 adult persons as an average. It seems that the members of SHGs are having an ideal of family.

Land Holdings

No doubt the rural poor mostly belong to assetless groups or have a marginal value of assets. An attempt is made in table 2 to explain the land ownship of sampled women, as the primary asset in rural areas is agricultural land.

The average size of the land holding is 0.81 acres, which shows that all the members are from marginal farmers category.

Loans Given

SHGs provide loans to their members for using loans as investment and to have access to income generating opportunities. The total amount of loan provided to SHG members is Rs. 20,12,099/-, to take up various economic activities. The average loan provided to each member works out to Rs. 9,960,09/-. 66 members (33%) received loans less than Rs. 5000, 75 members (37%) received loans between Rs. 10001-20000 and 8 members (4%) got loans between Rs. 20001-40,000 from the SHGs.

Table 1 : Distribution of Operational Holdings

Ownership of Land	*0-2.5 acress*		*2.5-500 acress*		*Total*		*Average in Landholding*
	No. of H.Hs.	*Land*	*No. of H.Hs*	*Land*	*No. of H.Hs*	*Land*	
Owned Land	155	121.24	–	–	155	121.24	0.60
Leased-out	–	–	–	–	–	–	–
Leased-in	47	42.20	–	–	47	42.20	0.21
Total Operational	202	163.44	–	–	202	163.44	0.81

Source : Primary Data

Purpose of Loan

In the field work, it was noticed that the members of the group took loans for different purposes. Since Karakambadi is an agro-dominant rural area, majority of the respondents availed of loans for raising crops and also for some non-farm activities as shown in Table 2.

The data in Table 2 shows that a total amount of Rs. 20,12,099 was provided to the sampled respondents as loan and majority of them received if for crop husbandry. They also received loan for allied activities like dairying and flower vending. It was found that the members also received loans for non-farm activities like tailoring, idly shop and cloth business.

Table 2 : Distribution of Loan - Purpose-wise

S.No.	*Purpose*	*No. of Households*	*Total Amount of Loan Received*	*Average Loan Amount per Household*
1.	Agriculture	79	9,40,649	11,907
2.	Dairying	72	7,02,000	9750
3.	Tailoring	14	98,000	7000
4.	Flower Vending	15	90,000	6000
5.	Snacks Idly Shop	02	11,450	5,725
6.	Cloth Business	20	1,70,000	8500
	Total	**202**	**20,12,099**	**9,960.09**

Source: Primary Data.

Employment Generation

No doubt any financial assistance, if utilized property, generates gainful employment opportunities in the rural economy. It was observed in the field survey that the sampled members also got gainful employment opportunities as shown in Table 3.

Table 3 : Impact on Employment Generation

(In person days)

S.No.	*Financing Activity*	*No. of Household*	*Total Employment Generated*	*Average Employment Generated*
1.	Agriculture	79	17,222	218
2.	Dairying	72	72,00	100
3.	Tailoring	14	3,360	240
4.	Flower Vending	15	2,700	180
5.	Idly Shop	02	600	300
6.	Cloth Business	20	6,000	300
	Total	**202**	**37,082**	**184**

Source: Primary Data

The Table 3 shows that on an average, the loans received generated 184 person days of employment per household. In all, 37,082 person days of employment was generated for 202 selected members. It was noted that non-farm activities generated higher no. of person days of employment in the sample village. Idly shop, cloth business and tailoring generated 300 each and 240 person days of employment. On the contrary, agriculture could generate 218 person days of employment on an average per household followed by 180 person days of employment by flower vending and 100 person days by dairying.

A look at this data makes us infer that the loans provided by SHGs are productive and efficient in generation of employment to rural farm and non-farm workers in general.

Generation of Income

There is a symbolic relationship between generation of income and employment opportunities and the potential of employment can be judged by the amount of income generated in any activity. Table 4 makes an attempt to explain the impact of loans provided on generation of income.

Table 4 : Impact on Generation of Income

S. No.	*Financing Activity*	*No. of Households*	*Total Income Generated (in Rs.)*	*Average Income Generated (in Rs.)*
1.	Agriculture	79	20,96,739	26541
2.	Dairying	72	11,66,400	16200
3.	Tailoring	14	2,01,600	14400
4.	Flower Vending	15	2,70,000	18000
5.	Idly Shop	02	10,000	5000
6.	Cloth Business	20	2,10,000	10500
	Total	**202**	**39,54,739**	**19,578**

Source: Primary Data

It seems the loans provided by SHGs had a favourable impact on generation of income in the village selected. On an average, each selected family could get an income Rs. 19,578, which is sufficient to bring the poor families above the poverty line. No doubt, the income generation caries from activity to activity and each activity has its own generate income. The data also reveals this fact. Income generated in the selected activities shows that it varies from Rs. 5000 per annum in the case of *idly* shop to Rs. 26541 in the case of agriculture. Highest amount of income generation is seen from agriculture. Flower-vending proved an efficient activity as it generated an average income of Rs. 18000 per household, followed by Rs. 16200 in dairying, and Rs. 14400 in the case of tailoring. The women members engaged in cloth business could receive an average income of Rs. 10500 each in the reference period.

Use of Income Generated

Generally it is assumed that whenever people are able to receive sufficient income, they usually invest it productively towards the improvement of the quality of their lives. It is true that income has a favourable effect on consumption expenditure in general and on education and health in particular. The opinions recorded from sampled beneficiaries also prove true in rural areas. Table 5 illustrates this favourable qualitative shift.

Table 5 : Use of Income Generated

S. No.	*Use of Income*	*No. of Households Reported*	*% to Total*
1.	Reinvested in the activity (agriculture)	79	39.11
2.	Spent on Household expenditure	23	11.38
3.	Spent of Education	41	20.30
4.	Spent of Health Care	32	15.84
5.	Spent on purchase of income yielding assets	27	13.37
	Total	**202**	**100.00**

Source: Primary Data

The opinions of sample respondents revealed that they productively made use of the income generated after receiving the loans. 39.11% of the respondents reinvested their income on agriculture, 20.30% of them revealed that a part of the income generated was utilized for educating their children, and 15.84% of them spent if on health care. 11.38 of the respondents told that income generated was spent for meeting the household expenditure and 13.37% reported that they have spent it on purchase of productive assets for them.

In general, the field experiences reveal that the micro-finance provided by SHGs is productive enough and had a favourable effect on employment and income generation. It is also observed that the credit extended to rural women also had a quality-improving effect on the families of sample respondents, because majority of the women beneficiaries utilized the income generated either for investing or improving the educational and health requirements. These expenditures, as we know, resulted in qualitative improvement of human resources.

Observations and Findings

The functional analysis of the study of SHGs undertaken in Karkambadi village reveals the following findings:

1. The number of SHGs is substantially increasing in Karkambadi village. These groups are mobilizing thrift deposits and receiving timely matching and revolving funds to generate employment activities to earn their livelihood.

2. The rural women have successfully demonstrated how to mobilize and manage tariff, appraise credit needs and enforce financial discipline.
3. This micro-level study of SHGs reveals that these groups generated awareness among rural women about Government development programmes.
4. The social outlook of the women has undergone a beneficial change and some degree of transformation of social outlook is found. On Social development, the women need further exposure. The changes that have occurred between "before" and "after" stages, are encouraging, but not adequate.
5. There was a sense of equality of status of women as participants, decision makers and beneficiaries in the democratic, economic, social spheres of life and sensitized the women members to take active part in socio-economic progress of rural areas.
6. The rate illiteracy can be further reduced through existing programmes. Formal education with focus on critical issues, needed for functional literacy should be imparted to the women groups so that they can manage their group affairs independently.
7. Periodical training at regulae intervals to group members of self-management issues is to be necessarily imparted with the help of experienced resource persons.
8. Change of leadership is a must for sharing the responsibilities by all members and generate leadership qualities in each member.
9. Anti-child labour measures should be made an integral part of these Self-Help efforts.

NATIONAL RURAL EMPLOYMENT GUARANTEE ACT

'Highlights'

Works and Their Execution

Permissible Works

The intention of the National Rural Employment Guarantee Act (NREGA) is to provide a basic employment guarantee in rural areas. The Act indicates the kings of works that may be taken up for this purpose. As per Schedule I of Act the focus of the Rural Employment Guarantee Scheme (REGS) shall be on the following works :

(*i*) Water conservation and water harvesting;

(*ii*) Drought proofing including afforestation and tree plantation;

(*iii*) Irrigation canals, including micro and minor irrigation works;

(*iv*) Provision of irrigation facility to land owned by households belonging to the SC/ST, or to land of the beneficiaries of land reforms, or to land of the beneficiaries under the Indira Awas Yojana;

(*v*) Renovation of traditional water bodies, including de-silting of tanks;

(*vi*) Land development;

(*vii*) Food-control and protection works, including drainage in waterlogged Areas;

(*viii*) Rural connectivity to provide all-weather access. The construction of roads may include culverts where necessary, and within the village area may be taken up along with drains;

(*ix*) Any other work that may be notified by the Central Government in consultation with the State Government.

Basic Implementation Principles

(*a*) Collaborative Partnership and Public Accountability

(*b*) Community Participation
(*c*) Role of Panchayats
(*d*) District Programme Coordinator and Programme Officer
(*e*) Coordination among Agencies
(*f*) Resource Support

Key Agencies and Their Respective Roles

Village Level

(*a*) Gram Sabha (GS)
(*b*) Gram Panchayat (GP)

Block Level

(*a*) Intermediate Panchayat (IP)
(*b*) Programme Officer (PO)

District Level

(*a*) District Panchayats
(*b*) District Programme Coordinator (DPC)
(*c*) Implementing Agencies
(*d*) Delegation of Powers

State Level

(*a*) State Employment Guarantee Council (SEGC)
(b) The State Government
(*c*) Employment Guarantee Commissioner

Central Level

(a) Central Employment Guarantee Council (CEGC)
(*b*) Ministry of Rural Development (MORD)

Funding

Central Government

- The entire cost of wages for unskilled manual workers.
- 75 per cent of the cost of material and wages for skilled and semi-skilled workers.
- Administrative expenses as may be determined by the Central Government. These will include, *inter alia,* the salary and allowances of Programme Officers and their support staff and work site facilities.
- Administrative expenses of the Central Employment Guarantee Council.

State Government

- 25 per cent of the cost of material and wages for skilled and semi-skilled workers.
- Unemployment allowance payable in case the State Government cannot provide wage employment within 15 days of application.
- Administrative expenses of the State Employment Guarantee Council.

PAYMENT OF WAGES AND UNEMPLOYMENT ALLOWANCE

Payment of Wages

Every person working under the Scheme shall be entitled to wages at the minimum wage rate fished by the State Government.

Equal wages shall be paid to both men and women workers.

It is recommended that wages should be paid on a weekly basis on pre-specified day of the week in each Gram Panchayat.

If workers are willing, a proportion of the wages may be earmarked and contributed to welfare schemes organized for the benefit of REGS workers such as health insurance, accident insurance, survivor benefits, maternity benefits and social security arrangements.

Unemployment Allowance

If a worker who has applied for work under NREGA is not provided employment within 15 days from the date on which work is requested, an unemployment allowance shall be payable by the State Government at the rate prescribed in the Act.

The payment of unemployment allowance shall be made no later than 15 days form the date on which it becomes due for payment.

Planning

Planning is critical to the successful implementation of the Rural Employment Guarantee Scheme (REGS). A key indicator of success is the timely generation of employment within 15 days while ensuring that the design and selection of works are such that good quality assets are developed. The need to act within a time limit necessitates advance planning. The basic aim of the planning process is to ensure that the District is prepared well in advance to offer productive employment on demand.

Chapter 11

Microfinance — Emerging Paradigms

India has about 24.5 million ha of wasteland and 16.6 million ha of fallow land. Sizeable part of these could be converted into cultivable land through appropriate crop selection, improved water-use efficiently adoption of watershed approach and development of irrigation potential. However, this requires substantial investments in the form of extending financial support to farmers in the form of cheap credit and subsidies.

However even today, the rural credit markets are dominated by moneylenders and some instances of bonded labour have been reported from some parts of the country. Farmers are often benefit of much needed funds for the adoption of modern agricultural practices, tools, fertilizers, insecticides etc. It is to address this short fall that the Union Budget (2004-05) has stressed on the need for doubling agricultural credit to within the next three years.

Over the past few years, **Micro-Finance,** through the formation of *Self Help Groups*—SHGs, has proved to be an effective channel for disbursement of credit in rural areas.

Factors Hampering Extension of Rural Credit

1. *High Transaction Costs* : The amount saved or required by people in rural areas is often miniscule and providing it through conventional banking system has not proved to be cost effective.
2. *Collateral Based Lending* : Conventional banking requires that some assets be pledged with the lender as a security, to ensure that the loan would be repaid. In the event of a default the lender has the right to amortise the asset to recover his losses. People in rural areas hardly own any assets (land, jewellery) and as such are left stranded by the conventional banking system.
3. *Availability of Credit in Remote Areas* : It is not possible to have a bank branch in every nook and corner of the country. As a result many areas, especially in the interiors are left without banking services.

Banks also provide loans for existing traditions for example need for food, medicine, social occasions, such as marriages and religious ceremonies. The conventional banking system does not recognise it and provide loans.

Here, ICICI Bank creating windows of opportunity–describes the Private Sector Investment in Rural Marketing Infrastructure. These points were discussed in Agriculture Summit 2006 at Vigyan Bhawan New Delhi 18 Oct. 2006.

AGRICULTURE MARKETING INFRASTRUCTURE : STATUS

Inefficiencies in the Present set up

- **Imperfect Markets**
  - ❑ Multiple Intermediaries (low realization for farmers) & high losses

- ❑ Underdeveloped Rural markets
  - – 85% of 27.294 rural periodic markets lack facilities for efficient trade
  - – 7.161 market yards/sub yards at primary level are ill equipped
- **Poor handling at farm gate & village level**
  - ❑ 7% of grains, 30% of Fruits & Vegetables and 10% of seed species go waste p.a.
- **Negligible value addition leading to low returns/high wastages**
  - ❑ Value addition to food production only 7%
  - ❑ Less than 2% of fruits and vegetables production is processed
  - ❑ Only 25% of produced food-grains utilize scientific storage.

An estimated Rs. 500.00 bn lost/go waste in the marketing chain.

NEED FOR AN EFFICIENT MARKETING SYSTEM

Efficient Marketing System

Optimizes resources;

Supports Value addition;

Provides right infrastructure;

Maximizes farm incomes;

Mitigates risks;

Increases employment;

Marketing Institutions

- Individual farmers
- Marketing Agencies
- Cooperatives
- Contract farming
- Processors
- Middlemen
- Agribusiness centers
- Agri Export Zones

Marketing Infrastructure

- Grading centers
- Warehouses
- Cold storages
- Processing facilities
- Transport facility
- Reefer vans
- Market yard

Marketing Services

- Quality labs

- Information kiosks
- Credit supply
- Weather forecast
- Extension machinery
- R & D/Tech back up

Intervention in marketing infrastructure can raise farm income* by 20– 30%

Necessitates Huge Investments in Agriculture Chain

Items	*Physical nos.*	*Xth Plan*	*XIth Plan*	*Rs. in Bn. Total*
Wholesale markets	3213	24.10	40.1	4.26
Modernization of markets	7293	22.60	37.66	60.26
Rural Periodic Markets	27294	10.24	17.06	27.30
Grading Centers	5786	4.00	7.50	11.50
Modernizing Labs	100	0.10	0.10	0.20
Agribusiness Centers	1500	0.50	1.00	10.50
Warehousing (mm tons)	51.8	59.65	62.00	121.65
Total		121.19	165.48	286.67

...with a dominating private sector investment

Items	*Public*	*Private*	*Total*
Rural Roads	740.00	—	740.00
Market Yards Development	60.00	—	60.00
Fruits & vegetables Markets	10.00	—	10.00
Rural Periodic Markets	21.00	—	21.00
Cleaning & grading - villages	19.00	1.00	20.00
Storage	27.00	27.00	54.00
Cold storage	68.00	202.00	270.00
Reefer vans	1.00	5.00	6.00
Agri Export Zones	2.00	4.00	6.00
Processing & Value adding	375.00	1125.00	1500.00
Total	1323.00	1364.00	2687.00

...but currently private sector investment less than optimum.

FACTORS IMPEDING PRIVATE INVESTMENT

Policy Related

- Complex regulatory framework/multiplicity of laws.
- Monopoly of State to Set up markets.
- Stringent control on storage & movement of agri-commodities.
- Cascading effect of multiple taxes from harvesting to marketing.

Others

- Dominance of unorganized sector.
- Lack of basic rural infrastructure like roads, transport, power etc.
- Lack of linkages between spot and future markets.
- Non transparent price setting.
- Lack of financial resources.

ICICI : FACILITATING EFFICIENT MARKETING SYSTEM

Through Creation of New Institutions

- Facilitating linkages (Sandhi)
- NCDEX/NCMSL Platform

Through Creating Local Optimas

Through Funding of Agriculture Marketing Infrastructure

- Rural Roads-Mandi Cess MPRRDA
- Storage Infrastructure

Through Funding/Creating Marketing Services

- Commodity based funding for farmers

CREATING MORE OPPORTUNITIES

PPP Models

- ❑ Shared infrastructure approach
- Allow private parties to use public infrastructure like extension machinery on charge.
- Management/utilization of government infra like markets, ware-houses by private sector.
- Part of public investment be outsourced to private entities with mega rural plants.
- Innovative funding mechanisms.
- Allow access of RIDF funds ot private parties for agriculture marketing infrastructure projects.
- Spare capacities of private parties be used for purposes such as storage, processing.
- Stipulations like 10% mandatory investment in wastelands in lieu of fiscal concessions/ contract farming lease rights : TDR kind of rights.

Facilitative Support of Government

- For large scale Agriculture Infrastructure projects – eg. Roads, macro irrigation, cargo facility.
- Least subsidy/viability gap funding.
- Strategic equity stake in large agriculture infrastructure projects.
- Bunding of assets & services for agriculture infrastructure projects.
- Allocate rights on non project revenue generating activities *e.g.* rights for real estate development near rural roads projects.

- For medium/small scale Agriculture Infrastructure projects – *e.g.* micro irrigation, private mandis.
- Partial Risk Guarantees/Agriculture risk/Innovation fund.
- First Loss Deficiency Guarantee (FLDG) based model.
- Providers of mezzanine risk.

REFORMS & ENABLERS – AGRICULTURE MARKETING

Regulatory Factors

- Amendment of APMC act.
- Amendments of ECA.
- Removal of excessive regulations on private trading.
- Facilitating Contract farming through model act.
- Warehouse receipts as negotiable instrument.
- Replace multiple laws with Unified food law.

Enabling Factors

- Allow Commodity Hedging for Banks.
- Agri infra funding as direct agri.
- Utilize mandi revenues for creating marketing infra.
- Promote private markets like Ryathu bazaarm Apni mandis.

For a strong agriculture economy with farmer focus.

BUILDING LIVELIHOOD FOR PRODUCERS : SANDHI

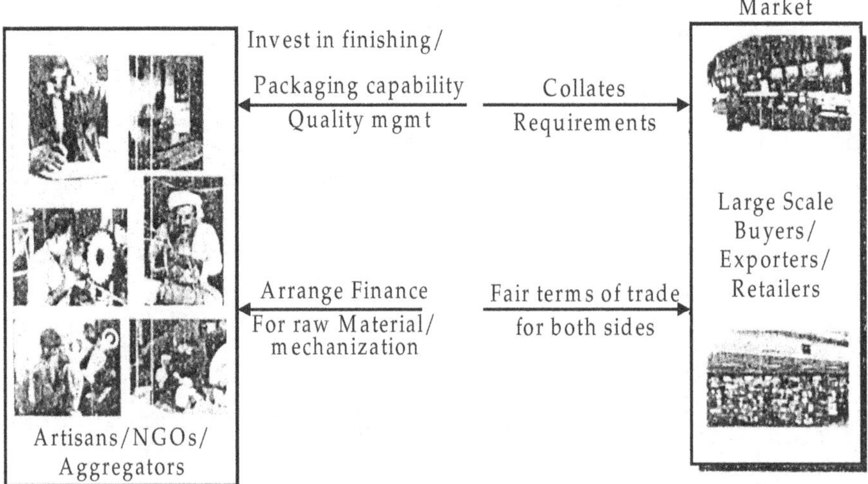

LEVERAGING OUR CORPORATE LINKAGES

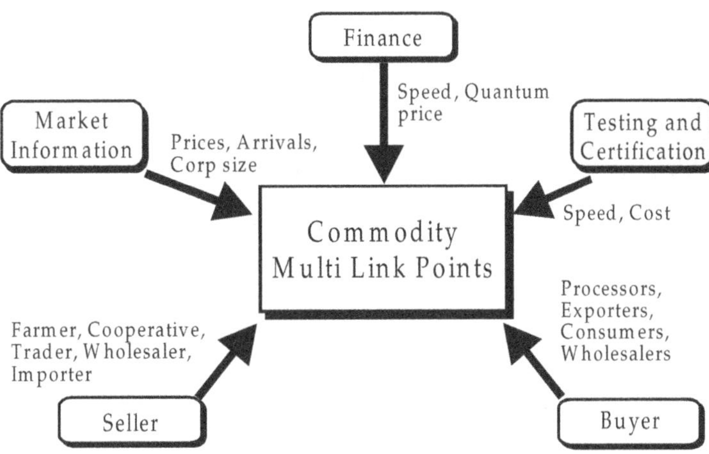

For Creation of accredited Warehouses performing more than mere storage function

INTERMEDIATION AT KEY TRANSACTION POINTS

- **Input Infrastructure :** Entire Supply chain intervention may not be desired by farmers.
- **Production & Harvesting :** Intervening in the entire supply chain not a feasible option in all locations/all crops.
- **Processing Infrastructure :** Real unlocking of values possible through professional intermediation at key transaction points in the value chain.
- **Distribution, Retailing & Export :** Ensures higher output value with greater sharing of benefits by the farmers (he pays only for what he desires with no common services cost being passed on).

Multiple local optimas may achieve more than generic global optima. *i.e.,*

INTERMEDIATION FOR PRICE DISCOVERY : SAFAL

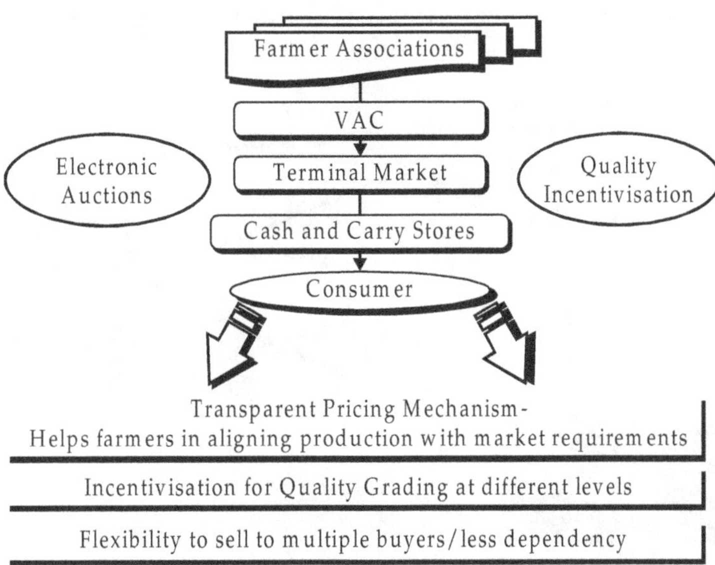

RURAL ROADS – LEVERAGING MANDI CESS

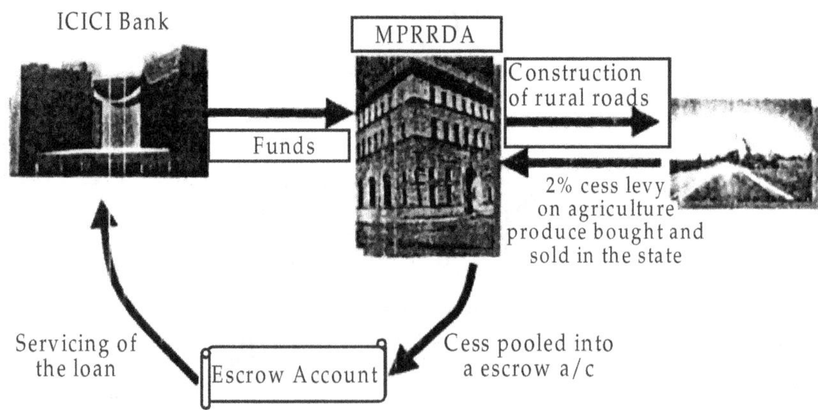

STORAGE INFRASTRUCTURE

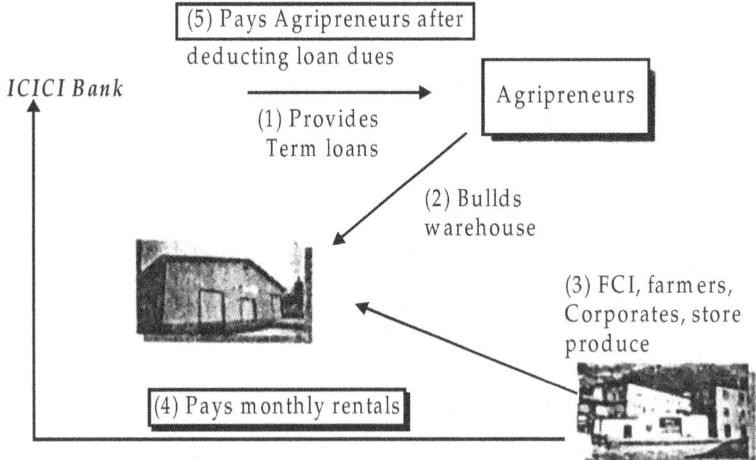

STRUCTURED FINANCING OF RURAL WAREHOUSES

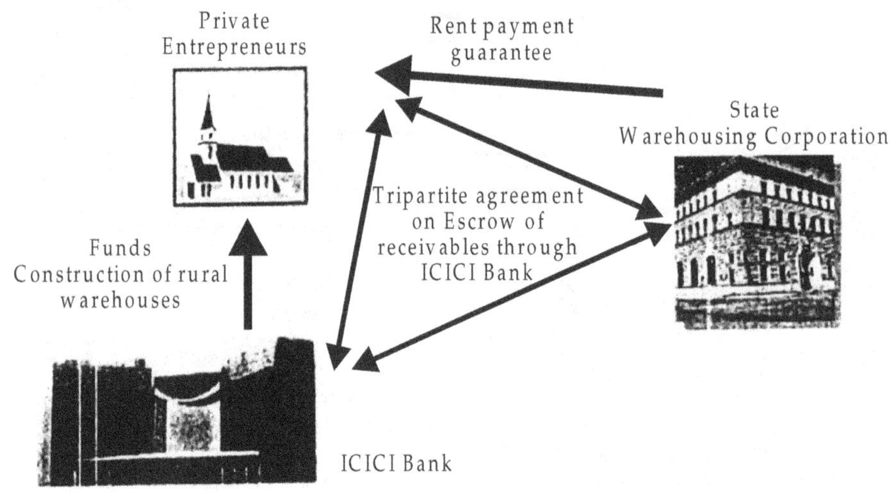

COMMODITY BASED FUNDING

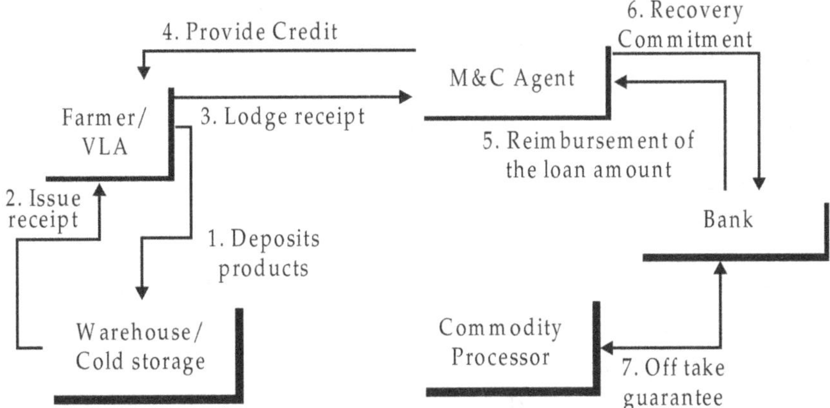

Commodities : Chilli, wheat, cumin, aniseed, mustard, cotton, groundnut, seeds, sunflower seeds

SUGGESTIONS FOR REFORMS – AGRICULTURE MARKETING

Amendment of APMC Act

- For facilitating purchase outside mandis : promote private investments.
- NCDEX NCMSL infrastructure to be declared as Deemed mandis.
- Permission to open private markets *e.g.,* Safal.

Market Efficiency

- Interlinking of mandis.
- Popularize www.agmarknet.nic.in and update information regularly.
- Private information kiosks set up with clear deliverables.

Reduction in post harvest losses/Logistics

- Aggregators could set up common infra facilities – collection centers, warehouses, primary processing units, transport facilities : SEWA.
- Price advantage through pooling.

SUGGESTIONS FOR REFORMS – AGRICULTURE MARKETING

Warehouse Receipts

- Negotiability/transferability issues.
- Needs to be treated as negotiable instrument.
- Private warehouse receipts not funded due to credibility and lack of appropriate systems – NCDEX accredited warehouses to be promoted.
- Warehouse Receipts based advance needs to be delinked from crop loan.

SUGGESTIONS FOR REFORMS – AGRICULTURE MARKETING

Infrastructure

- Allow infrastructure funding as direct agriculture to boost agri investment.
  - ❑ Rural Roads, Rural electrification.

- ❑ Irrigation (micro/macro).
- ❑ Post harvest infrastructure, Food processing infrastructure.

- RIDF funds could be deployed to fund state wise strategic options study for creation of basic infrastructure like roads – funding to be linked to the same.
- Integrated air/cargo facilities suited to the movement of agri commodities to be set up in PPP.
- Provision of basic infrastructure facilities through public private partnership model.
- Last Support Model.

ENABLERS FOR GROWTH – AGRICULTURE MARKETING

- NCDEX could place view terminals across the country for price discovery.
- Address connectivity issue.
- Subliksha kind of models could work for rural areas.
- Rural entrepreneurs must be encouraged to set up such centers.
- Allow FDI in Retailing.
- Rural Hyper markets on lines of Spandana could be set up across the country.
- Special support could be given for infrastructure creation by MFI/NGO/ Aggregators.

Today our farmers cannot afford to ignore the global market. There is a huge global opportunity that awaits us provided we are willing to change the way we look at farming. Our farmers will have to look outside the price support system that creates supply driven cultivation and switch to crops that deliver better returns in the global market. But the farmers in unlikely to get all this knowledge of what sells in the global market all by himself. And for this he needs to partner with the corporate sector and public sector agencies that can tell him what varieties to grow, how to grow them and then preserve them or process them so that it is delivered in the form that the global consumer wants.

The private sector and the public sector including research agencies such as ICAR and farm universities that can deliver global competitiveness to our farm sector and in the process raise productivity and incomes of our farmers. But we need reforms to make these partnerships work and remove all kinds of roadblocks that hinder them whether it is the APMC act or lack of a cold chain or warehousing systems or processing facilities that add value to crops to create what consumers want.

To capture global markets :

- ❑ We need a stable exim policy on farm products. Sometimes we allow exports and sometimes we ban them. This ad hocism drives away buyers as India cannot be trusted as a reliable supplier. We also need stability in our approach to imports and here allow great play for market forces subject to adequate tarrif protection for our farmers.
- ❑ We must create adequate infrastructure to support farm trade, that is both imports and exports. This would include the cold chain at ports and airports, apart form increased handling capacity at ports. I would urge both our ministers to support our demand of 100 per depreciation benefit on cold chain investments by the private sector to accelerate investments in this area.
- ❑ We should accelerate the creation of crop and agro-climatic zone-specific export zones and processing zones.

Friends I must complement The Minister of Food processing Mr. Sahay for the proactive manner in which he is promoting the cause of value addition through processing. But we need his support for uniformity of VAT rules on food processing industries. We also need his support in faster implementation of the integrated modern food processing law that is in parliament.

We also need the support of both our ministers present today to promote FDI in food retailing as we need global knowledge of how to create efficient supply chains that link the farmer to the consumer both at home and abroad. Large foreign players in food retailing would enable our farmers to reap higher returns for their produce through higher exports.

India has the potential to become the world's food basket. We have all the agro-climatic zones and can grow virtually every crop. All we need is the right knowledge, a facilitatory framework for taking these crops to the market and the creation of the infrastructure that supports quick and efficient movement in this produce. As we stand on the threshold of change, there is a lot more that is to be done.

The money depositedly in SHG groups in utilized for existing traditions for *e.g.*, need for food, medicine, social occasions such as marriages and religious ceremonies etc. The conventional banking system doesn't recognize and provide for such needs. This would result in diversion of funds availed for productive activities to meeting contingencies such as sickness or marriages etc; as a result the desired activity, would remain underfinanced and fail. These are the micro finances which can be used.

What is a Self Help Group?

A Self Help Group can be defined as "*.... a small economically homogenous and affinity group for rural poor voluntarily coming together ...*" The salient features of a SHG are as follows :

1. Group members save small amounts regularly and voluntarily agree to contribute to a common group fund.
2. Within the group they have simple and responsive rules.
3. Decision-making is collective and is based on consensus. Conflicts, if any are resolved democratically.
4. Loans are provided for production as well as consumption purposes.
5. Loans are provided without collateral and interest rates charged are comparable to those of the market *i.e.*, banks, cooperatives etc. However, these interest rates are lower than those charged by the moneylender.

Operations within a SHG

Membership of a SHG smoothens an individual's conduct and makes him more reliable both as a borrower and as a saver. It is also felt that the approach for poverty alleviation should be self-help rather than support, hence the term *Self Help Group.* Usually meetings are held at weekly intervals along with the bank/NGO and the amount saved is deposited with the bank/NGO, in the name of the group. Although the account is operated by the office-bearers of the group, all the operations *viz.*, savings and loans are carried out in the name of the group. The bank/NGO in this case is called *Self Help Group Promoting Institution* — SHPI. The objective of the regular meetings and savings is to *train* the group. This is usually carried out for a pre-determined period *e.g.*, 6 months or a year. During this period the group is also required to frame its buy-laws etc.

The loan amount and the accumulated group savings are used for on-lending within the group. The interest rates charged and other terms and conditions, which are applicable for on lending within the group, are left to the discretion of the group. The formulation of terms and conditions, decision regarding interest rates, deposition of savings and decision regarding the grant of loan and its actual disbursement are carried out at group meetings. This highlights the democratic nature of group operations. The group is also empowered to fine errant members and charged penal rates of interest. In India, the group size usually varies between 10 and 20. The groups may be exclusively male, female or mixed.

Above all the uniqueness of the group lies in the fact that it can speedily disburse loans for both production as well as consumption needs. Thus it is an exclusive instrument designed specially to meet the credit needs of the rural people.

Last but not the least, the formation of SHG leads to consolidation of accounts. Instead of maintaining accounts of all group members, only a single account of the group needs to be operated. Effectively the banking services *viz.* savings, collecting and paying interest on deposits, loan appraisal and lending are taken up by the group. The SHG thus starts functioning as a micro banking unit.

Dynamics of a SHG

When a SHG is formed, care is taken to ensure that it is as *homogenous* as possible *i.e.,* group members have the same socio-economic background, profession, age, region, level of aspirations etc. This minimizes the chances of conflict later. The ensuing synergy contributes substantially to the success of the group. Further because all the members are well acquainted with each other and because they all possess the knowledge for an activity under progress, they can effectively monitor each other (because of their homogeneity and closeness). Thus diversion of funds or any other attempt at cheating becomes next to impossible. Since the SHG concept already provides for consumption loans to take care of emergencies as and when they arise, the causative pressure for diversion of funds is also removed. Last but not the least, loans are provided on the basis of trust. The members are genuinely grateful for being given a chance to improve their lot and try to make the most of it.

All the members are collectively responsible to the bank/NGO for the repayment of the loan. The entire loan amount can be only paid off from the group. Thus an members exercise vigilance over the loanees and ensure that they make the repayments on time. This vigilance or compulsion is called peer–*pressure* or *group pressure.* It has proven itself to be so effective that it is regarded as a substitute for collateral.

As long as the group functions effectively and loan repayments are on time, it enjoys absolute freedom with regard to the choice of group activity, internal byelaws, admission of new members (subject to a maximum of 20) etc. The activities financed by the group may be individual or team based. The bank/NGO can not coerce the group to take up a particular activity. This is because it is stressed that members are themselves aware of their capabilities and local conditions. Also any sort of external intervention would go against the purpose of making people self reliant. However, the group cannot lend to anyone outside the group *i.e.,* all the lending has to be strictly *intra-group lending.* Otherwise the SHG will itself become a moneylender! The emphasis on intra-group lending is valid because the group will not be able to exercise peer pressure and monitor a non-member.

Formats for Lending to SHGs

Model–1 : Here the bank itself takes the initiative in the formation of SHG. Once the SHG is formed and starts functioning properly it can be given loans for enhancing the group corpus as per the existing schemes. The corpus can be used for on lending among the members. The group has to meet regularly with the bank staff and show continuous and profitable activity. The bank on the basis of the accumulated savings provides the credit. Thus the savings act as partial collateral in this case. This model is being actively followed in many parts of Uttar Pradesh and Uttaranchal.

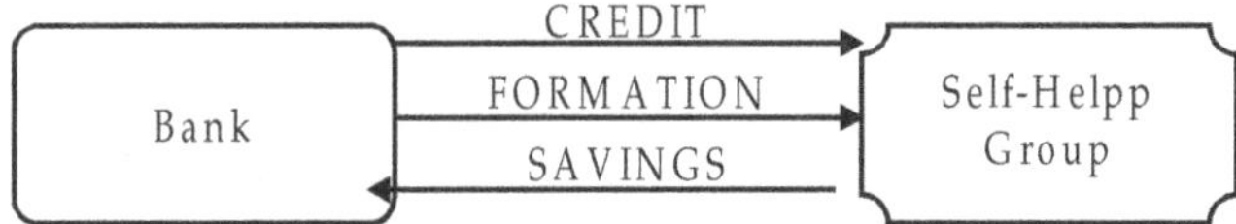

Model–II : In Model II the groups are formed by a NGO/SHPI. Other than this it is similar to Model I. The savings of the group may be kept by the bank as partial collateral and the credit for the group corpus is linked to the amount of savings. The SHPI may include farmers' clubs (Aligarh Model), *Vikas Volunteer Vahini* etc. Instead of the bank, the SHPI bears the responsibility of the group meeting and saving regularly.

As in the case of Aligarh Model, farmers themselves may help in the formation of new SHGs after realizing the benefits therein. In such cases the favourable experiences of farmers act as *positive word of mouth publicity.* Many NGOs also actively participate in the formation of SHGs, out of a spirit to serve the society and the nation. It is infact the missionary zeal of NGOs in south India that has led to the huge success of micro credit programs.

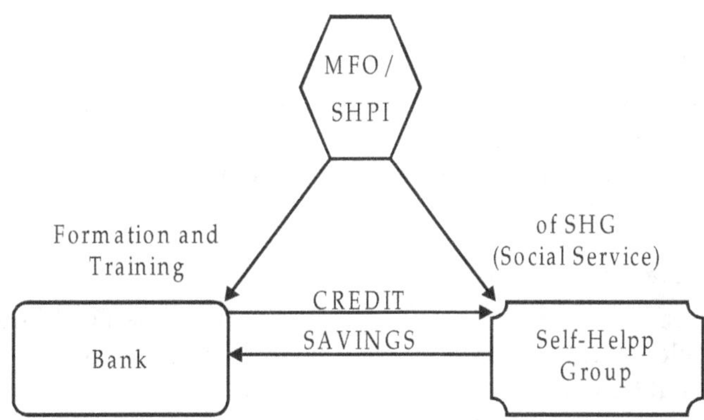

Model–III : In this case, the NGO acts as a financial intermediary. It forms and trains the SHGs. It also lends to them on behalf of the bank. The NGO is supplied credit at rates, which are cheaper compared to the rates at which the banks lend directly to the SHG under Models I & II (about 1.5% to 2% cheaper). However, the NGO itself lends to the SHG at the same rates as the banks do as in Model I & II. In other words, the SHG gets the funds at the same rate irrespective of the model in as their full time occupation. It is important role. In model II the motive is philanthropy but in model III the motive is strictly business. This is thew reason why such NGOs are also referred to as *Micro Finance Organizations.* In a way MFOs act like commission agents working for the bank. Since the difference in interest rate is fixed, the amount of money earned by the MFO varies directly with the number of SHGs formed and the loans disbursed to them. Thus in order to achieve break-even the MFO has to lend out a minimum amount of money. This break even point essentially depends on the operating costs of the MFO and other factors. The break even point could vary depending on various local factors such as transport linkages available, level of development in the area, quality of staff employed and general economic condition of the area.

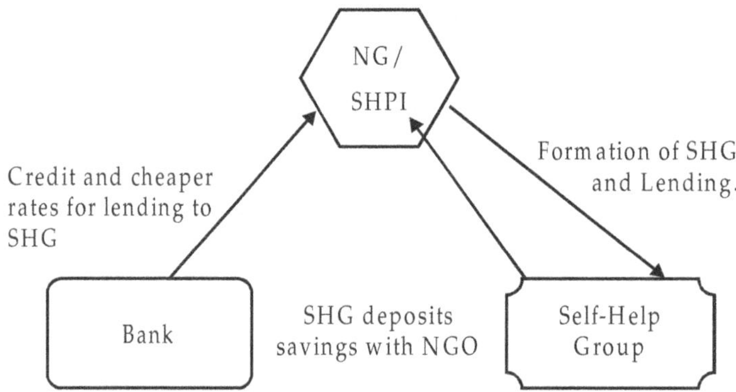

The Need for Insurance Services

SEWA or the Self-Employed Women's Association was founded in Ahmedabad by Elaben R Bhatt in 1972. It too offers micro-financial services.

In Andhra Pradesh, BASIX is carrying out exemplary work in extending rural credit. *Bhartiya Samruddhi Investments and Consulting Services Ltd.* (BASICS Ltd.), is the holding company, through which equity and debt investments are made in the group companies. *Bhartiya Samruddhi Finance Ltd. (Samruddhi),* a RBI registered NBFC, provides micro credit, retail insurance and technical assistance to borrowers. BASIX too has realized the importance of insurance in protection the people from falling into debt traps.

Providing Insurance services in rural areas is important because a single *economic stress event i.e.,* natural calamity or sickness & death can wipe out all the assets of a poor family and force it to turn to the moneylenders. Therefore, lending must be coupled with insurance service in rural areas and agricultural lending.

Initiatives of the Private Sector in Microfinance

BASIX has also secured loans worth Rs. 26.3 crores from the Ford Foundation – USA, the Swiss Agency for Development and Cooperation–USA, Shorebank Corporation–USA, Cordaids–The Netherlands, and DID Canada. In India BASIX has been lent Rs. 12 crores from Global Trust Bank Ltd., SIDBI, ICICI bank and the HDFC. This can be considered to be a **working example of Model III of bank – SHG linkage** format.

Another such tie-up was completed between ICICI Bank and Micro credit Foundation of India (MFI), founded and run by Dr. K.M. Thiagarajan, former Chairman of Bank of Madura. Under the arrangement, MFI would facilitate formation of self-help groups. Later, MFI would provide finance for the income generating activities. In all these areas, ICICI Bank would back MFI with the funds needed. MFI is already involved with around 10,000 SHGs in Tamil Nadu. The Foundation expects the number to go up to 50,000 in the next three years. These groups are expected to need about Rs. 1,250 crore, which ICICI Bank as per different, agreed formulae. While MFI is likely to get funds from ICICI Bank at around 5 per cent, the ultimate borrower would pay around 14 per cent (because of the high operational costs.

In addition to this, the ICICI Bank bought Samruddhi's crop loans to the extent of Rs. 42.1 million. The buy-out is of Samruddhi's crop loans (to Joint Liability Banking, on behalf of ICICI Bank executed this transaction on Nov. 20, 2003 at Mumbai.

Last but not the least, HLL's Project Shakti owes its existence to Micro-finance. The seed capital and working capital required by women and the who become HLL's direct to home distributors in rural areas can be provided easily through loans from the SHGs.

Conclusion

The role of the SHGs in providing financial services to the agricultural sector and rural areas can't be overemphasized. As it is, variants of group lending methodology are functioning successfully in many countries, especially Bangladesh. However, lending needs to be coupled with insurance and other services such as training and marketing support, government subsidies etc.

Earlier the rural banking system was plagued with the problem of huge Non Performing Assets and losses. Of the many Regional Rural Banks, many were saddled with huge losses and were on the Urge of closure. Even now many banks are contemplating closing down their loss making rural branches. Against this backdrop, Microfinance has ensured up to 90% *on time repayment* while the Non Performing Assets are close to nil. At the same time it has helped in the empowerment of women and poverty alleviation in rural areas.

Critics may dismiss the participation of public sector banks in Microfinance as a compulsion owing to the fact that they are owned by the government. But the initiatives of the private sector (through NGOs – Model III) confirms the viability and feasibility of microfinance as a legitimate and profit making business. There is no doubt that Microfinance, as practiced through the SHG concept will play a major role in the development of rural areas and thereby of the whole country.

Chapter 12
Rural Finance

Rural finance has concurrent relevance for all, the developmental agencies/institutions, academicians, policy makers and corporate sector. It interests developmental agencies and policy makers as they have been entrusted with the task of taking rural people and economy to higher and higher standards both in terms of per capita income and status of living or in terms of human resources development index. It interests academicians as there exists many unresolved areas and potential for developing new theories and conducting research exists in it. Corporates find for their customers/consumers (existing and potential) lies in the rural masses.

Since widespread recognition, in mid 1980s, of failure of the old paradigm of directed agricultural credit with subsidized interest rates, rural and agricultural finance kept a low organizations in rural as well as agricultural finance. This was also reflected in the current budget of the Union Government of India, doubling of credit flow to agriculture in next three years. The principal motivating factors which have contributed towards creating the renewed interest are decline in formal rural and agricultural credit supply; role of rural finance for agricultural and economic growth, food security and poverty reduction is quite visible.

Decline in Formal Rural Credit

A few developing countries mainly in North Africa, Middle East and South East Asia such as Egypt, China, India and Pakistan continue with their state owned subsidized rural banking infrastructure. In other countries, with the dismantling of government and donor support to subsidized rural and agricultural financing has taken place, this has led to the considerable decline in formal rural and agricultural credit. It is still left to be assessed what amount of this decline has been compensated by agribusiness houses, trade credit by traders, micro finance, credit groups and informal savings. In most of the developing countries it has been observed that commercial banks have not entered rural and agricultural credit market on a substantial scale, actually after liberalization some banks have closed their branches in rural areas.

Rural Finance as Engine of Economic Growth

Agriculture as such is a declining sector in course of development world over but in many developing countries it is still a leading economic sector, the main exporter and major employer, especially for poor and women. Improved financial markets accelerate agricultural and rural growth. Financial services lend a helping hand to household in maintaining food security and smoothing consumption, thereby safeguarding and increasing labour productivity.

Since, agriculture yields a strong forward and backward multiplier effect on whole economy, the economic growth in agriculture-especially in sub sectors that directly or indirectly benefit small land holders, tenants, and wage labours is a key precondition for overall economic growth and poverty reduction.

Increasing number of micro-finance institutions and their considerable achievements in reaching a relatively large number of relatively poor women and men are coming up. Successful microfinance institutions operate partially from rural areas with much of their lending for non farm enterprise.

The stakeholders in the field of rural finance comprise farmers, artisans, landless agricultural labour, landless labours, traders, private moneylenders, banks, cooperatives, NGOs, SHGs, governments post-offices, insurance providers etc. Rural finance includes interplay of all these institutions/individuals in terms of lending and deposit relationship.

Many a times people take rural finance and agricultural finance to be synonymous. Therefore, at the very outset it should be understood that agricultural finance is a subset of rural finance. Agricultural finance involves the credit or finance generated solely for the purpose of agriculture. Where as rural finance involves besides agriculture finance, rural-credit; micro-finance; insurance-crops, livestock and infrastructural development.

Rural indebtedness is deep in the country and its growing pressure has fettered the growth of rural economy. It has manifested as a serious threat to economic, political and social life of rural India. In 1951, the National Income Committee put the total rural debt at Rs. 1913.8 crores of the Indian union, though Bhawani Sen estimated Rs. 2,000 crore a safer estimate.

The credit taken by rural masses can be divided into two categories :

1. Credit taken for productive purpose or productive credit
2. Credit taken for consumption purpose or consumption credit.

Rural people require credit for both the purposes as rural economy is mainly farm dependent economy with no fixed monthly or daily income. In 1895, Nicholson observed that 1.3 per cent of registered loans in Madras were due to land improvement, and M.L. Darling recorded that only less than 5 per cent of the debt in Punjab was caused by land improvement.

Per cent of then existing loans was taken for unproductive purposes and same trend was witnessed in the case of Bombay and Bengal. Rural credit survey committee reported that out of all loans about 56.3 per cent was unproductive marriages, birth days, and other social obligations.

Causes of Rural Indebtedness

There are numerous factors which lead to rural population getting indebted leading finally to debt trap. To list a few of these would be :

- Poverty of Rural Masses
- Brought Forward Debts
- Population Explosion
- Fragmented Land Holdings
- Unpredictable Monsoon
- Heavy Livestock Mortality Rate
- Illiteracy
- Litigations
- High Interest Rates
- Money Lenders
- Low Waiting Period.

Rural people suffer from poverty, which force them to borrow money both for production and consumption purpose. "Chronic insufficiency of the farmers income and the consequent tendency of consumption to out run production". Poverty, decay and misery have yet not been removed from the rural areas of the country and even landholding farmers can not live without extra earning in the form of wages.

- Micro-Finance
- Rural Credit
- Agricultural Finance
- Insurance
- Institutional Framework
- Policy

Micro-Finance

Micro-finance refers to that part of financial sector which responds to the financial demands of low income households. This is a happy innovative development. It has benefited an increasing number of low income people and micro small entrepreneurs. Micro-finance institutions in the urban areas have been very successful and have shown a remarkable success in rural areas also. New lending technologies strategies have been designed for low-income clients. Micro credit services, can be broadly divided into four groups :

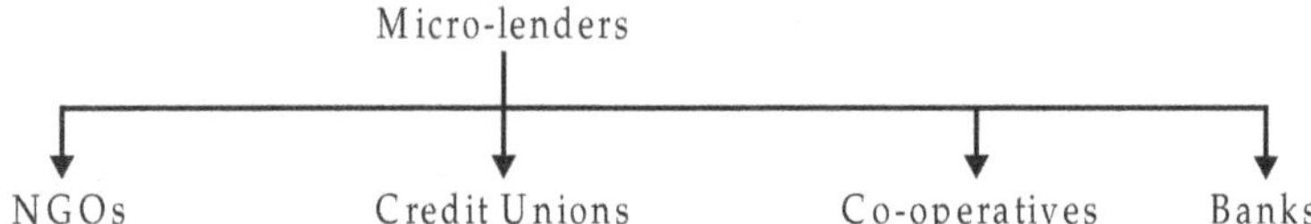

The majority of the micro credit programmes world wide are being operated by NGOs. These include national as well as international organizations. National NGOs operate their programmes through affiliated local agencies. They are committed to work for poor people. NGOs have the comparative advantage as they are familiar with the household livelihood strategies and the financial situation of their target population. They are well knit with local communities having good access to the population. However, NGOs do not possess the required professional expertise to effectively and efficiently execute the programmes, therefore, they have encountered many challenges in the administration of credit programmes. In fact, they will have to undergo a substantial transformation if they intend to become specialized financial services providers. They will have to alter their public image. Instead of serving "beneficiaries", they must establish contractual relationships with clients.

Since the mid-1980s, a number of NGOs have established themselves as specialized microfinance institutions. Some of them have discontinued their social services, while others have created separate affiliated/sister organizations to provide financial services. Specialized NGO micro lenders have initiated the design of innovative micro credit technologies. Even though in a small period of time significant advancement has taken place in the field of micro finance, still the majority of the NGOs serve only a few hundred or a few thousand clients. Most provide loans and usually have only one or two loans products. Although some require mandatory savings deposits from their clients that form part of their loan collateral, just a few mobilize voluntary savings. There has been interest in operating savings deposit facilities as a means to mobilize loanable funds and to enhance their customer services. Since, NGOs are generally restricted from accepting deposits; they fall outside the purview of banking regulation and supervision system. Due to this restriction many NGOs have got motivated to transform themselves into regulated involved in micro finance freedom to expand their range of financial services. They can also access financial markets for raising additional loanable resources. Bancosol in Bolivia was the first NGO to achieve the status of a regulated financial institution. BRAC in Bangladesh and K-Rep in Kenya are also in the process of obtaining bank licenses.

Credit Unions and Co-operatives

Generally, credit unions are meant to serve people with difficulties in accessing commercial banks for credit. They are more formal in their structure than NGOs, sometimes having regional or even national network. The centralized financial management helps them in reallocation of surplus (liquidity) funds between fund deficit member credit unions. Normally credit unions and co-operatives limit their services to members, whose savings provide funds for their lending operations. Providing services only to members has an inherent advantage of better screening of prospective borrowers and appraise, monitor and recover loans. As in the case of informal savings and credit groups, members are self-selected, and peer pressure is exerted to attain full and timely loan repayment. Even best of management theories suggest that social pressure and superior information on member clients are effective mechanisms of control. This functions as long as members know each other and the scope of the financial operations remains manageable.

Despite their advantages, credit unions and co-operatives face notable challenges. The provision of financial services is restricted to members and thus limits their outreach and growth potential. Because loanable funds are generally limited to the mobilized member savings, the credit union is restricted in its efforts to satisfy the amounts than were applied for. They are only available after outstanding loans have been reimbursed. This restricts borrowing opportunities and the effectiveness of the loans. These co-operatives and credit unions lack in professional management and if borrowers dominate the policy making, they have propensity of setting lower interest rates. Lobbying amongst members can lead to poor quality of the loan portfolio. The drawbacks cause the poor potential for loan portfolio growth. The credit unions and co-operatives which are not infected from the above problem have demonstrated considerable success to prove themselves as viable institutional model for providing micro credit. Modified and adaptable regulatory and supervisory framework for credit unions coupled with technical assistance and services from international credit union organizations are key factors towards strengthening the performance of credit unions.

Banks

The involving of commercial banks in microfinance is quite recently. There are a variety of strategies in serving low-income clients, who are normally perceived as unbankable. Micro credit granted indirectly or directly. The credit flow in indirect lending adopt channel; as given below :

Commercial Banks ⟶ NGOs ⟶ SHGs ⟶ Members of SHGs

Indirect ways in which commercial banks lend to small clients include the so-called *linkage* programmes with NGOs or other intermediary organizations. In these cases, banks provide loanable resources and the intermediary organizations on-lends the resources to members of self-help groups for micro enterprise activities. In these arrangements, banks have limited contacts with the final borrowers. They are not the NGO for all aspects of loan appraisal, loan monitoring and loan recovery.

While this model has increased the access of low-income clients to bank loans, it has proven to be only moderately successful in the provision of sustainable banking services. The bank has few incentives to develop appropriate and cost effective credit technologies. It relies on a number of organizations, each with different objectives and performance standards. The direct linkage service to low income clients is referred to as *down-scaling.* It implies the creation of a specialized micro credit department in the bank. This development is particularly attractive in view of the outreach and the financial expertise contained in commercial banks.

Table 1 : SHG Bank Linkage Programme

(Bank Loan Rs. crore)

Year	*Number of SHGs financed*	*Growth (%)*	*Cumulative Number of SHGs*	*Amount*	*Growth (%)*	*Cumulative*
1992-99	32995	–	32995	57	–	57
1999-00	81780	148	114775	136	138	193
2000-01	149050	82	263825	288	112	481
2001-02	197653	33	461478	545	89	1026
2002-03	255882	29	717360	1022	87	2049
2003-04	3617731	41	1079091	1855	81	3904

Banks should change its stubornness. In these cases, it may be preferable to create a new microfinance institution that has a clear corporate mission and set of objectives. Banks will have a poor reputation due to failed directed credit programmes, or when their operations have been undermined by government interference.

Agricultural Finance

The agrarian history from Uttaranchal to Arkansas is that an essential of agriculture is credit. According to an old proverb "Credit supports farmers as the hangman's rope supports the hanged." The farmers in developing countries can not expect their credit needs to come from savings. Their income from farm operations is insufficient to meet any thing else than minimum necessities of life. This causes them dependant on out side finance.

The studies conducted have shown a positive relationship between agricultural growth and availability of credit. The finance for Agriculture can be divided into two heads broadly speakings :

- *Short-term :* for the purpose of input expenses
- *Medium/Long-term :* for development of fixed assets

The need for agricultural finance can hardly be overemphasized where its productivity is still low due to financial constraints. According to Husband and Dockery, "Finance is necessary for any economic activity connected with agriculture or manufacturing industry. It may be said to be the circulatory system of the economic body, making possible the needed cooperation among the various units of activity."

Table 2 : Disbursement of Agricultural Credit 1978–2003

(Rs. Crores)

Item	*1979-80*	*1985-86*	*1993-94*	*1995-96*	*1997-98*	*1998-99*	*1999-00*	*2000-01*	*2001-02*	*2002-03*
A. Co-operative Banks										
Short term	1,300	2,787	7,839	8,331	10,845	12,571	14,845	16,564,	21,542	
Medium term & Long term	125	1,087	2,278	2,148	3,190	3,386	3,518	4,220	5,538	
Total	275	3,874	10,117	10,479	14,085	15,957	18,363	20,784	27,080	
B. Commercial Bank & RRBs	850	3,131	6,377	11,553	15,831	18,443	24,773	27,711	31,964	
Grand Total	2,550	7,005	16,494	22,032	31,956	36,860	46,268	52,714	64,000	

Source : Economic Survey, Govt. of India 2002-03

The table 2 reveals that the volume of agricultural credit has increased from Rs. 2,250 crore in 1979-89 to Rs. 82,073 crore in 2002-03.

The share of commercial banks alone in total institutional credit is nearly 48 per cent, followed by co-operative banks with a share of 46 per cent. Regional rural banks account for just 6 per cent of total credit disbursement.

Sources of Agricultural Finance

The short term and medium term loans of Indian agriculturists is met through loans borrowed from semi-formal and informal lenders and long term requirements are largely met by formal lenders. The table below presents the types of rural lenders that can be found in developing countries.

Table 3 : Typology of Rural Lenders

1. **Formal lenders**
   - Agricultural development banks
   - Rural branches of commercial banks
   - Co-operative banks
   - Rural banks-community banks
2. **Semi-formal lenders**
   - Credit unions
   - Co-operatives
   - Village or semi-formal community banks
   - NGOs
3. **Informal lenders**
   - Relatives and friends
   - Moneylenders
   - Rotating savings and credit associations
4. **Interlinked Credit Arrangements**
   - Input suppliers/Crop buyers
   - Processing industries

Role of Money Lender

Despite rapid development of banks in rural areas, moneylenders still remains the preferred source of credit as they are approachable directly and easily. Moneylenders can be clubbed in two categories : (*i*) Professional moneylenders, and (*ii*) Agriculturist moneylenders. According to All India Credit Survey, 1951-52 and Reserve Bank Survey, 1961-62, moneylenders accounted for nearly 70 per cent and 49 per cent respectively of total rural credit. They enjoy nearly a monopoly in rural credit. These moneylenders accept immovable property by way of security but they lend on personal security also. All India Credit Survey Committee stated that "One of the noteworthy aspects which is common to both professional and agriculturist moneylenders is that about four-fifth of the debt owed to either class is unsecured." It is exorbitantly high in terms of interest.

Problems

The agricultural finance sector is plagued with problems of limited coverage, complicated procedure, lack of co-ordination among various lending institutions, misutilization of loans, regional imbalance, inefficient administrations, poor recovery, lack of saving etc. Certain measures were taken by government in 1998-99 to improve credit flow to agriculture.

One the most critical components in raising the income of poor is credit. Providing adequate credit to the rural poor has become a problem due to the complex nature of the rural society. Situation further worsens when the credit is not put to productive use. Some of the major problems of rural credit system are :

(*i*) Lack of desired results in terms of direction, quantum, and quality of the credit flow;

(*ii*) Afflicted by high over dues, bad debts, loan defaults, unreliability, low profitability, over-burdening of staff, declining control and deteriorating customer services;

(*iii*) Structural deficiencies have contributed towards information imperfection and inefficiencies like high transaction costs and low recycling of credit;

(*iv*) Lack of performance motivators;

(*v*) Viable functioning of credit disbursing units has been seriously hampered due to directed credit programme and subsidized lending.

Table 4 : Debt Owed to different agencies by Rural Households (All India)

Credit Agency	*Cultivators*				*Non-Cultivators*			
	1951	*1961*	*1971*	*1981*	*1951*	*1961*	*1971*	*1981*
Government	3.3	2.6	7.1	3.9	1.5	0.6	3.4	4.5
Co-operatives	3.1	15.5	22.0	29.8	1.5	5.3	6.0	13.9
Commercial Banks	0.9	0.6	2.4	28.8	2.0	1.4	0.8	17.3
Landlords	1.5	0.6	8.1	3.7	4.9	1.2	12.6	8.4
Agricultural Money-lenders	24.9	36.0	23.0	8.3	24.8	23.0	23.8	11.4
Professional Money-lenders	44.8	13.2	13.1	7.8	38.0	10.6	18.7	13.4
Traders	5.5	8.8	8.4	3.1	9.9	16.4	10.9	5.8
Relatives/Friends	14.2	8.8	13.1	8.7	15.5	8.6	19.0	14.4
Others	1.8	13.9	2.8	5.9	1.9	32.8	4.8	10.9
Total	100.0	100.0	100.0	100.0	100.0	100.0	100.0	100.0

Source :

(*i*) Report of the All-India Rural Credit Review committee, RBI, Mumbai 1969.

(*ii*) All-India Debt and Investment Survey 1981-82.

(*iii*) RBI Bulletin, June 1986, p. 477.

The all India Rural credit survey in the year 1951-52 collected data on socio-economic conditions of farmers so that it can be helpful for central as well as State Governments and Reserve Bank of India in formulation of an integrated scheme of rural credit. In the year 1961-62, All India Rural Debt and Investment survey was conducted to arrive at a reliable and statistically valid estimates of debts, investment and other related features of rural households. Following the nationalization of 14 commercial banks in 1969, commercial banks took up task of providing impetus to priority sector. To take a fresh look at the condition of farmers, during 1971-72, the all India debt and Investment survey was conducted covering both rural and urban households of the economy. The economic power and rural indebtedness have surveyed. The survey showed out of the total assets of Rs. 88,409 crore, cultivator households, which was 72.3 per cent of rural households accounted for 93.6 per cent of the total with an average of Rs. 14,694 per household. The proportion of agricultural labour households to total rural households may be taken as an index of poverty and inequality in rural areas. This proportion was found to be 14.6 per cent at the national level. The non-cultivated class had an average of Rs. 2000 per household. Among the non-cultivators, 53 per cent were agricultural labourers with an average asset value of Rs. 1,139 per household and 8.7 per cent were artisans with an average asset value of Rs. 2,367 per household. According to All India Debt and Investment survey of 1981–82, the dependence of rural households for cash debt on non-institutional agencies had come down from 93 per cent in 1950-51 to as low as 39 per cent in 1981. In past land lords, professional and agricultural money-lender were enjoying monopoly by disbursing credits up to the extent of 75 per cent. But it came down significantly by 1981 to the level of 17 per cent.

Rural Insurance

It is one of the most promising times for insurance companies to enter the rural areas. Insurance opportunity in rural areas comprises :

1. People life insurance
2. Live stock insurance
3. Agricultural/crop insurance

People Life Insurance

Prior to liberalization and entry of private sector insurance companies, insurance was mainly used as a tax saving instruments. It was mainly limited to working persons whose incomes were enough to lend them in the income tax relief. As under income tax act some relief has been provided on insurance premium paid, there was always a sharp rise in the number of insurance done in the month of February & March every year. As the competition started becoming intense with the opening up of insurance sector, companies are advertising heavily to spread the awareness in rural masses also. Only to some extent they achieved success in making a dent.

Live Stock Insurance

Live stock constitutes a good chunk of rural people's assets. High mortality rate in live-stock has been identified as one of critical variable contributing towards the increasing rural indebtedness. Various insurance companies are already having policies framed for the insurance of livestock. But still a lot of effort remains to be done. Still farmers do not go for insurance of their live stock except for live-stock purchased on bank-loan as it is mandatory. This means insurance companies should act promptly and spread awareness in rural masses. For corporate this may turn into a gold-mine.

Agricultural/Crop Insurance

Majority of the farmers are totally dependent on their crops and if something happens to their crops they are nowhere. Therefore, it is but natural that there should be some crop insurance scheme in vogue. Now in case the crop has been insured, another factor that can take toll on the farmer's income is fluctuating price-levels. That means there should be some insurance scheme for assured farm income also. Governments every where in the world are providing such facilities.

The government in India had initially formulated a comprehensive crop insurance Scheme which got replaced by scheme 'National Agricultural Insurance Scheme (NAIS) from Rabi 1999-2000 season. The scheme is available to all farmers irrespective of their loanee status and size of land holding. This scheme covers all food crops, oilseeds and annual commercial/horticultural crops for which historical yield data is available. The premium rate differs from crop-to-crop ranging from 1.5 per cent to 3.5 per cent of sum issued while actuarial rates are being charged in case of commercial/horticulture crops. Small and marginal farmers are entitled to a subsidy of 50 per cent. Remaining 50 per cent is shared by central and state government in the ratio of 50 : 50. At present this scheme has been implemented in following 23 states and 2 union territories.

1. Andhra Pradesh
2. Assam
3. Bihar
4. Goa
5. Gujarat
6. Himachal Pradesh
7. Karnataka
8. Kerala
9. Maharashtra
10. Madhya Pradesh
11. Meghalaya
12. Tamil Nadu
13. Uttar Pradesh
14. West Bengal
15. Sikkim
16. Chattisgarh
17. Jharkhand
18. Tripura
19. Orissa
20. Jammu & Kashmir
21. Uttaranchal
22. Haryana
23. Rajasthan
24. Pondicherry
25. Andaman and Nicobar Islands

Table 5 : Performance of the National Agricultural Insurance Scheme

Sl. No.	*Particulars*	*Rabi*	*Kharif*	*Rabi*	*Karif*	*Rabi*	*Kharif*	*Total*
		2000-01	*2001*	*2001-02*	*2002*	*2002-03*	*2003*	
1.	Farmers Covered (in lakh)	20.92	86.96	19.55	97.65	23.27	79.69	328.04
2.	Sum Insured (Rs in crore)	1602.68	7502.46	1497.51	9429.44	1837.53	8110.13	29979.75
3.	Insurance Charges (Rs. in crore)	27.79	261.62	30.15	325.38	38.50	283.14	966.58
4.	Area coverage (in lakh ha)	31.11	128.88	31.46	155.22	40.38	122.89	509.94
5.	Total Claims (Rs in crore)	59.49	492.63	64.66	1821.79	193.03	191.92	2823.52
6.	Claims Paid (Rs. in crore)	59.25	492.32	64.37	1754.28	114.15	12.02	2496.39

To cope up with the twin problems, the Department of agriculture and co-operation has formulated the Farm Income Insurance Scheme (FIIS). This scheme targets two most critical components of farmers income that is yield and price. This scheme was conceived to provide protection to farmers by integrating the mechanism of insuring production as well as market risks. It the actual income of the farmers falls short of the guranteed income (average yield × minimum support price) of the farmers, they would be eligible for the compensation to the extent of indemnity from Agriculture Insurance Company of India Ltd. (AICI). Initially this scheme would cover paddy and wheat only would be mandatory for any farmer availing corp loan. Initial this scheme has been taken up on pilot basis in Rabi 2003-04 in 18 districts of 12 states for paddy & wheat. Based on the feedback from this pilot project necessary fine tuning of the scheme would be done.

Rural Financial Institutions (RFI)

Development of rural areas is a matter of serious concern world wide. To address this issue, besides numerous other things, sound rural financial institutions are required. Nearly every country, at least developing, has a rural banking system, which has been utilized for raising and chanalising the millions of rupees to these rural areas. Even today they act as a dominant tool for rural development affecting the lives of billions of rural poor around the world.

Before going to broad patterns concerning rural financial institutions, let use see why we need RFIS. We need them due to following reasons :

1. Farmers are poor

2. Farm credit needs creed
3. Rural development should be promoted by government, and
4. Rural development needs supply leading finance as a pre-requisite.

Now all the points argue for a cheap or subsidized rural credit. But any institution doing this will not be able to face the music of commercial criteria of normal banking. This led to the establishment of specialized RFIs in many of the developing nations.

The World Bank has identified the following constraints in the working of RFIs (BMZ 1997) :

- The RFI suffered from inflexibility of objective within an inappropriate policy and legal framework and lack of institutional autonomy;
- Subsidized or regulated interest rates on loans have blocked the emergence of new credit markets;
- Target group access to credit has been severally restricted and loans at concessional rates have mostly gone to larger enterprises;
- Domestic resources mobilization and national financial market integration were seriously neglected;
- The accumulation of disclosed and undisclosed losses due to defaults, inadequate risk management and excessive inflation has eroded the capital base;
- Government-owned RFIs have acted as administrative agencies in channelising funds from donors/central banks/government;
- Personal loans for operational efficiency have been absent;
- Due to lack of autonomy, loans have been unprofitably allocated or misused.
- Monitoring and evaluation by donors/funding agencies have been inadequate.

Since the failure of old paradigm of directed agricultural credit with subsidized interest rates in mid 1980's a paradigm shift has been witnessed from 'directed credit paradigm' to new 'financial market paradigm'. It is believed that financial markets forces would be helpful in allocation of resources more efficiently between surplus and deficit unit of the economy.

Summing Up

Following nationalization of major commercial banks in 1969, the inflow of credit for rural development had been steadily increasing. Even the analysis of credit flow for the last four fiscal years recorded a compound annual growth of about 18 per cent. During 2003-04 the total credit flow to agriculture and allied sectors is estimated to have increased to Rs. 80,000 crore. The trend is encouraging but not adequate. Realizing this, the latest Union Budget (2004-05) aims doubling of credit flow to agriculture in next three years.

Prior to banks nationalization 1969 rural financing was the sole domain of exploitative private money-lenders. Cooperatives were the only institution providing small proportion of rural credit in the country. Since 1970s the share of institutional credit increased significantly. NGOs and SHGs are the new entrants in rural financial activities.

Even though the Government of India as also Governments in various states have taken various pro – rural policies in order to encourage flow of institutional rural credit to curtail the dominance of the private money-lenders, the private money-lenders still continue to play a significant role in rural financing. Their interest rates are exorbitant and the payment schedules exploitative. For want of proper understanding of rural psyche, various institutional sources of rural financing like banks, cooperatives etc. are yet to make significant impact to make institutional credit the main component of total rural credit. Further, there are wrong notions that agricultural credit is rural credit and lending is financing. The other aspects of these are totally untouched For instance, majority of rural families are not land holders or farmers. They include land less agricultural labour, rural artisans, petty traders etc. The credit flowing to this section which is more than half of the rural population cannot be treated as agricultural credit. Unfortunately, this section continues to remain out of focus in rural financing. Mobilization of small savings from rural people also continues to be ignored in view of less or limited quantum thereof. Other aspects of rural financing which remained out of focus or received only little attention are micro financing, consumption credit, lending for social purposes etc. Micro credit which has become a recent buzz word is also to play a significant role in rural financing. Thus there is a need to consider rural financing under holistic manner rather than taking it as only agricultural loaning. Issues like directions, quantum and quality of rural credit; problem of over dues; debt relief to farmers; low profitability; deteriorating law and order situation in rural areas; high transaction cost; and subsidized lending are the issues which require attention of scholars and institution.

REFERENCES

Adams, D.W., and J. Von Pischke (eds). 1984. Undermining rural development with cheap credit. Boulder, Colo., U.S.A. : Westview Press.

Adams, D.W. 1998. The conundrum of successful credit projects in floundering rural financial markets. *Economic Development and Cultural Change* 36 (2) : 335–367.

Adams, D.W. 1998. The decline in debt directing : An unfinished agenda. Paper presented at the Second Annual Seminar on New Development Finance, Goethe University of Frankfurt am Main, Germany.

Alderman, H., and C.H. Paxson. 1992. Do the poor insure? A synthesis of the literature on risk and consumption in developing countries. Discussion Paper No. 169, Woodrow Wilson School of Public and International Affairs, Princeton University, Princeton, N.J., U.S.A.

Ali, I.S., S. Malik, and M. Zeller. 1994. The rural financial sector in Egypt – PBDAC's future role and policy options. Final report submitted to the Ministry of Agriculture and Land Reclamation, Government of the Arab Republic of Egypt, and U.S. Agency for International Development. Washington, D.C. : International Food Policy Research Institute. May 1994.

Baker, J.L. 2000. Evaluating the Impact of Development Projects on Poverty. A Handbook for Practitioners. The World Bank, Washington, D.C.

Branch, B., and A.C. Evans, 1999. Credit unions : Effective Vehicles for microfinance delivery. World Council of Credit Unions, Madison, Wisconsin.

Braun, J. von., H. Bouis, S. Kumar, and R. Pandya-Lorch, 1992. Improving food security of the poor : Concept, policy and programs. Washington, D.C. : International Food Policy Research Institute.

Braverman, A. and L. Gausch. 1989. Institutional analysis of credit cooperatives. In : Bardhan, P. (ed.) : The economic theory of agrarian institutions. Oxford : Clarendon Press. BRI, and John F. Kennedy School, Harvard University. 2001. BRI Microbanking Services : Development impact and future growth potential. Mimeograph, Jakarta.

Chao–Beroff, R. 1999. Self reliant village banks, Mali. Consultative Group to Assist the Poorest, Working Group on Savings Mobilization.

Christen, R.P., E. Rhyne, R.C. Vogel, and C. McKean. 1995. Maximizing the outreach of microenterprise finance : An analysis of successful microfinance programs. Program and Operations Assessment Report No. 10. Washington, D.C. : U.S. Agency for International Development.

Churchill, C.F. (ed.). 2000. The MicroBanking Bulletin, Vol. No. 4, February 2000. Calmeadow, Washington, D.C. (and more recent issues of the Microbanking Bulletin).

Churchill, C.F., D. Liber, M.J. McCord, and J. Roth. 2003. Making insurance work for microfinance institutions : A technical guide for developing and delivering microinsurance. International Labour Organization (ILO), Geneva : Switzerland.

Conning, J. 1999. Outreach, sustainability and leverage in monitored and peer-monitored lending. *Journal of Development Economics* 60 : 51– 77.

Cuevas, C. 1999. Credit unions in Latin America : Recent performance and emerging challenges. Sustainable Banking for the Poor. Washington, D.C.

Deaton, A. 1992. Understanding consumption. Oxford : Clarendon Press.

Diagne, A., and M. Zeller. 2001. Access to credit and its impact on welfare in Malawi. Research Report No. 116. Washington, D.C. : International Food Policy Research Institute.

Dror, D., and A. Preker (eds). 2002. Social Re-insurance : A new approach to sustaining community health financing. Washington, D.C.

Economic Survey 2003-04, Government of India.

Evans, A.C. 2001. Strengthening Credit Unions in Sri Lanka : Dispelling the middle class myth. Research Monograph Series No. 19. World Council of Credit Unions, Madison, Wisconsin.

FAO. 1998. Agricultural Finance Revisited : Why? Food and Agricultural Organisation (FAO) and Deutsche Gesellschaft für Technische Zusammenarbeit (GTZ).

Fleisig, H., and N. Pena. 2003. Legal and regulatory requirements for effective rural financial markets. Conference paper.

Fernando, N.A., and R.L. Meyer, 2002. ASA – The Ford Model of Microfinance. In : Finance for the Poor, Asian Development Bank, Vol. 3, No. 2, June 2002, pp. 1–3.

Fruman, C. 1998. Mali : Self–Managed village savings and loan banks (CVECA Pays Dogon, Mali. In : Sustainable Banking with the Poor, Africa Series. The World Bank, Washington, D.C.

Gittinger, J.P. 1982. *Economic Analysis of Agricultural Projects,* (Economic Development Institute of the World Bank), The Johns Hopkins University Press, Baltimore and London.

Gons, N., B. Branch, and M. Cifuentes. 2001. The road to Jinotega. Research Monograph Series No. 18, World Council of Credit Unions. Madison, Wisconsin.

Gonzalez–Vega, C. 2003. Depending rural financial markets : Macroeconomic, policy and political dimensions. Conference paper.

Gonzalez–Vega, C., and D. H. Graham. 1995. State–owned agricultural development banks : Lessons and opportunities for microfinance. Economics and Sociology Occasional Paper No. 2245, The Ohio State University, Columbus.

Gurgand, M., G. Pederson, and J. Yaron. 1994. Outreach and sustainability of six rural financial institutions in Sub-Saharan Africa. Discussion Paper 248. Washington, D.C. : World Bank.

Heidhues, F. 1995. Rural financial markets - an important tool to fight poverty. Quarterly Journal of International Agriculture 34(2) : (105-108)

Heidhues, F., J.R. Davis, and G. Schrieder, 1998. Agricultural transformation and implications for designing rural financial policies in Romania. European Review of Agricultural Economics, Vol. 25, pp. 351-372.

Henry, C., M. Sharma, C. Lapenu, and M. Zeller. Assessing the relative poverty of microfinance clients : A CGAP operational tool. Book published by the Consultative Group to Assist the Poorest (CGAP), Washington, D.C.

Huppi, M., and G. Feder. 1990. Cooperatives in rural lending. World Bank Research Observer, Vol. 5.

Hollis, M., and A. Sweetman. 1998. Microcredit : What can we learn from the past ? *World Development* Vol. 26, No. 10, pp. 1875-1891.

IFAD, 2002. IFAD Finance Policy. International Fund for Agricultural Development, Rome, 2000.

Kankhoje, D.P., Institutional Changes in Indian Agriculture, NCAP, New Delhi, 2003.

Klein, B., R.L. Meyer, A. Hanning, J. Burnett, and M. Fiebig. 1999. Better practices in agricultural lending. Food and Agricultural Organization (FAO) and Deutsche Gesellschaft für Technische Zusammenarbeit (GTZ).

Krahnen, J.P., and R.H. Schmidt. 1994. Development finance as institution-building : A new approach to poverty-oriented banking. Boulder, Colorado : West view Press : Kropp, E. 1989. Linking self-help groups and banks in developing countries. Eschborn, Germany : German Agency for Technical Cooperation (GTZ).

Krüger, A.O., M. Schiff, and A. Valdes. 1991. The political economy of agricultural pricing policy. Baltimore, MD : John Hopkins University Press for the International Food Policy Research Institute.

Lapenu, C., and Zeller, M., 2002. Distribution, growth, and performance of the microfinance institutions in Africa, Asia and Latin America : A Recent Inventory. *Savings and Development,* No. 1 - XXVI, pp. 87-111. Appeared previously as Lapenu, C., and M. Zeller. 2001 under the same title as Discussion Paper No. 114, Food Consumption and Nutrition Division, International Food Policy Research Institute, Washington, D.C., July.

Lehki, R.K. and Singh, Joginder, Agricultural Economics, 1996, Kalyani Publishers, New Delhi.

Mellor, J.W. 1996. *Economics of Agricultural Development.* Ithaca, N.Y. : Cornell University Press.

Meyer, R.L., and G. Nagarajan. 2000. Rural financial markets in Asia : Policies, paradigms, and performance. Asian Development Bank and Oxford University Press. Mosley, P. 1999. Micro-Macro Linkages in Financial markets : The impact of financial liberalization on access to credit in four Africa countries. World Development.

Navajas, S., and M. Schreiner. 1998. Apex organisations and the growth of micro-finance in Bolivia. Occasional paper No. 2500, Rural Finance Program, Ohio State University.

Morduch, J. 1995. Income smoothing and consumption smoothing. *Journal of Economic Perspectives 9 (3) :* 103-114.

Morduch, J. 1999*a*. The microfinance promise. *Journal of Economic Literature* 37(4) : 1569-1614.

Morduch, J. 1999*b*. The Role of Subsides in Microfinance : Evidence from the Grameen Bank," *Journal of Development Economics 60 (1) : 220-248.*

Satyasundaram, I., Rural Development 1997, Himalaya Publishing House, New Delhi.

Chapter 13
Fisheries Co-operatives

Need for the organism of co-operatives.

→ Fisherman & food & fishing is a seasonal activity.

→ Because of their illiteracy, poorly, back risky nature of this profession no any agency to finance fishermen.

→ They are to obtain the finance from intermediaries whose interest is only to buy fish at cheapest price.

→ Fishermen to forced to obtain the loan during the offseasons and clear them during the fishing season. It is very difficult to solve this problem of this fishermen community.

→ The only alternative for the fishermen is to farm a fisheries co-operative to obtain the services like credit, transport marketing & supply of various inputs, members of the co-operate can obtain services at reduced load and got high monitory return.

Indian Fishery Co-operative Management

Fishery → Commercial practice or business practice of aquatic organism or economic importance for welfare of human being.

Fish → Cold blooded aquatic animal popularise by fins.

Fishery co-operative was suggested by Royal Agricultural Commission for providing solid structural base in the development of Indian fishery.

Agricultural growth rate – 4.6%

Fishermen growth rate – 10%

1944 → Fishery sub-committee constituted by agri planing committee suggest that direct and indirect help must be provided to Indian Fishery Industry as in Japan.

co-operative marketing of Japan – 75%

co-operative marketing of India – 15%

In 1946 co-operative, planning committee suggested that the financial support for the development of fisherman community must be provided through the co-operative societies. Thus co-operative society must infrastructural helps to the members.

→ The red movement was started after the independence as a result of inclosure of fisheries development in the 5 year plans.

→ Over the earth the fishery structure it into multifunctional units at primary level central and regional level.

How is a cooperative different from other organizations → Different types of business organization such as :

1. Individual traders,
2. Sale traders,
3. Partnership firm,
4. Joint stock company,

are distinct from other composition function goals.

(*i*) The sole trader organism his business with his own capital, his aim is earning profit a sole trader can establish his business without any rigid, legal formalities and he alone decide the policies and he alone get the entire property.

(*ii*) **Partnership Firm** – The objective of a partnership firm is to share the profits b/w partner, at least 2 person and not more than 20 are to form a partnership firm. In a partnership firms the powers of members cannot be derivated. Investment is not compulsory and sharing of profits depending on the forms & conditions, laid down at the start of business and may differ from partner to partner.

(*iii*) **Joint stock Comp** – A J.S.C. is registered under the law the continues to an existence irrespective of the death or origination of any member, the main aim of setting to secure profits for its members, the profits earn by a member in a proportional to the shares held by him.

Cooperative Society → Co-operatives an essentially, a union of person and honourship is generally unlimited, the control of a co-operative is democrate and is based on the principle one member one, these are voluntary organisation and open to members including the weaker sections of the community who use the services.

"Co-operative solve the economic problems of the members with group action".

Principles → The co-operative organisation was born in England in 1844 with the starting of a consumer store by the "Rochdale poineers".

The cope principles as they accepted today had their origin in the rules adopted by the Rochdale Pioneer.

→ In order to serve the economic activities in different economic system the international co-operative allownce formulated the following co-operative principles.

1. Membership of a co-operative society should be voluntery and available without artificial restrictions of any social political or religious problems,
2. Co-operative societies are decorative organizations, their affairs should be administered by persons elected or appointed by the other members all members her equal rights of voting. In societies other than private societies the administration should be conducted in a democratic banks in a suitable form,
3. Share capital should receive only a strictly limited rate of interest,
4. Surplus or saving should be distributed equally among the members, this may be done by decision of the members as follows :
   (*a*) By provision for development of the business to their property,
   (*b*) By provision for common services,
   (*c*) By distribution among the members in proportion to their transaction with the society.
5. All co-operative should make provision for the education of means officers, and employes of the society.
6. All co-operative organizations in order to best serve the interest of their members and

their community should actively co-operate in every practical may with other co-opeatives at local, national, and international levels.

Structure and Activities of Co-operatives → The structure of co-operative counts of primary societies at the base or at the village level with individuals as members. These prim societies are affiliated to the district or regional federation, at state level these are apex societies to in the federations are affiliated. These are national federation serving the interest of co-operatives at the national level. As regards the activities the primary societies and to concentrate on production while other high level societies undertake services and supplies pooling of catches (stockry, processing and marketing).

→ National Insertive value the link b/w co-operative networks and the govt.

Structure of fisheries co-operatives in India → There are 4 type system of fisheries co-operative network in India.

1. National level federation
2. State level federation
3. Central/Regional/district co-operative societies.
4. Primary co-operative societies at grass root level.

Fun and objectives of National Level Federation :

1. In coordinates various activities of fisheries across all the states.
2. It should have a research or advisory survive all.
3. Research activities should be field oriented.

It should limit training to the members and ranges of state and central co-operative societies.

Fun and Objectives of state lend federation → It should maintain level b/w fisheries depart and state govt. to secure necessary association to the central co-operative societies.

It should an part training to the members of state level federation and to the member of central co-operative society.

It should arrange finance to the central co-operative societies and primary co-operative societies.

Objective and Fun of Central Co-operative Society

- ❑ It should provide necessory services to primary co-operative society.
- ❑ It should bring the latest technology. Its the knowledge members of P.C.S.
- ❑ It should arrange finance for P.C.Ss.
- ❑ It should co-ordinate various activities b/w state federation & primary co-operative societies.
- ❑ It should arrange marketing of produce of primary co-operative society.

Fun and Objective of Pri co-operative Society

Fishermen populations 14.28 lakhs

Nor of P.F.C.S. 12000 (Primary fisheries cooperative societies)

Nor of C.C.Ss. 108 (Central Cooperative Societies)

State level federation – 17

National federation – 01

National federation of fisherman co-operative (Fish co-operative fed)

Add- 7, Sarita Vihar, Institutional Areal, New Delhi-44

Present President of National Federation is Shri Bindeshwar Sahni.

Fish Cop Fed. : Thus *Fish Cop Fed* was *registered on 26 Feb., 1980 under Maharashtra co-operative society net under the friendship of Late Shri J.C. Barway.*

The head quarter was shifted from Mumbai to New Delhi 1982.

Objective of Fish Cop Fed : To facilitate, coordinate and promote. Fishery Industry in country through co-operates and for this purpose to understates organise and develop production, processing, storage and marketing of fish & fish products and to manufacture and to distribute machinery implements and other inputs required by the fishing industry.

Fish Objectives of Primary co-operative society.

1. It should provide service to individual members
2. It should arrange necessary finance assistance.
3. It should procure now unit in place of old existing unit.
4. It should bring the latest technology in notice of member

FISH COP FED

Activities of Fish Cop Fed

1. To undertake, purchase, sale and supply of fish and fish products, valuating and processing spare parts of native engines and other fisheries inputs.
2. To act as an insurance agent.
3. To organise consultancy work.
4. To undertake manufacture of fishing vessels engines and other fishery requisites.
5. To set up storage unit including cold storage.
6. To maintain transport units of its own or in collaboration.
7. To subscribe to the share capital or co-operative institution as well as other publish and joint sector enterprises.
8. To arrange for training to employs.
9. To establish processing units for processing and preservation of fish and fish products.
10. To undertake grading, packing and standardisation of fish and fish products.
11. To advance loan.

Insurance Scheme of Federation

An accident insurance scheme was formulated for providing economic security to the families of fisherman.

The scheme was approved by govt of India with the premium subsidy of 50%

Now initially accident coverage of rupees 15,000 against accidental depth and rupees 7,500 against partial/permanent disability was available against an annual premium of rupees 12. A discount of 25% was allowed by the govt thus bringing the effective rate of premium rupees.

At present of 24 hrs accident coverage of against accidental death and rupees 25,000 against partial/permanent disability is available against an annual premium of rupees 14.

Types of Fish Cooperatives

1. Fish Marketing co-operatives.

2. Credit & Supply co-operatives.
3. Fish transport societies.
4. Vessel ownership societies.

The fishery co-operative can also be found at three different levels in a state.

1. Village
2. Distt.
3. State

At the village level the produce/primary societies in engaged in production. At the district level the untrue society and concern with procurement and supply of inputs and at the state level the body of fishermens co-operatives federation limited looks after the management of different co-operatives in the district.

Arranging supply of input and procuring finance for development of co-operatives on all most all states these cooperatives at three different level and non functional and dormant, above all, the national level three exist the national federation of fishermans co-operaives (Fish Cop Fed) which co-ordinates with central government ministry of Agriculture and co-operation, state government directorate of fisheries. FFDAs, financial Institutions, N.C.U.J., MCDC and etc.

An analysis of working pattern of fishery co-operative reveals that they suffer from many management difficulties and structural defects and thus are not economically viable and strong.

Co-operatives and Insurance → Fishery as well as fishering is a hazardous profession thus are several risk in surround the lives of fishermen, their houses, house hold properties, tools, boats, and the trade, such accidental Haggards may happen due to natural or man-made factors. Insurance comes in where there is a risk or haggard, fisheries insurance practices are advanced an developed countries look Japan & Korea where Insurance is a part of fisheries organisation mostly in co-operative sector.

1. The package of Insurance offered provide for Insurance for Individual and familiar.
   (*a*) Personal accident Insurance.
   (*b*) Medical or Hospitalisation Insurance.
   (*c*) Insurance of houses and house hold properties.
   (*d*) Loss of income due to insured causes.
2. Insurance of fish in hatcheries, ponds.
3. Insurance of ponds.
4. Insurance of travelers, mechanised or non-mech. boats.
5. Insurance of net, instrument and gear.
6. Insurance of cold storage.
7. Insurance of fish in refrigerated vehicles and slops.
8. Welfare Insurance for fishermen and their families.

Insurance Scheme for fishermen and fisheries in India.

The bank is that an India the fishermen live in a poverty condition and are socially at a lower level they are ignorant and illustrate, they need therefore to be organised in co-operatives and these co-operative among the other things do awaken them to realise their economic interest including Insurance benefit.

In India the burden of looking after the fishermen and fish formers welfare has fallen on or owned by the Fish Cop Fed.

The fisheries cop fed, central govt. agency moniter, central, and manage the Insurance schemes.

The central govt. in co-ordination with the state govt. share and pay the premium, the fishermen are therefore not direct Interest with Insurance Industry.

The United India Insurance comp to has about thousand offices in the country spread over even to the remotes to area collaborates with the Fish cop fed.

The following Insurance schemes are in operation in India

1. Janta personal Accident Insurance
2. Pond fish Insurance scheme.
3. Fish Part Insurance scheme.
4. Fish Insurance, fingerling culture.
5. Prawn Insurance

Janta Personnel Accident Insurance → The fishermen and who are fishery co-operative members are covered with the Janta Per A.I.

The planning to extend this scheme to all fishermen and women, it is a 24 hrs cover (but not only an cases of accidents while fishing of any type of accident while fishing or other wise, the rate of premium is Rs. 9 only per annum. It is shared by central and state govt. on 50 + 50% bases.

One interesting feature of this insurance scheme is claim procedure has been simplified, the insurance comp pays the claim check with in 2 weeks, but in no case later than 1 month.

Rs. 15,000 is pays to the members of family of insure in case of his or her accidental death and Rs. 7,500 paid to the insured members of he/she becomes physically disable.

The insurance scheme has been well received in states for *e.g.*, In 24,222 fisherman in 1987–88 and about 25,000 fishermen on 1988-89 has been protected by this insurance scheme, however this scheme is being implemented on phases due to identification and administrative difficulties.

Up → 24,222 in 1987-88

25,000 in 1988-89

Pond Fish Insurance Scheme : Under this scheme insurance cover is provided to the fish farmers and the fishermen engaged in inland fish prod against the total loss of seed or fish by death resulting from accident or disease on the tanks or ponds. The Insurance gives 24 hr cover during the period of Insurance.

The Insurance Comp pays the insured value of fish at the time of happening of loss, this Insurance scheme is also extend to the financing institutions like the National co-operative land Development Bank (LDB) to recover and realize the loans advanced by them, further hatcheres own by state govt. FFDA or the state fisheries development co-operative are also covered by, this scheme.

The details of insurance cover, method of insurance rate of insurance cover, method of insurance rate of premium, settlement of claim and other terms and conditions has to be checked out :

Fish Pond Insurance Scheme : Ponds have been classified under there categories according to the nature of constant.

1. Brick and Cement
2. Brick and mud
3. Mud only.

The insurance cover includes insurance against fire, flood, strick, risks and damages have been happening to the ponds due to calamities mentioned above and therefore, the Fish cop fed, LDB and United India. I.C. have been decided to insure ponds or tanks against such calamities.

The premium rates are as follows :

LD Br. National Cooperative Land Development Bank.

Category of Pond Premium rate 11000 fisher

	Category of Pond	Premium rate
1.	Brick and cement	Rs. 1.2
2.	Brick and mud	Rs. 1.3
3.	Mud	Rs. 4.3

Fish Insurance for Fingurling Culture : The scheme is applicable to fingurlings culture in tanks or ponds or other still F.W. Projects only. It does not cover B.W. or marine fisheries, the policy covers total loss of fingerlings due to any accident or disease occurring during the period of Insurance. The Insurance covers the risk like strike, earth quake, explosion, damage by road vehicles, flood, cyclone, storm etc. or pollution, summer kill due to temperature beyond 40°C.

Prawn Insurance : It provides Insurance cover against total loss of prawn or seed on lateleries own by state govt. FFDAs, state fisheries cooperations or MPEDA.

It provide Insurance cover to these engaged in B.W. prawn farming against total loss of seedlings or juveniles or prawn of all spp. culture/raised in B.W. after being transformed into the form.

The risk covered are summer kill, pollution, poisoning earth quake, explosion, storm, cyclone, flood, volcanic, eruption, damage due to road vehicles, shell disease, viral disease, and parasited attack.

Fisheries Cooperative in India : Fishermans are economically down treateden and socially at a lower level, even after mechanization of fishing the fishermen do not have bigger boats and modern implement for offshore fishing so they limit their fishery to inland fishing only. It is subsistance fishing of a shall scale fishing. The fishermen are ignorant of fish market, fish prices so the middleman exploit them thus their Income is small, if the fisherman can organize fishery cooperatives there socioeconomic problem can be solve to a great extent by raising the production. of fish and organising a marketing system and Insuring fair prices, their economic Independence can be guaranteed. Following types of fishery co-operative have been assisting in India.

A. Fish Marketing co-operative : The main objective of this co-operative are :

1. To provide better monitary return for a resonable price for produce and eliminate middle man in distribution of fish.
2. To establish of regulate fish prices.
3. To process and preserve surplus fish for the lean season.

B. Credit and Supply Co-operative : Main day are

1. To provide loans to members for purchasing fishery equipments, net, engines, boat, ice, rope, prames paints etc.
2. To directly provide or supply ice, fuel, nylon net boat and other equipment.
3. To develop quality fish sold, fertilize and manuring the ponds.
4. To link the credit and supply co-operative with marketing processing and prod. Cooperative so as to guarantee real economic benefits to the fisherman by eliminate the middle man and money lander.
5. The loan can be easily recovered on easy Instalment.

6. The credit and supply co-operative are now a days helping in mechanization of fishing and arranging subsidy loan from the govt. to their members.

C. Fish Transport Societies : Such societies eliminate not only the middle man but also the marketing societies while marketing the fish, the members of transport societies can take their fish produce directly to the market in a vehicle of society and sale the fish directly and release the money without any delay.

D. Vessel Ownership Society : Boats and other equipments and costly and may be beyond the means of fisherman more than one fisherman, may jointly purchase boats and use it or the society can purchase it and issue it to means jointly.

Such societies can manufactur vessels, repair equipments and crafts and provide for cold storage.

Fisheries Co-operatives in Asia

Fishery co-operative big and small many in nor have been existing in ask for several pears, national level organization and govt organization at different level have been dealing with them.

1. *Govt. Incentives for fishery cooperatives :* There are department of fisheries and of co-operative to provide advice and guidance for fisheries development through govt help and support.

   Govt. stabilizes fisheries colleges, cooperative colleges, training center to give greater knowledge to co-operative member and others on the subject like betterment of members management training manage of fund educational courses etc.
2. *The International Co-operative Alliance (ICA) :* Headquaters are at Tokyo with the support of Japanese govt is engaged in overseas develop assistance (ODA) programe for advanced fisheries co-operatives. The ICA and ODA jointly conducted fisheries seminars and lectures in many Asian countries such as in Shri Lanka in 1986-87 and in 1988 at Lucknow in India.

In India there are existing the national federation of fisherman co-operative (Fish cop fed). The ICA fisheries such committee for Asian region is Lead quartered in New Delhi and members countries must actively participate in its programe.

Fishery Co-operative in other Countries

1. Japan : There is a good network of fishery co-operative in 1976 there were 5682 cooperatives out of which 999 were inland fish cooperative they were grouped into 29 federations, there is a exclusively a law mode for fishery co-operative, it is made as fisheries co-operative association low of 1948. It have been ammounded in 1975 (FCA law 1948).

Features of Japanese Fishery Co-operatives

1. They can start functioning even without a working capital.
2. They don't depend upon share capital from member.
3. They raise money from lease rights on inland or other waters.
4. For registering as fish co-operative society qualifications are 2/3rd household with in a local distt engaged in fishing for at least 90 days a year.

Foundation of Japanese Co-operative

1. Providing loan or advancing funds.
2. Awaiting banking facilities for receiving deposit.
3. Supplying goods or services needed for business or providing equipments.

4. Transporting, processing storing and selling of clothes.
5. Providing disaster relief on sea.
6. Management of fishery.
7. Educating and providing general information of fishery.
8. Welfare activities.

Fish cop fed : Fish cop fed was registered in the year 1980 however it started functioning in year 1982. Head quarter of Fish cop fed in New Delhi.

Role of Banks in Fisheries

The landing towards the fisheries sector by the commercial bank started only after the nationalisation. In 1969, the banks are landing towards fisheries sector under and broad categories.

1. Financing for mechanized boat.
2. Financing for deep sea fishing vessel.
3. Financing for inland fisheries development such as fish culture and establishment of fish hatchery.
4. Financing for B.W. fisheries such as Shrimp farming and shrimp seed hatchery.

Bank have been financing fisheries sector since 1971 and has played a dynamic role in assisting the fisherman community and fisheries industry.

Aspects to be Considered while Financing

While financing weather it is working capital loan of short term loan or long term loan the following aspects are to be considered.

(*i*) *Experience of the party, in the live :* To know weather party is experienced in the live and can take the activity successfully

(*ii*) *Project Report from applicant :* Normally the applicant don't give any project report, only in few cases project report is receive, the project will be appraised for the technical feasibility and financial viability.

(*iii*) *Suitability of the Area :* In the case of low and B.W. fisheries suitability of the area is to be considerate.

Nationalisation in 1969 and Starts Financing in 1971

(*iv*) *Proper infrastructural facilities :* The success of any prg. depend on development of infrastructural facilities such as ice plant, cold storage, market yard etc.

(*v*) *Feasibility Certificate :* In the case of marine fisheries at is required the produce certificate from the depart of fisheries. This is in order to regulate the fishing afforts in the areas.

(*vi*) *Forward and Backward linkages :* F and B linkages essential in the success of prog. such as proper road, transportation facilities dying yards etc.

(*vii*) *The tie up arrangement of recovery :* It is better that the recovery of loan is linked with the marketing so that their will not be any problem for loan recovery.

(*viii*) *Subsidy assistance :* If there is subsidy assistance the burden on the party will be less.

Scope for Financing Marine Fisheries : The scope for financing sector is lowest.

The fishing activities at present are restricted to a arrow belt of Inshore water, the inshore fishing resources are being exploited by the traditional craft mechanised transless, motorized gill nets and purse seins.

The CPUE has declaimed making the units less profitable a point of saturation has already being reached in the inshore fisheries as bank is mainly concepted towards inshore fishing the recovery is not satisfactory. This is due to the following reasons :

(*i*) *No support prize for fish* : Thus is not a deficit mode of catch sometimes it is in pluntry when the catchery plunty the price goes down or rise versa.

Fishes are highly commodity it has to be sold for a low prize. It thus is no demand, thus is at present no arrangement for fixing regular price for fish except prawn while has got high export value.

(*ii*) *Step Increal* : In the cost of Diesal and Spare parts

(*iii*) Poor fish catch.

(*iv*) Poor marketing system

(*v*) Intra group disputes case of group loan

(*vi*) Enhancement in Interest rate

(*vii*) Lack of timely fellow of action.

A. Suggestions for improvement in loan Recovery

(*i*) Introduction of multipurpose vessels

(*ii*) Reduction in capital cost as by using cheap material in boat construction

(*iii*) Better design of boat to save fuel

(*iv*) Decleration of fishing gone for different types of boats may be mode by govt.

(*v*) Marketing co-operatives should be strengthen to take up marketing activities.

(*vi*) Legislation must be framed and inforced strictly to avoid fishing during breading season.

(*vii*) To fix longer repayment period.

B. Scope for Financing Inland Fisheries : Bank has financed sizeable amount for this sector particular the state of A.P.

There are systematic aquaculam development has taken place, the recovery from this sector is better than the maintained boat sector (Man sector) the main problem in the sector are

1. *Natural Lagards* : There is a problem of cyclone followed by flood, once in 3 yrs. around the east coast of India.
2. *Problem of Disease.*

Suggestions for Improvement

(*i*) Timely follow up of the amount.

(*ii*) Insurance policy to settle the claim with in a man period of 6 months.

(*iii*) First formers are to be property educated.

Scope for financing B.W. Fisheries ® Banks, have been provided huge loan to big private companies for undertaking shrimp faring and shrim hatcherks, there is good scope in the sector problem of this sector are :

Problems of Sector

(*i*) Problem of disease.

(*ii*) Export market restricted to Japan and USA.

(*iii*) Price of shrimp is using continuously in international market as more and more countries are producting shrimp.

In order to sustain this sector it is necessory to seract the market in other countries also. Practice of new candidate spp. adoption of in term is culture systems and skill oriented training to farmers for better job performance are necessary for developing this sector.

Suggestions for Improvement : It is clear that the bank landing towards marine capture fisheries is limited where as the landing towards is unlimited.

Banking System

Introduction to Currency : Gold coins were introduced during the period of Gupta's 390 A.D. to 550 A.D.

Rupees was minted in India during the period of Sher Shah Suri around 1542 A.D.

It was silver coin weighing around 179 gms and it replaced gold coin

Paper currency in India was introduced in 1882 by the British govt.

All notes above 1 rupee are issued by RBI and bear the signature of RBI Governor, where as the one rupee bear the sign of the secretory ministry of finance.

RBI into existence on April Ist 1935 as private own bank with only 5% share of Govt. of India.

On Ist January 1949 RBI become a state own bank on account of private share holding by the Govt.

Function of RBI Regulate issue of banks note above 1 rupee undertakes distribution of all currency notes and coins on behalf of Govt.

Act as a banker to the Govt. of India and state govt. and co-operative banks.

Formulate and administer the monitory policy.

Maintain exchange value of rupee.

Represent India at the International monitory fund (IMF).

Note : No personal account are maintained in RBI.

SBI : It is the largest public bank of India and was credit after the nationalization of Imperial Bank of India on 1955, in terms of Branches 10,836. It is largest in the world.

Nationalization of Banks : Govt. of India nationalized 14 banks on July 1969 and 6 more banks on 15 April 1986. Ist 14 bank are :

1. SBI
2. Union Bank of India
3. Bank of Baroda
4. Bank of Maharashra
5. P.N.B.
6. Indian Bank
7. Indian Overseas Bank
8. Central Bank of India
9. Canara Bank
10. Syndicate Bank
11. United commercial bank
12. Allahabad Bank

13. United Bank of India
14. Dena Bank

IInd phase of Nationalization

1. Andhra Bank
2. Corporation Bank
3. New Bank of India
4. Oriental Bank of Commerce
5. Punjab and Sindh Bank
6. Vijaya Bank

New Bank of India is merged with PNB leaving 19 national banks.

Financial Requirement for Fisheries

1. Financial Sources is an essential and basic input for developing fisheries in India. Poor fisherman of fish was exploited by the money lander. Middle man, by giving then credit at high interest rate and also banking their catches at non profitable price.
2. Fishermen needed credit for their basic needs previlance of non Institutional credit arrangement in fisheries sector has contributed to the segmentation of rural financial markets.
3. Fisheries is a major area of Investment, credit landing organisation therefore providing Investment credit long, mid and short term loans equally participation consultancy services, technical and financial monitoring promotion and guidance for developing the fisheries sector.
4. Fisheries is now being considered as an Industry since it has capacity to convert waste land into productive use create job for unemployed earn lively hood for fisherman, boost up country food prod and foreign exchange by earning through sea food export so financial input is one of the key factor for developing fisheries industry and for blue revolution.
5. Finance resources for fisheries are available both external and resonance utilization, technology transfer, foreign consultancy overseage training, study tools, project preparation and monitoring studies. Sides on abroad 15 either in the form of grand or soft loans from international agencies like IDA, IMF, world bank, IFAAD, VNDP, SIDA and Asian Dena Bank, most of these organisation support developmental activities through govt. top banking organization on long term arrangement of concessional rate of interest.

Financial assistance with in the country is available from different financial institutions and banks.

E.g., NABARD (National Bankers for Agriculture and R.D.)

IDBI	Industrial Deve Bank of India
IFCI	Industrial Financial Corporation of India
ICICI	Industrial Credit and Investment Company of India
SFC	State Finance Corporation of India
SIDBI	Small Industry Deve Bank of India
NCDC	National Co-operative Development Corporation.

Central Co-operative Banks : The C.C.B. are Independent banks, the working area of these banks is limited to one distt only there are two types of C.C. Bank.

1. Pure
2. Mixed

Those banks the membership of in is confined to co-operative organization only included in pure type while those banks the membership of in is open to co-operative organization as well as individual are included in mixed type.

The pure type of banks can be seen in Kerala, Mumbai, Orissa, etc. while the mixed type can be seen in A.P., Assam Chennai a Mysoore.

The pure type of banks is based on spict co-operative principle while the mixed type does not adhare to any such type of principals.

The main fun of these banks is to finance the credit societies they carry on commercial banking activities like acceptance of deposit, giving loans receiving fided deposit, gold, goods and documents of individuals collection of bills, checks etc.

They also act as balancing centers malcing available temporary excess funds of primary society to another which is a need of them.

Co-operative bank originated in India with inactment of co-operative credit societies act of 1904 in provided for the formation of co-operative credit societies, a new act was passed in 1912 in provided for the establishment of co-operative central banks a union of primary-credit societies.

The chief sum of these banks are :

1. Attracting deposit from non agriculturist.
2. Using excess funds of some societies temporarily to makeup for shortage in others.
3. To superwise and guide the affiliated societies.

Co-operative banking in India is federal in its nature, at the lower level there are primary credit societies then there are central union or central co-operative banks and act the top there are provincial co-operative banks or state co-operative banks or Apex bank.

The primary societies may be coopaired with joint bank their main fun is landing money to villagers on carrier ferms, much of their works is done by members themselves they have their own funds supplemented by funds drawn from central co-operative banks.

The central co-operative bank obtain the funds from share capital deposit, loans from apex bank and where Apex bank do not exist from the Reserve Bank of India.

The Apex banks obtain their funds from share capital deposits loans from commercial banks. R.B.I. and the Govt.

Rural Banks

Institutional finance is an important prerequistic for rural development in the existing delivery system there are and channels for institutional credit in rural areas. There are :

1. Commercial Bank (C.B.).
2. Regional Rural Bank (R.R.B.).
3. Short term co-operative Institutions viz primary agricultural credit societies (PACS) and their Apex body at the distt level *viz.*, District Central cooperative bank.
4. Long Term Cooperative Credit Organisation.

NABARD : National Bank for Agricultural and Rural Development.

It is devoted to Agricultural and Rural areas and financial resources are most important for fisheries sector. NABARD has played role in both market and layland sector resulting in the phenomenal growth in fish production.

Credit for this change could be shared by Govt. of India and state Fisheries department and research institutes banking sectors. That's why fisheries now being recognise as a "major area of investment". Fisheries is now getting Lones as an Industry since it has the capacity to convert wasteland into production and foreign exchange earn through sea food export.

NABARD was established in July 1982 through an act of parliament by merging Agricultural Credit Development (ACD) of Reserve Bank of India and Agricultural Refinance Development Cooperation (CARDC).

It is an Apex financing institution for providing refinancing facilities to all scheduled banks for multipurpose, multiterm and multidevelopment oriented credit project for Agri and rural develop of the country.

System is one of the prime areas among national bank refinance operations. Bank credit facilities are available to all Commercial Banks (CBS) Regional Rural Banks (RRBs), Land Develop Banks (LDBs) State Cooperative Banks (SCBs).

NABARD loans can be given on long term or mid term basis to all the banks but short term or cooperative loan is available only to RRBs or SCBC. Among India group of entrepreneur co-operative societies, fish formers federations, state fisheries development cooperatives and corporate groups can avail loan for fisheries development.

Role of NABARD in B.W. Sector

The bank supported cultural activities in about 7,000 ha for developing coastal aqua by strimp farming techno. In fisheries sector I research projects with financial assistance of rupees 91 lakh have been supported for standardizing field level technologies field of mariculture.

Role of NABARD in Mariculture

Now NABARD refinance assistance is available for all subsector of fisheries *i.e.,* projects covering distant water and coastal fisheries and fisheries projects covering aquaculture.

NABARD also support for setting up infrastructure like fish processing plants. Its plants, cold storage, hatcheries, market yards etc.

NABARD has so far finance were than 5408 fisheries schemes with total re-finance of rupees 583.43 crores.

To the field of mariculture bank have servisioned a self lone assistance to Tamilnadu pearl culture joint venture company found by southern petrochemical Industries cooperation.

One research and deve project to demonstrate pilot scale operations of edible oyster culture was sanctioned for supplementation by CMFRI with a total grant of rupees 6.44 lakhs.

As an outcome of such afforts the bank has so far senctioned refinance assistance for two mariculture practices.

Marketing of aqua products is the performance of all business activities involved in flow each aqua proof and services from the initial aqua production until they are enhance of consumer.

Marketing is the name given to management process responsible for finding out what consumer needs and supplying there as efficiently and profitably as possible.

Marketing starts from fish farm and ends with the satisfaction of consumer fish marketing is neither a mechanical nor an automatical operation. It can be a complex process. Where the product is changed in the form such as from fish to fish cake or fish sausags etc. and the food undergone many buying and selling transection before reaching to the ultimate consumer.

Marketing has a central role in the management of fish farmer decision made on any aspect of fish farm management in affect the consumer and producer are part of marketing and must be considered from the marketing point of view.

A market may also be described as an element for organising and facilitating business activity and for answering following basic economic questions.

What to produce : Size and quality and the price this will depends upon the liking of the consumer.

How much to produce : How much of each species and size to be produced at different periods, it will depend upon marketing demand.

How to distribute : It will depend upon the marketing system in the area.

Marketing is a productive task it creates utility *i.e.,* marketing makes goods and services useful.

The marketing process is the movement of proof, it is a series of action that takes place on a sequence.

Marketing Functions : Can be defined as major specialized activity an accomplishing the marketing process.

1. Exchange Junction ® Buying and Selling.
2. Physical Fun : Storage of product, transportation and handling and processing.

Facilitating

(*i*) Standardisation
(*ii*) By financing
(*iii*) Risk bearing
(*iv*) Market intelligence

Exchange Function : Are there activities to are involved in the transfer to fish, they are present at the point at in prize determination is existed buying is simply assembling of various species of fish and fish products.

Selling fun involve a series of action caller merchanting means ownership, advertising and promotional efforts are also apart of selling, this process include, packaging displaying and advertising.

Physical Function : Are those activities that involve Landling movement and physical change of actual commodity. It provides storage transportation handling and processing.

(*a*) *Storage fun :* It is considered with making fish available at desired time. It may be done in refrigerator of storage houses.

(*b*) *Transportation :* Is concern with making fish available at proper place, this process is very sophisticated in developed countries.

Size may be transported live, in tanks and processed in large refrigerated trucks.

But in developing countries fish may be transported in 10 box or plastic bags.

(*c*) Handling and processing function. It involves all those basic activities in are required for freezing, drying, canning, smoking and salting.

Marketing Channel

Fish Marketing : It was used to denote supply, buying and selling of first at landing center but present system of fish marketing is not only to melt the demand of product but also creating demand in the market. Changing in the form of sale are to consumer liking and need of consumer would result in more profit however processing the fresh fish until it reaches to the consumer is one of the essential requirement for using its market, new products have to be designed by developing a packing and brand name and fix price of product to get a return. Constent advertising has to be done to aware public about new products.

Traditional Fish Marketing System : All the traditional methods of fish marketing are based on some customs they have been uncharged and unequipped over decayed the retailer is namely a women to sale the fish, the fish is usually is a whole fish not cutted or there is no scientific procedure for fish marketing by retailer and there is no maintenance of fish quality.

Modern fish Marketing System : For keeping fish fresh for long time fish has to be preserved by proper means for enhancing marketing of fish and fish products, some times fish is in its original form after its capture may not be immediately developed by the consumer but when preserved into other forms would find a good market.

Micro and Macro Level Fishery Trend

(*i*) **Micro Level :** In this method of marketing small fish markets are involved who arrange the distribution of fish from landing center to main consuming center.

(*ii*) **Macro Level :** Sea foods of new varieties are processed at production center and distributed by some organization, state fishery department, fishery co-operative societies and other marketing agencies. Frozen products are transported in refrigerated trucks at a regular interval of time to each retail marketing center.

Market Intermediaries : There are six capitaries of intermediates providing a link b/w fish provides and consumer there are :

1. Option Auctional.
2. Purchasing Commission agent.
3. Whole saller.
4. Selling Commission agent.
5. Retailer.
6. Bendors.

1. **Auctional :** Is an Important Intermediator in marketing chain except in the cases their family member sale the fish as a retailer they conduct the auction of fish landed on the beaches taking commission from fisherman they also give working capital loans.
2. **Purchasing Commission Agent :** C.A. operate at all lensing centers they collect the fish from whole sale at large landing center or consuming centers. Individually after the auction the CA makers payment to the fisherman or to the auctional, they also assist the whole saller on packing and transportation of fish to their respective centers they provide loan to fisherman through bank.
3. **Whole Seller :** They directly process the fish from fisherman or through the commission agent from different landing centers and sale them to retailers, they usually finance fisherman for boat, and to make social obligation directly or through C.A.
4. **Selling Commission Agent :** The category constitute the most Important link in whole sale market they are predominant at consuming center they are engaged in receiving fish from whole saller for supply to the retailers, they directly finance fisherman to purchase coat net etc. and also for melting various other requirements of fishermen.

5. **Retailer :** They are either had load wanders or cycle renders. They carry first to the consuming centers for retail marketing retail wanders buy fish from C.A., whole seller or fishermens.

Fisherman Share in consumer rupees : There is no uniformity in the price even for quality fish at a given point of time or location as the fish move through the intermediator to consumer the price varies with place or location of the market and time of selling etc. The producers price is equivalent to the amount given by the first intermediates or auctional, the amount received by last intermediator is the consumer price, the different b/w the price in consumers pays and the fishermen actually receive is dependent on factors such as nor of intermediators and distances b/w landing and consuming center, usually the fisherman get maximum share of consumer price in a distinguishing claim involving only one intermediate.

The Cooperatives

Earlier the cooperatives were treated as parasitic/depending organizations. It is time for the cooperatives are coming up and standing on their feet and they have to be like professional organizations. It is a well-known fact own Cooperatives have a great role in rural development. The cooperatives are innation-building in a different form. They are as professional organizations.

The Cooperatives are facing cut-throat competition as the market is open and terms are dictated not only by internal factors but also external factors. At this juncture, professionals are highly indispensable to manage the affairs of the organizations. The thump rule is "professionals will flourish and non-professionals perish". And Cooperatives are no exception to this rule.

Professionalism has many folds at many levels and is also handled in many different ways. The professionalism, we think of, for agricultural credit cooperative society might be different from that of an urban cooperative bank or housing/diary cooperative society. One might find that what an organization sees as being professional, others would see differently. This can cause considerable confusion for some other sector of cooperative, trying to define professionalism in their own business. But, the specific point to be kept in mind is that the core definition of professionalism is always the same.

Professionalism is the expertness characteristic of a person/organisation with a conscientious awareness of the role, image, skills, knowledge and commitment to quality and client-oriented service. In other words, Professionalism is a focussed, accountable, confident, competent, motivation towards a particular goal, with respect for hierarchy and humanity, with less of emotion.

The Secretary/President of a cooperative society leave out the outbursts and emotional thrills that accompany stressful situations and success. They maintain focus, a sense of urgency, and accept responsibility on a path towards a specific goal of cooperative society. In the process, they maintain respect for their people in the society. The Secretary must understand that most of the members of the society are moderately educated, in weaker section societies. He should understand that emotion varies wildly between individuals. Emotional responses have no place. Moreover, this wastes his valuable time in moving towards goal. He should understand that a business situation has a purpose and a goal, there is advantage in dealing with professional situations without emotion; it provides a common foundation from which professional relationships can flourish.

When a co-operative society is called a professional organization?

There are many parameters to make an organization professional. For co-operative societies, following are certain areas wherein co-operatives have to practice so as to become professional organizations. Adhering to one or two areas listed out below is not enough to become a professional co-operative. Integration of many parameters is highly indispensable. Hence, people in the cooperatives a various levels have to stick on to practice the following parameters to make the co-operatives a professional one.

Act and Rules

To become a professional organization, a Cooperative has to :

1. Adhere to various provisions to the Cooperative Act and rules. Special care is needed at the time of elections. The society should not endeavour any undemocratic steps. Society should not allow membership just for the sake of elections, *i.e.,* registering more members just before the election period.
2. Conduct regular meetings like Management Committee (MC), General Body (GB) etc., for taking decisions to the advantage of the Society as well as members as a whole and not merely for the a few members/directors. It should shun the decisions with vested interests.
3. Adhere to not only Cooperative Act and rules but also other allied laws applicable to the cooperatives.

Bye-Laws

1. When the cooperatives prepare its own bye-laws, they have to identify the needs, urgencies and interests of the members, and then set its objectives. It should not simply copy the bye-laws prepared by a similar society functioning elsewhere.
2. Amend the bye-laws after adhering to the act and rules. The amendments should be in the best interest of the members as well as society as a whole, and certainly not for the vested interest of a few people.

Financial Code

1. Cooperatives should not indulge in any financial irregularities, malpractices, etc.
2. Cooperatives should maintain all books of accounts in a neat and crystal clear manner as per statute and get it audited in due course of time.
3. Societies should timely pay taxes.
4. Cooperatives must settle dues if any, to the various parties timely if due.
5. Societies have to invest the Society funds judiciously with maximum returns and not in stock market or any other kind if investment for their personal gains as done by some of the urban banks.

Knowledge Economy : A Development Perspective

The ancient belief that Knowledge is Wealth assumes prominence in the contemporary world, which is rapidly changing from an industrial economy to a knowledge-driven economy. It provides ample opportunities for India to gain much in promoting economic progress and social welfare. Dr. R.A. Mashelkar, the Director of Council of Scientific & industrial Research, has stated that *India* will be among the world's three biggest economies by the year 2050 and we will be there much earlier if we can master the economics of knowledge.

For the last 200 years or so, the neo-classical economics has recognized land, labour and capital as the given factors of production. This has been changing over time. Information and knowledge are replacing capital and energy as the prospective wealth-creating assets. In addition, technological developments in the 20th century have transformed the majority of wealth creating work from *physical* based to *knowledge* based. Technology and knowledge are now the key factor of production.

The economists have termed this as *the expansion of the production frontier.* The source of technology is science and science is rooted in knowledge (Mashelkar, 1999).

The history of nations clearly shows that their progress from poverty to prosperity over centuries was conditioned by a leading sector, which formed the engine of their economic growth. In Britain, it was textiles; in the U.S., the railways; in Sweden, the timber and timber products, and; in Denmark, milk and milk products. Over the last 50 years or so, India has been striving to find its leading sector. Knowledge sector has now emerged as the leading sector of the Indian economy.

The economy is rapidly shifting along several dimensions from manufacturing to service to capital resources to knowledge resources.

The sectoral composition of the Gross Domestic Product (GDP) changes with economic development. As the economy develops, the predominance of agriculture is reduced by the increasing importance of manufacturing and subsequently services. As such, the rates of economic growth tend to increase. The transition is now occurring globally and is reflected in the unprecedented growth of the services sector, especially in the fields of financial services, information and communication technology, insurance, education and health. The economy of their primacy is termed as Knowledge Economy. The dimension of the New Economy is nothing but the realm of knowledge and the knowledge economy provides the pathway for future progress and prosperity.

The knowledge economy is an economy of a *work in progress* which requires significant investment in harnessing skills, technology and learning. In a knowledge driven economy, scientific knowledge and information from the major sources for creating value. There should be rapid changes in technology, wherein greater use of information and communication technology help the growth of knowledge-intensive industries with increased networking and enhanced social engineering.

These are in contrast with the earlier economies such as agrarian economy where agriculture forms the key elements and industrial economy in which manufactures help to generate wealth. In the emerging knowledge economy, as much as, if not more than labour and capital, knowledge becomes the *basics* in creating wealth and improving the quality of life.

The emerging knowledge economy has these four goals for India's future development *viz.*, the innovation goal, the economic goal, the environment goal and the social goal. The environment goal is set to accelerate knowledge creation and development of human capital, social capital and learning system and networks in order to enhance India's capacity to innovate. The economic goal is to increase the contribution that knowledge makes to the creation and value of new and improved products, processes, systems and services in order to enhance competitiveness of Indian enterprises. The environment goal is to increase knowledge of the environment and of the biological, physical, social, economic and cultural factors that affect it in order to establish and maintain a healthy environment that sustains nature and people, and the social goal aims.

If we wish to transform India into a major knowledge-based economy and also face the challenges posed by the WTO regime, protection of traditional knowledge, which has the potential to create wealth and social good, is crucial for the nation. The protection of our biodiversity resources takes us to a tremendous responsibility of strengthening the Intellectual Property Rights. Every endeavour has to be made to meet this challenge. Our ancient knowledge and culture should also be protected against unjustifiable attacks placed by multi-national corporations.

The Rau's IAS (Quoted as example)

The Vision

Rau's IAS Study Circle was established as a top ranking institute nearly 50 years ago, solely with the aim of helping serious students achieve success in Civil Services Exam by providing the highest quality coaching. The method, content and teaching standards established by the Study Circle have become synonymous with success in the minds of civil service students.

Be Sure, we have no branches or associates any where in India. Our name which has become a legend among students for the highest standards in teaching, and hence has been copied by a lot of centres across India, but it can never be equalled.

It is a well known fact that Rau's is the most trusted and recommended name all over the country for IAS, PCS and Judicial Services Coaching.

- As the President of the cooperative, he should know what is exactly needed by and wanted by the members. He must possess a service mind-set and be of a helping nature not only to the members but humanity at large.
- The people in the cooperative society should keep work area clean and surroundings orderly. A professional never keeps the work area dirty.
- The personal in the cooperative society should look decent and speak well. They consistently are well-groomed and appropriately well-dressed.
- The personnel of the cooperatives have to learn every aspect of their job. They should always update their knowledge by way of taking part in seminars, workshops, training programmes organized by professional bodies. They are supposed to believe in modern education/technology, but at the same time maintain the values, principles and philosophy of the cooperatives.
- The President/Secretary should be optimistic. He should not get upset. He should consistently be positive, constructive in nature, pragmatic and more strategic in his approach.
- A professional secretary of a cooperative society should be enthusiastic, cheerful, contended. He keeps no scope for anger, hostility, fear etc.
- To become a professional, one should become member in an association, which is relating to our job/profession. Likewise, a cooperative society should be member of a union or federation at district level/state level/national level. The cooperative should function in such a way that even the federation gets inspired.
- A professional cooperators always tries to manage the scarce resources to the best advantage of the society for optimum returns.

The cooperative in India, even though they have completed hundred years (1904–2004), still wait to demonstrate their actual competence at fullest capacity. The reasons might be many, such as Board of Directors blame the Cooperative Department, in turn the Department blames the actions of the Management of societies, and both of them blame the employees of the societies, saying that they lack the competence in timely executing the orders, are unable to run the business on sound lines, lack vision, lack proper goal setting, etc. However, it is the time for the cooperative to change their mindset and plan to grow vertically with forward and backward linkages simultaneously so as to survive in the competitive era. It is possible that the cooperatives can grow vertically, only when the employees of the societies possess competence. Besides this, the Board of Management too bring changes in decision-making processes with creativity and innovativeness, considering changes that are taking place in the environment, government policies. Initiative to move up and keep up the professional standards in their business with intensively trained staff and bring in awareness among members that they are the owners, and part and parcel of the cooperative society.

It is high time the cooperative aligned and focussed their systems and procedures towards the changing needs of their members. It is possible only when the cooperatives take up the performance measurement at Board of Directors' level, employees level seriously.

There should be adherence to high values and ethical code. The values and ethics at any cost should not be diluted or practiced at the surface level; rather these have to be deep-rooted to emerge as market leaders.

1. Increasing productivity by following management methods if possible, such as Business Process Re-engineering (BPR), Kaizen (Continuous improvement), and zero defective goods/service delivery, TQM, Benchmarking.
2. Reducing wastage and increasing productivity, growth, generating profits (as least modest).
3. Bringing out new and value added products by understanding desires of members.

Cooperative Values

A cooperative society is treated as professional when :

1. It practices the values, ethics, philosophy,
2. Adhers to reformulated cooperative principles and values framed by the International Cooperative Alliance (ICA).

How people in the cooperatives can become professionals

It is very difficult to say as to how people in the cooperatives can become professionals. It requires a lot of practice and should gain from experience. The people ought to know certain business, behavioural, technical tips and management techniques required to get shaped as professional. Let us discuss what happens of a *Military Band* that has obligations as it was hired. These obligations include :

- Presenting the best performance by the individual with collective abilities by every one and every time.
- Arriving at given time (early would be better) and be ready to perform, practicing repeatedly to work through the rough spots.
- To be aware of maintaining the visual experience of their music, uniform, colourful and impressive dress, minimal clutter of on stage performance (no instrument cases, coats, unwanted items, etc.)
- Commitment to work or play the melodious and captivating band, even if part of the concert may not be the most exciting, challenging, and delightful.
- Maintaining quality performance and rehearsal-discipline, which shows that the Band is much serious about their music.

Now, it is very clear as to how tough it was to complete obligations. As they are professionals, the band could do it well because of hard practice, rehearsals, and concentration and become perfect professionals. In Cooperatives too, professionalism is quite possible. Best examples are Amul, IFFCO, KRIBHCO, etc. However, following are certain requisites/qualities that are required for people of cooperatives to become better professionals.

- First of all, the people of the cooperative institution have to question themselves – "Are they professionals?"
- A good cooperator/professionals cooperator produces a high-quality product or provides good service. He is a good leader, inspires his followers and remains ahead in the market.
- To become a professional, the employees should avoid shrinking from difficult assignments. They should stick to it and see that it is accomplished.

They should complete the projects at the earliest. They should never pile up the unfinished work. They should consistently accomplish tasks and responsibilities.

- A person is called a professional when he is focussed and clear-headed. He should have clarity in his actions and never distract from them.

- A professional cooperator persists until the objective is achieved. He ensures that the objectives are for the advantage of members, society and humanity at large. He always evaluates the objectives of the Society, which are set to be achieved.
- Generally, a professional does not commit mistakes. Even if it happens accidentally, he tries to learn from the mistakes and ensures that they do not occur again.
- A professional handles financials and accounts very carefully. He is the trustee of the organization and will never desert members of the cooperative.
- A professional Secretary/President of a cooperative never makes false promises but will definitely boost up the confidence.
- A professional believes in value-based management and good business practices.

Strategic Code

1. A cooperative Society (even a small Primary Agricultural Cooperative Society) should set its mission statement, vision statement objectives and goals.
2. The cooperative needs to indicate its activities clearly, *viz.*, what to do, how to do, when to do, who has to do and why to do.
3. Stress needed on business development plans and other allied plans, etc.
4. Always stick to fundamental rules as well as to the strategic code. Evaluate time-to-time alignment with related organizations according to the changing environment for the benefit of the Society as well as members.
5. Evaluate the objective every year and remove unachievable objectives to have a clear-cut road-map of the Society.
6. Cooperatives have to churn the concepts such as "organization learning" and "learning organization" to formulate strategies to reduce weaknesses and increase strengths to avail of the opportunities.

HRD Code

1. The Society should follow the strategies for development of human resources, fair wages, knowledge, skills, competencies, attitudes increasing values and ethics among employees.
2. Organize in house and outdoor educational, training and developmental programmes. Deputing its human resources to the professional institutes in cooperatives, such as Vaikunta bhai Mehta National Institute of Cooperative Management, Pune, IRMA Anand, and Institute of Cooperative Management located in various parts of the country.
3. Treating human being as an important asset, measure performance of the personnel correctly every year, carrot to performers, sometimes stick to non-performers.
4. Making Total Quality People (TQP) in the Society.
5. Maintaining good and cordial relations with the employees. Increasing zeal and morale, job satisfaction among employees.

Members-Driven and Member-Focused

1. The cooperatives should focus at the needs of the members, assessing and reassessing their needs every year and explaining to the members about the status and changing activities of the Society.
2. Getting valuable guidance from members.
3. Maintaining member satisfaction and security.
4. Member development through member education programmes.

Technology Upgradation

To be professional :

1. The cooperatives have to update its technology and introduce modern technology.
2. Maintain Management Information Systems (MIS) to facilitate the best decisions.
3. Computerize the various operations.
4. Change methods of office procedures and communication systems.
5. Be inclined towards research, innovations and developing its creativity.

Business Development and Productivity

Cooperatives will become professional when they aim at :

Professionalism has many levels in the organization and is handled in many different ways in cooperatives. The cooperatives in India, even though they have completed hundred years (1904–2004), still wait to demonstrate their actual competence at fullest capacity. However, it is the time for the cooperatives to change their mindset and plan to grow vertically with forward and backward linkages simultaneously so as to survive in the competitive era.

SHG is always considered an important institution for improving the life of women in various economic and social components such as health, education, human rights, water and sanitation, etc. However, in reality it does not happen without any specific inputs or direction.

Even if it happens, it is very slow. To make these SHGs a success, an elaborate planned field work was done.

Identification/Selection of Beneficiaries

For the purpose of awareness building regarding benefits of entrepreneurship and self-help groups, an information workshop was conducted. This was done through public meetings, bronchures, posters, visual aids and skits.

SHGs are considered as one of the most significant tools to adopt participatory approach for the economic empowerment of women. It is an important institution for improving the life of women on various social components. The basic objective of an SHG is that it acts as the forum for members to provide space and support to each other. SHGs comprise of very poor people who do not have access to formal financial Institutions. It enables its members to learn to cooperate and work in a group environment.

To identify the beneficiaries, a survey was conducted at Janata Colony and on the basis of this survey, these potential groups were identified. A market survey was conducted in the market to seek out the products for which demand is there.

Criteria for Selection of Potential Entrepreneurs

- Residence in concerned area for more than three years.
- Matching entrepreneurs (potential) with the enterprises.
- Assessment of the commitment, interest and involvement of men/women towards self employment.
- Determining the latent business competence existing in men/women.

Formation of SHGs

Rules and Regulations

Number of members was from 5 to 10.

Not more than one member from a single household.

- Criteria and procedure for approving individual members loan applications.
- Procedure for electing group officers, length of term.
- Attendance at meeting.
- Criteria and procedure for expulsion from the group.

It was decided that members must observe the group's rules and regulations which are determined partly by the implementing agency and partly by group members themselves.

Objectives

- Saving
- Income generation and gradually becoming self reliant.
- Internal and external lending.
- Discussions on common problems.
- Social development.

Training

Groups were given training for development for skills like :

- Record keeping.
- Awareness of different related documents like promissory note, loan sanction letter etc.

Credit Delivery

For credit delivery, a sanctioning committee was formed with following, members in the panel – representative of the bank, a member of the selection committee, member from the local community (Sarpanch/Pradhan/BDO, etc.), appointee from the Commonwealth Youth Programme (CYP) Asia Centre and two senior members from the NSS unit of PEC. The initial sanctioning limit was :

For Individuals

A maximum sum of Rs. 5000/- (Rupees Five Thousand Only) as their first loan.

For SHGs

A maximum of four times the money saved by them in the incubating period or a maximum of Rs. 40,000/- (Rupees Forty Thousand Only) whichever is less.

Monitoring : was done to evaluate the functioning of SHGs and to develop future strategies.

The different purposes of monitoring were to evaluate :

- Are the meetings being held regularly?
- Are the rules and regulation being followed?
- What is being done with savings?
- Role of leader in relation to the group?
- Are the members active?
- Whether the money is being utilized for the purpose for which it was taken?
- Whether the recovery of loan is on time?
- Reason, if any, for non-repayment of loans, identification of problems?

Monitoring Machinery

Monitoring was done by field inspectors, NSS volunteers, village heads, banks and CYP, various documents which were monitored were minute book, saving register, loan register, individual passbook, individual and expenditure book, inventory register. Group monitored at its level, the attendance and loan repayments of its members. Group imposed fines, penalties to defaulters and offered incentives for those who worked efficiently.

Tools for Monitoring and Evaluation

A tool for monthly, quarterly and annual monitoring was developed to evaluate the functioning of these SHGs.

Monthly Report

It was from implementing agency to lead agency and management advisory board.

- No beneficiaries defaulted.
- Repayments overdue.
- Loans outstanding

Quarterly Report

- Detailed financial statement.
- Performance indicators.
- Narrative report – achievement of goals.

Annual Report

- Detailed financial statement.
- Narrative focusing on outcomes of programme.

Successful working of these three SHGs has given enormous benefits. Organized working of the women through these SHGs has increased the income of the families involved. Most of them are now able to repay their old debts and started asset building. The existing enterprises of beneficiaries are better managed now. Success of these SHGs not only improved the economic status of the women concerned but there is also a drastic change in their social status.

State Level Marketing Reforms : A Status Report

- The Need for Reforms.
- Creating an Enabling Environment.
- Public Private Participation – Some Case Studies.
- Inter-State Barriers to Trade.
- The Way Forward.

Introduction

- The recent boom in India economy (GDP growth of > 8%) has been led by industry and services sectors.
- Agriculture, the dominant sector has remained, more or less, stagnant (2.3% growth).
- Agriculture contributes 24%, to the GDP but over 63% of the population is dependent on it for livelihood.
- Marketing is the key not only to catalyze agricultural development but to foster inclusive growth.

- An efficient marketing system can :
  - Reduce post-harvest losses
  - Enhance formers' realisation
  - Reduce consumer prices
  - Promote grading and food safety practices
  - Induce demand-driven production
  - Enable higher value addition
  - Facilitate exports

EXISTING SUPPLY CHAIN

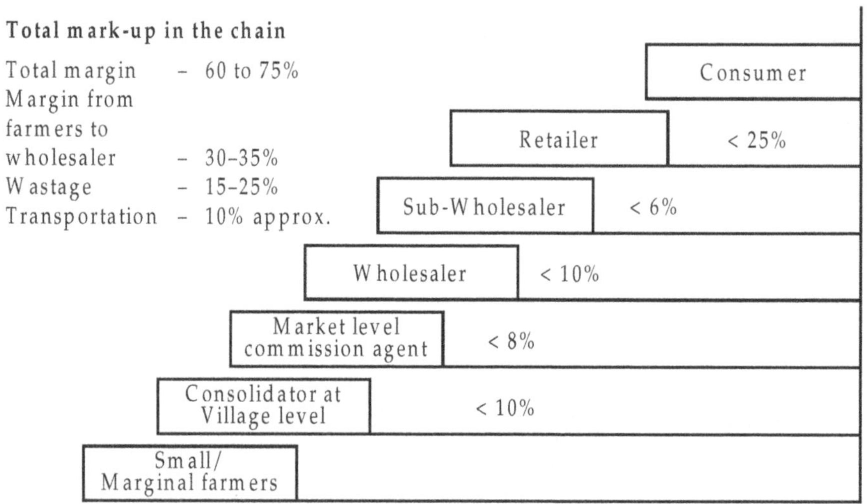

Farmers get 25-30% of consumer price

WHAT AN EFFICIENT MARKETING SYSTEM CAN DO

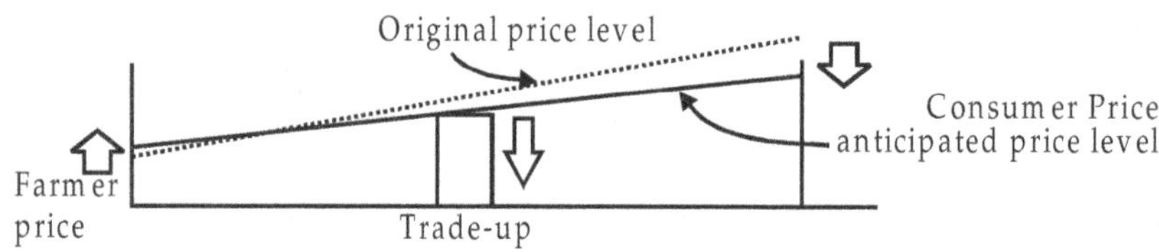

Reduce intermediaries
increase farmers' realisation and lower consumer prices

PRESENT POSITION

- APMC Acts prevalent in most States.
  - 7,521 regulated market.
  - 27,294 rural periodic markets (village heats).
- Area served varies widely (from 114 sq km in Punjab to 11,215 sq km in Meghalaya) –average 459 sq. km.
- APMC act creates a State monopoly in marketing of agricultural produce.

- Most regulated markets lack critical infrastructure.
    - Estimated cost of developing infrastructure Rs. 12,230 Crs.
- High transaction costs (Punjab–11.5%, Haryana 10.58%).
- Lack of grading and standardization.
- Undue influence of Commission Agents (Arhtias).

Marketing Reforms critical to development of Agriculture.

RECOMMENDATIONS OF NCF

- Legal Issues
    - State APMC act needs to be amended.
    - Harmonization of inter state and centre-state tax.
- Marketing Issues
    - Auction system should be more transparent.
    - Professionalization of the exiting regulated markets.
    - Encouraging private sector investment.
    - Allowing direct marketing.
- Financial Issues
    - Credit availability for eliminating distress sale.
    - Need to encourage secondary markets for distress sale.
    - Future trading in commodities is important
- Contract farming
    - The Government should work out a farmer centric 'Code of Conduct' for Contract Farming arrangements, which will form the basis of all contract farming agreements.

THE MODEL ACT

- The State Agricultural Product Marketing [Development & Regulation] Act, 2003 – main features :
    - Direct Marketing by producers.
    - Establishment of markets by private bodies/co-ops or PPP.
    - Separate markets for special commodities.
    - Promotion of Contract Farming.
    - Prohibition of commission agency in transactions with producers.
    - Role of APMC in promoting alternative marketing system, contract farming etc.
    - Wider role of State Marketing Boards in training & extension.
    - Constitution of Standards Bureau at State Level.

REFORMS UNDERTAKEN BY STATES

- State/UTs where APMC Act has been amended to permit Direct Marketing; Contract Farming and establishment of markets in Private/Coop Sectors :
    - Madhya Pradesh
    - Himachal Pradesh
    - Punjab
    - Sikkim

  - Nagaland
  - Andhra Pradesh
  - Rajasthan
  - Chattisgarh
  - Orissa
  - Maharashtra
  - Arunachal Pradesh

MADHYA PRADESH

- Agriculture Marketing Reforms.
- Single License
- Contract Farming
- Establishment of soil-testing laboratory in Mandi yard.
- Soil sample collection centers in mandis tied up with Kisan Bandhus.
- Farmer Road Fund.
- Agricultural Research & Infrastructure Development Fund.
- E-Agricultural Marketing – 'EKVI'

HIMACHAL PRADESH

- The Himachal Pradesh Agricultural & Horticultural Product Marketing (Development & Regulation) Act, 2005.
  - Setting up of private market yards other than committee.
  - Direct sale/procurement from farmers field.
  - Contract farming.
  - Public – private partnerships for market management.
  - Market fee.
  - Setting up of separate Agriculture Produce Marketing.
    Standard Bureau for grading and standardization of the produce.

SIKKIM

- All provisions amended.
  - Setting up of private market yards other than committee.
  - Direct sale/procurement from farmers field.
  - Contract farming.
  - Public–private partnership for market management.
  - Setting up of separate agriculture produce marketing standard bureau for grading and standardization of the produce.

PUNJAB

- The state has amended all the provisions including.
  - Setting up of private market yards other than APMC markets.
  - Allowing direct sale/procurement from farmers. No fee on wheat, maize, fruit and vegetable purchased by processing industries through contract farming.

- 2 years exemption for new units of fruit and vegetable processing even without contract farming.
- No change and provision on promoting public–private partnership and market fee.

MAHARASHTRA

- The state has not amended some of the provisions of the act and has added chapter wherein.
  - The director may grant license to any person to establish a private market.
  - License for Direct marketing.
- There is no provision for contract farming but the government is considering to issue an ordinance in the matter.

States Where Partial Reforms Undertaken

- Direct Marketing.
  - Haryana, Karnataka, NCT of Delhi, Chandigarh, UP
- Contract Farming.
  - Haryana, Gujarat
- Market in Private/Coop Sector
  - Karnataka (Only NDDB)

REFORMS INITIATED

- States/UTs where administrative action is initiated for the reforms.
  - Assam
  - Mizoram
  - Tripura
  - Meghalaya
  - J & K
  - Uttaranchal
  - Goa
  - West Bengal
  - Pondicherry

JAMMU & KASHMIR

- Reforms for Agriculture Marketing
  - Establishment of Terminal Market to Operate on 'Hub and Spoke' format.
  - 40 Satellite/Terminal markets and 22 Apni Mandies are being set up in the State.
  - Kissan Ghar at New Delhi consisting of 24 rooms has been commissioned.
  - Toll Tax on export of fruits exempted.
  - Market Intervention Scheme (MIS) for procurement of C–grade apple and Garden Fresh Scheme for marketing of premium quality apples have been introduced.

KARNATAKA AND WEST BENGAL

- Karnataka

- APMC act has not been amended but has allowed NDDB and Raithra Santhe (farmers' Market) to set up private markets and perform direct sales.

- West Bengal
  - The state has not amended its APMC act but contract farming in the name of participatory farming is practiced informally in the state.

STATES IN WHICH MAJOR REFORMS NOT REQUIRED

- States/UTs where there is no APMC Act and hence not requiring reforms
  - Kerala
  - Manipur
  - Andaman & Nicobar Islands
  - Daman & Diu
  - Lakshadweep
- States/UTs where APMC Act already provides for the reforms.
  - Tamil Nadu
- Only state where the APMC Act has been withdrawn
  - Bihar

OTHER MARKETING REGULATIONS AND REFORMS INITIATIVES

- Recent steps towards deregulation of domestic markets.
  - Essential Commodity Act has been substantially pruned-now restricted to only 15 entries.
- Restrictions on storage and movement of certain food items reimposed recently.
  - Amendment of Milk and Milk Products (Order (MMPO), 1992.
  - Integrated Food Law (Food Safety and Standard Bill, 2005) passed by Parliament.
  - Warehousing (Development and Regulation) Bill 2005 has been placed before the Parliament.

PUBLIC PRIVATE PARTNERSHIP

WHY PUBLIC PRIVATE PARTICIPATION?

- Massive investment needed to provide critical Agricultural Marketing Infrastructure.
  - Rs. 12,234 crs, for regulated markets.
- Constraints to Public Investment :
  - Gross Fixed Capital Formation in Agriculture as a % of GDP has come down to 1.6%.
  - Choice between competing needs.
  - Delivery mechanism
- PPP ensures
  - Efficient resource utilization and commercial insight.
  - Better management and operation.
  - Overcomes last mile delivery hurdles.
  - Achieves targeted results.

SOME CASE STUDIES IN PPP

- Modernisation of Wholesale Markets.

  - Safal Market in Karnataka.
  - Several new Terminal Markets coming up in other states following amendments in APMC Acts.
- Marketing & Distribution
  - ITC E–Chaupal
  - Haryali Kisaan Bazaar
  - Mahindra Subh Labh
  - Cargill Farm Gate Business
  - Tata Kisan Sansar
- Commodity Exchanges & Future Markets
  - National Commodity & Derivatives Exchange Limited (NCDEX).
  - Multi Commodity exchange Limited (MCX).

MODERNIZATION OF WHOLESALE MARKETS

- NDDB has set up a modern wholesale market for Fruits and Vegetables near Bangalore facilitated by an amendment of the Karnataka APMC Act.
- Its main features are :
  - Electronic Auction for transparency in price setting.
  - Assistance in post harvest care at Collection Centers.
  - Packing and grading.
  - Direct contract between growers' cooperatives and buyers without intermediatories.
  - Prompt delivery and payment.
  - Add on facilities such as mechanized handling, Ripening Chambers etc.

GRADING OF PRODUCE BY THE FARMER

- Collection centers set up near production areas.
- Extension and QC services provided.
- Transparent price setting mechanism.
- State Govt. facilitated the project by amending APMC Act.

NEW MARKETS COMING UP

- Flower Auction Market in Mumbai.
- New Wholesale Market in Thane.
- Wholesale Markets in Solan and Parwano in Himachal Pradesh.
- Expression of Interest invited for 8 new Wholesale Markets by GoI.
- Rs. 100 crores provided for new markets in the National Horticulture Mission.

PRIVATE SECTOR INITIATIVES IN AGRICULTURAL MARKETING & DISTRIBUTION

- ITC 'e-chaupal'.
- Haryali Kisaan Bazaar.
- Mahindra Shubh Labh.
- Cargill Farm Gate Business.
- Tata Kisan Sansar.

SAATHI IS A "SOLUTIONS AND ACCESS" PLATFORM MANAGED BY CARGILL

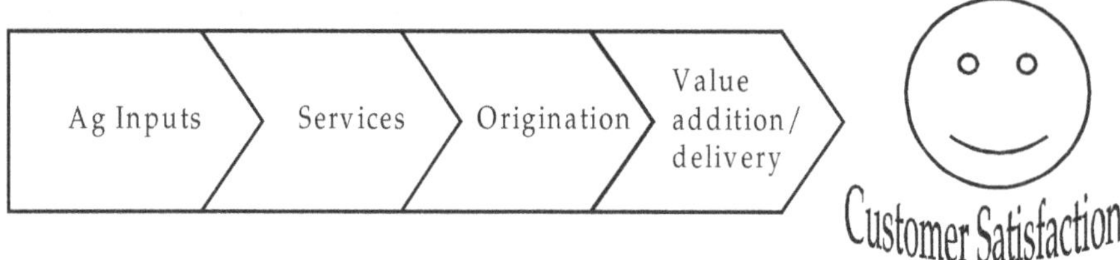

- Customer and Farmers get connected with each other to fulfill their need by cutting across non value adding layers
- Cargill with its experience and expertise creates, captures and delivers desired value to fulfill your needs.

COMMODITY EXCHANGES

- The Forwards Contracts (Regulation) Act, 1952 has been amended to permit futures trading in (54) agricultural commodities.
- It is expected to reduce the exposure of farmers to the risk of price fluctuation.
- 24 recognised associations are active and the value of future traded exceeds Rs. 13.87 lakh Crs (2005).
- Agricultural commodities account for 62%.
- However, initially these exchanges are being used mainly by traders and few big farmers.
- There is a need of creating greater awareness, wider access and deeper understanding so that small farmers can better manager their risk.

BARRIERS TO INTER-STATE TRADE

- Marketing of agricultural products are constrained by several barriers to Inter-State trade.
- Fiscal
  - Vat (Sales/Purchase Tax)
  - CST
  - State Excise
  - Motor Vehicles/Passenger/Goods Tax
  - Market Fees
  - Entry Tax/Octroi/Toll Tax
- Regulatory
  - Essential Commodity Act
  - PFA Act
  - Physical Check posts at borders
- Environmental
  - Forest Act
  - Environ (Protection) Act.
- Procedural/Technical
  - Standards/norms

- Multiple licensing
- Language/forms

TRADE BARRIERS–CASE OF AGRI PRODUCTS

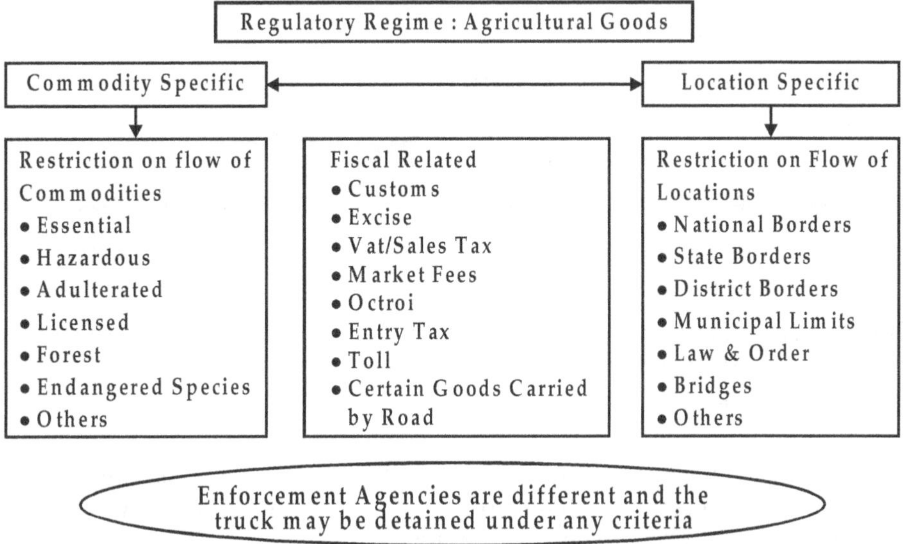

TOWARDS AN INDIAN COMMON MARKET

"Constitution of India conceived of India as a unified, common market. Large economies have advantages of scale, scope and ease of conducting transactions. However, over the years, we have managed to fragment the national common market by erecting a large number of financial and physical barriers which choke inter-state commerce. When the whole world is moving towards dismantling barriers to trade and promoting trade facilitation, it is imperative that states too realise that these policies – which may be driven by revenue or development concerns – will not pay off in the long run"

– Dr. Man Mohan Singh at Chandigarh, 20th Sept. 2006

MILES TO GO..

- Agricultural Marketing Reforms are critical to development of agriculture with inclusive growth.
- The reform process has been initiated, but its progress has been slow and tardy.
- In spite of enabling legislations in many states, private sector participation has not been forthcoming in creating marketing infrastructure.
- The condition of periodic markets (village haats) is deplorable – they lack minimum hygiene facilities.
- Greater efforts should be made to promote successful PPP models in agricultural marketing.
- Barriers to inter-state trade must be removed to create an unified Indian Common Market for agricultural products.

Chapter 14

Basic Concepts

Inventory : It includes tools, standard supply items, raw materials, goods in process and finished goods having economic value.

Inventory Management : The balancing of a set of costs that increases with larger inventory holdings with a set of costs that decreases with larger order size. In other words, it is the procedure and the body of knowledge, which can help us in planning to maintain an optimum level of the idle resource.

Inventory Control : Inventory planning is generally based on the information concerning the part usage and also on factors that are likely to crop up in the future, then only the control process starts. Thus, inventory control and planning goes together.

Methods

1. Economic Order Quantity (EOQ)
2. Cost Comparison Approach

EOQ method is mostly practiced for the determination of optimum ordering quantity.

Economic Order Quantity (EOQ) : It is the optimum (Least Count) quantity of inventory that should be ordered.

Inventory of any item consist of :

(*a*) Working stock : which depends on pattern of inflow and outflow.

(*b*) Safety stock : is designed to guard against unexpectedly high demand, delays in receiving shipment or both.

Different Costs

(A) **Ordering Cost :** It is the cost incurred to get the materials into the inventory of an organisation. *e.g.* Advertisement, salaries, etc.

(B) **Carrying Cost :** Costs incurred for maintaining for given level of inventory are called carrying costs. *e.g.* Capital investment cost, storage cost, etc.

Ordering Cost	*Carrying Cost*
Requisitioning	Warehousing
Order placing	Handling
Transportation	Insurance
Receiving, inspecting and storing	Depreciation
Clerical and staff	Obsolescence

Objective of determining the EOQ is to minimize the sum of Ordering Cost and Carrying Cost.

Determination of the Optimal Order Quantity

To avoid problems that may arise from carrying too much or too little inventory, a business must determine the optimal quantity of a product to purchase each time an order is placed.

(*iv*) Savings

	Institutional	*Personal*
Monthly : ______________	Rs. : __________	Rs. : ________
Annual : ______________	Rs. : __________	Rs. : ________

(*v*) Loans

Pending Amount : Rs. ___________

Rs. ___________

Rs. ___________

Problems in Paying : ____

____

(*vi*) Affiliations to Institutions

Political : Yes ☐ No ☐

Social : Yes ☐ No ☐

Specify : ____

Co-operative Society : Yes ☐ No ☐

Specify : ____

(*vii*) Information known to them on modern technologies

____

____

(*viii*) Other Information

____

____

Place : _____________ Name of Surveyor : _____________

Date : _____________ Signature : _____________

Ordering cost decreases the no. of orders per year are decreased. On the other hand decreasing the no. of orders per year increases the inventory involved and therefore annual carrying cost.

For determining the EOQ there are two approaches.

1. Trial and Error approach - by the help of table.
2. Order formula approach - by the help of formula.

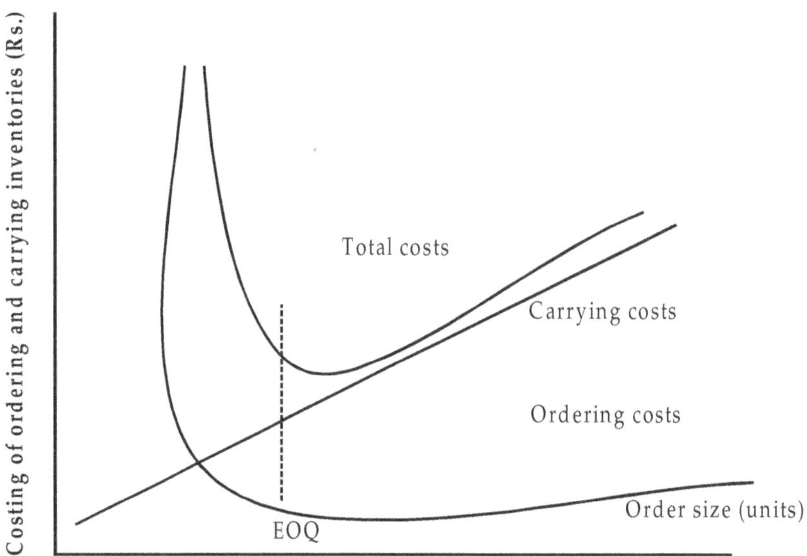

Fig. : Carrying costs increases steadily as order size increases; ordering costs on the other hand decline with larger order sizes. The sum of these two curves, and the lowest point on that curve is the optimal order size, or EOQ

1. Trial and Error Approach

The following table gives :

No. of Orders/Yr.	*Order size (Rs.)*	*Average inventory*	*Annual carrying cost*	*Annual ordering cost @ Rs. 25/- order*	*Total cost*
1	30,000	15,000	3,000	25	3,025
3	10,000	5,000	1,000	75	1,075
5	6,000	3,000	600	125	725
10	3,000	1,500	300	250	550 (Minimum)
15	2,000	1,000	200	375	575
20	1,500	750	150	500	650
25	1,200	600	120	625	745
30	1,000	500	100	750	850

Note : Average inventory value - is equal to half of the maximum inventory (order size).

Annual carrying cost - 20% of the average inventory value. Total cost = Annual carrying cost + Ordering cost

Economic order quantity or EOQ is that quantity where the total cost is minimum.

By observing the above table it is clear that the total cost 550 is minimum value. Therefore, it is clear that the order should be in 10 lots and the EOQ will be 3,000, *i.e.*, for this economic order size the investment on the inventories will be minimum.

2. Formula Approach

$$EOQ = \sqrt{\frac{2.4 \times 0}{C}}$$

Here EOQ = Economic ordering quantity or the optimal quantity to be ordered each time an order is placed.

A = Order size

O = Ordering cost per order

C = Carrying cost expressed as per centage of inventory value.

By Using this Formula

$$EOQ = \sqrt{\frac{2 \times 30,000 \times 25}{0.2}}$$

$$= 2738$$

$$\simeq 3,000$$

Therefore, EOQ will be 3,000 to keep the inventories at the lowest.

Assumptions of EOQ Model

1. Sales can be forecasted perfectly.
2. Sales are evenly distributed throughout the year.
3. Orders are received with no delays whatever.

Example

Following information is given by a fish processing company about the number of orders and order size. Find out the Economic Order Quantity. Represent it graphically.

Given Carrying Cost is 10 per cent of any inventory value. Order Cost is Rs. 50/order.

No. of Orders/Years	1	3	4	5	8	10	20	25
Order Size	49,000	30,000	15,000	10,000	6,000	4,000	2,000	1,000

Make a table and also find out the total cost.

Solution

No. of Orders/Yr.	*Order size (A)*	*Average inventory*	*Annual (C) carrying cost*	*Annual (O) ordering cost @ Rs 25/ order*	*Total cost (O + C)*
1					
3					
4					
5					
8					
12					
20					
25					

EOQ will be 6,000.

i.e., order size of 6,000 in 8 lot will be *EOQ*

By Formula $EOQ = \frac{2 \times A \times 0}{C}$

Conclusion

An inventory management system should include an area approach before a micro approach involving units to be effective.

By the application of inventory management, a firm can maintain a smooth, continuous supply of product throughout the year as per worked demand. By using models like EOQ, optimum order quantity can be found and to avoid the profile loss due to large investment on inventories and storage and cost of production interruption caused by inadequate inventories.

A system must be designed to fit the present situation and capabilities must be enhanced to incorporate system improvement.

Chapter 15
Money

Barter System : Exchange of goods and services with goods and services and money was not involved. (In ancient times).

Difficulties in Barter System

1. Lack of double coincidence of want. (Suppose a person has cow and wants to purchase goat for this to happen he must find a man who has goats and wants to purchase cow which was difficult).
2. Lack of common measure of value. (*i.e.* lack of any common scale in which we could count any thing in its terms *e.g.* now every thing can be expressed in terms of money *i.e.* rupees).
3. Lack of difficulty of sub-divisions (*e.g.* half cows).
4. Lack of store value (especially for perishable goods which have shorter life) *i.e.* stored goods were sold for a lesser price.

Evolution of Money

Ist Stage : Barter unit of value.

2nd Era or Stage : Standard commodity money (*e.g.* ivory sea shell cows etc.).

3rd Era or Stage : Metallic money (especially gold, silver copper).

4th Era or Stage : Convertable paper currency.

5th Era or Stage : Inconvertable paper currency (*i.e.* at present both coins and paper currency).

6th Era or Stage : Bank money or bank credit.

Def. of Money

Ist Def. (By Robertson) : Anything which is widely accepted in payments for goods or in discharge of other kinds of obligation is money.

2nd Def. (By Crowther) : Anything that is generally acceptable as a means of exchange and that at the same time acts as measure and store of value.

The Major Functions of Money

1. Medium of exchange.
2. Standard measure of value (other can be converted in terms of money *e.g.* pen can be converted to Rs.)
3. Store value *i.e.* over a period of time its value being unchanged or does not decrease.

4. Deferred Payment (paying after a certain period - only possible if stored value is maintained).

Institutions which deal with Money : (Main institution is bank).

∴ Bank is an institution which deals in money.

(**Function :** It draws surplus money from non-uses and lend money to those who need it at that particular moments).

Functions of a Bank : *(Commercial Banks)*

1. Receiving deposits (in different forms and schemes) (for different period of times, it encourages deposition by giving security and interest).

   46 – 90 days

   91 – 180 days

   181 – 365 days slabs with diff. amount of interest.

   1 – 3 years

   3 –
2. **Advancing loans :** (Rate of interest on loan is higher than interest on deposits.)
3. **Discounting Bills :** (Type of loan for a shorter period.)

Types or Kinds of Banks

1. **Commercial Banks**

   (*a*) Schedule Banks

   (*b*) Non-schedule Banks

   (*a*) ***Schedule Commercial Banks :*** Those banks capital reserve is more than 5 lakh as they are scheduled by the R.B.I.

   1990 Data (figures) 278 = total banks
   279 = schedule C.B.
   4 = non-schedule C.B.

   Out of 274, 224 are in public sector. Out of 224, 196 are regional rural banks.

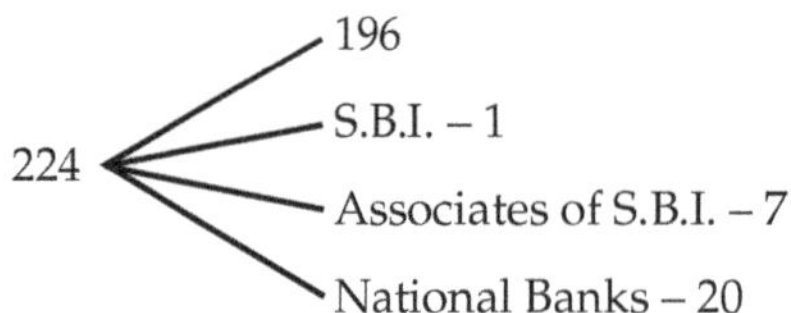

   (14 and 6) in different phases

   14 in 1969 (19th July)

   6 in 1980 (15th April)
2. Exchange Banks or Export and Import bank or EXIME bank. (Conversion or issuing of foreign currency).
3. Industrial Banks *e.g.*

   (*i*) Industrial Finance corp. of India.

   (*ii*) National Industrial Development Corporation. (give loans for medium and long terms to industries).
4. **Agricultural Banks :** (Banks operating in Co-operative sector and providing loans for agricultural) *example*

(*i*) District co-operative bank.
(*ii*) P.A.C.S. Primary Agricultural credit Society.
(*iii*) State Co-operative Bank.
(*iv*) Land Development Bank (LDB).

5. **Saving Banks** : (For small savings) *e.g.* post office also takes small savings.
6. **Central Bank :** (Apex body of all banks) In India it is R.B.I.

NABARD (National Bank for Agricultural and Rural Development)

Central Bank : (Apex body) in banking bodies. It stands at the Apex of the banking system of a country. It carries out monetary policy (*i.e.* related to money) or policies of the government and helps in maintaining a general economic stability of a country. In India the central bank is R.B.I. established in 1935.

Main Functions of a Central Bank or R.B.I.

1. Functions as a note issuing agent (Government) Govt. of India issues the unit currency *i.e.* 1 rupee note, others by R.B.I (governor R.B.I. signature).
2. Banker of the government.
3. **Banker of the Banks :** (Schedule banks are required to deposit a fixed per centage of there total deposit with R.B.I.) called minimum reserve ratio. Per centage is fixed by R.B.I.
4. **Lender of the last resort :** (Gives loan to other schedule banks and government as the last means of source).
5. **Controller of Credit :** (Exercises control over the amount of money in circulation with people). By increasing loan and deposit rate. Reserve Bank of India (R.B.I.).
6. Custodian of the nations foreign currency.
7. An agent of development.

Chapter 16

Cost Concept

Cost : Production cost can be classified as following :

1. Money cost and opportunity cost.
2. Fixed cost and variable cost.
3. Total, average and marginal cost.

Money Cost : Those costs which a producer spends to attracts different factors of production towards production process.

Money cost is of following types :

1. **Explicit Cost :** These are paid out cost *i.e.* payment made for productive resources, purchased or hired by the producer or expenditure made on those factors which are not owned by the producer.

   It includes :

   (*i*) **Production Cost :** Such as raw material, wages, interest, rent etc.

   (*ii*) **Selling Cost :** *i.e.* advertisement, sales promotion activities and commissioning (agents and brokers).

   (*iii*) **Others :** Depreciation (cost of replacement of wear and tear units.) Expenditure on maintenance *e.g.* oiling, lubricating etc. Charges on safeguarding (security purposes).

2. **Implicit Cost :** These are cost of self-owned or self-employed resources for which producer doesn't pay anything. (It is measured on basis of how much you would have payed if you had not processed it.)

3. **Normal Profit :** It is that minimum profit which induces the entrepreneur to be in the production process. The entrepreneur must be sure of normal profit if he is to continue in the business.

Opportunity Cost

A	B	C	Crops
1000	1200	800	to grow on one hectare

Opportunity cost of taking B is 1000 (*i.e.* return of A). Over normal profit, super normal profit.

Different : Opportunity cost means not the efforts or sacrifices undergone or made but the most attractive alternative foregone (or left) or the next best choice sacrifice.

In other words, the amount of money which any particular unit could earn in its best paid alternative used is called opportunity cost. Sometimes it is also called as transfer earnings.

Short-run period and Long-run period : Short period is a period of time within which the firm can change or vary its output by changing the amount of variable factors such as labour, raw material etc.

In short-run fixed factors such as heavy equipment such as tractors, building, top management personnel cannot be changed or varied. So the short-run is the period of time in which the variable factors can be changed, while fixed factors remain the same.

Long-run Period : It is a period of time during which all factors variable as well as fixed can be changed, varied or adjusted.

Fixed Cost : (Supplementary or overhead cost). Those cost which in short-run do not change with change in output, it includes all costs which an entreprenuer has to bear even when there is no production. Such as interest on capital invested, in land building, machinery, salary of permanent staff etc.

Variable Cost (Prime Cost) : Those cost which vary or change with the variation of output in short-run, they are direct cost and include items such as price of the raw material, wages paid to labours, etc.

Total, average and marginal cost :

Total Cost : When variable and fixed costs are added together they are called total cost.

T.C. = fixed cost + variable cost

T.C. = F.C. + V.C.

Average Cost : (Per unit cost production) cost of producing a single unit.

$$A.C. = \frac{\text{Total cost}}{\text{No. of units produced}}$$

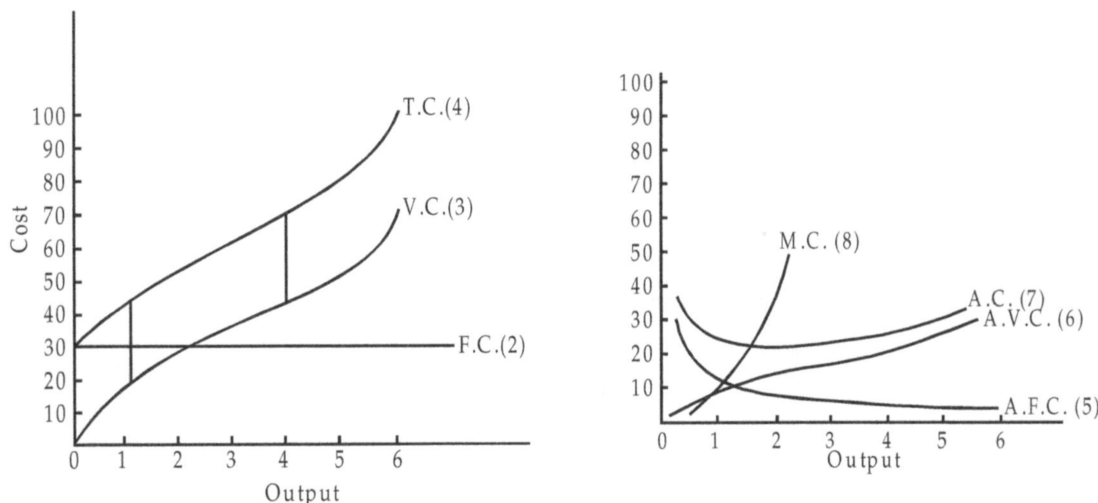

Marginal Cost : It can be defined as the addition made to the total cost by the production of one additional unit of production (output).

$$M.C. = \frac{\Delta \text{ T.C.}}{\Delta \text{ output}}$$

$$\text{Marginal cost} = \frac{\text{change in total cost}}{\text{change in no. of units of output}}$$

11 – 105

$$\text{M.C.} = \frac{105\text{–}98}{11\text{–}10} = 7$$

Example :

1 *No. of units produced*	2 *Fixed cost*	3 *Variable cost*	4 *Total cost*	5 *Average fixed cost*	6 *Average variable cost*	7 *Average cost*	8 *Marginal cost*
0	30 (units)	0	30	–(∞)	1	–(∞)	–
1	30	10	40	30	10	40	10
2	30	18	48	15	9	24	8
3	30	24	54	10	8	18	6
4	30	32	62	7.5	8	15.5	8
5	30	50	80	6	10	16	18
6	30	72	102	5	12	17	22

Why difference between A.C. and A.V.C in initial stages is more but becomes less in later stages?

Initially in total cost the proportion of F.C. is more that of V.C. but in later stages the proportion of V.C. becomes more than that of F.C.

In beginning (*i.e.* at 0 no. of unit produced) T.C. = 30 but F.C. = 30 and V.C. = 0 at 1 no. of unit produced T.C. = 40 but F.X. = 30 and V.C. = 10 at 6th unit of production T.C. = 102 but F.C. = 30 and V.C. = 72 ∴ the distance between the two decreases.

Price and concept of different types of markets

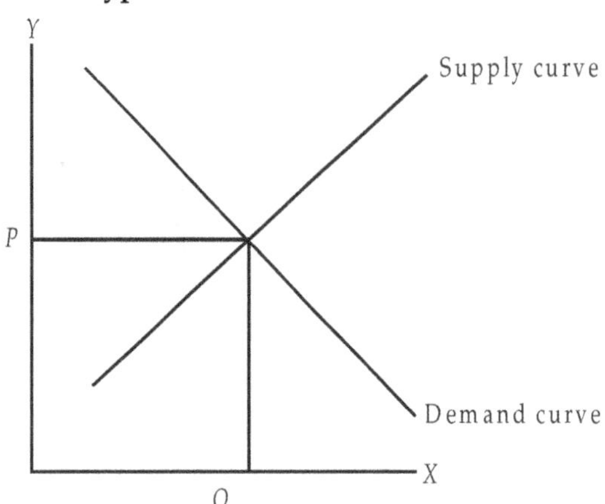

There are two types of prices :

1. Normal price
2. Market price

1. **Market Price :** It is determined by the equilibrium between demand and supply in a market period or in a very short period. The market period is a period in which the maximum that can be supplied is limited by the existing stock. Supply cannot be adjusted according to the demand. Supply cannot be increased or decreased but is fixed from demand.

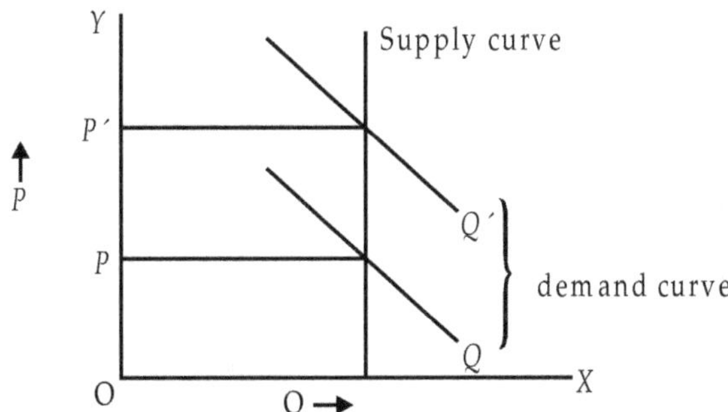

2. **Normal Price :** It is determined by the equilibrium between demand and supply in a long period. Long period is a period of time in which a firm or firms can adjust the supply according to the demand.

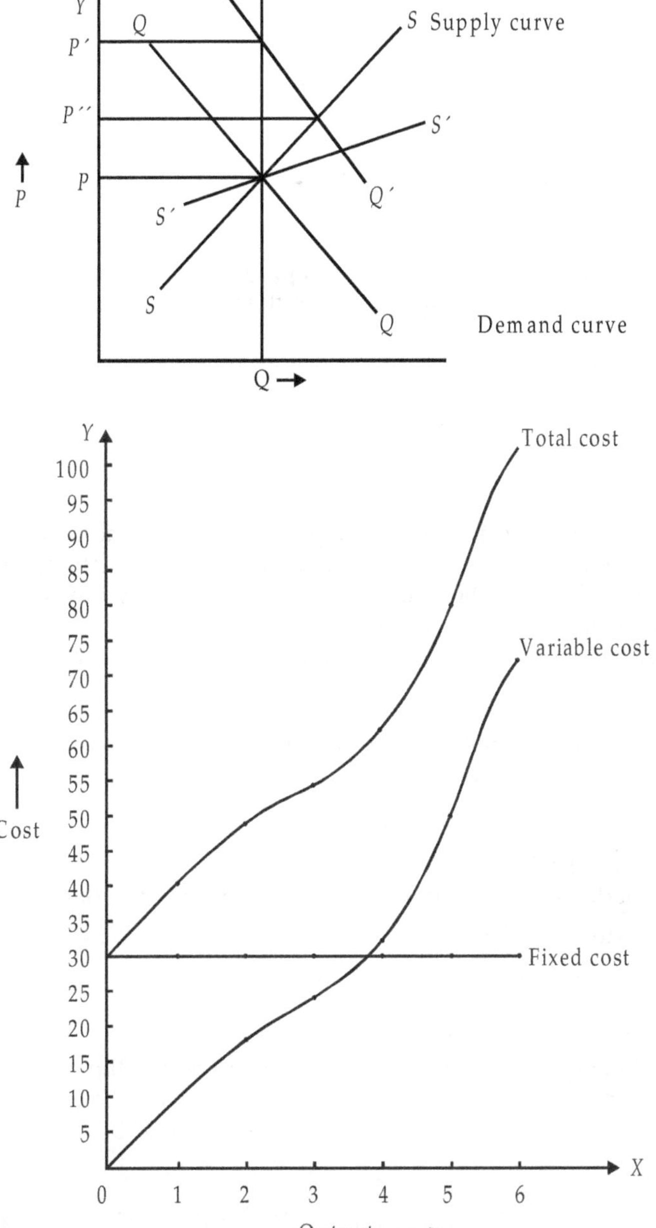

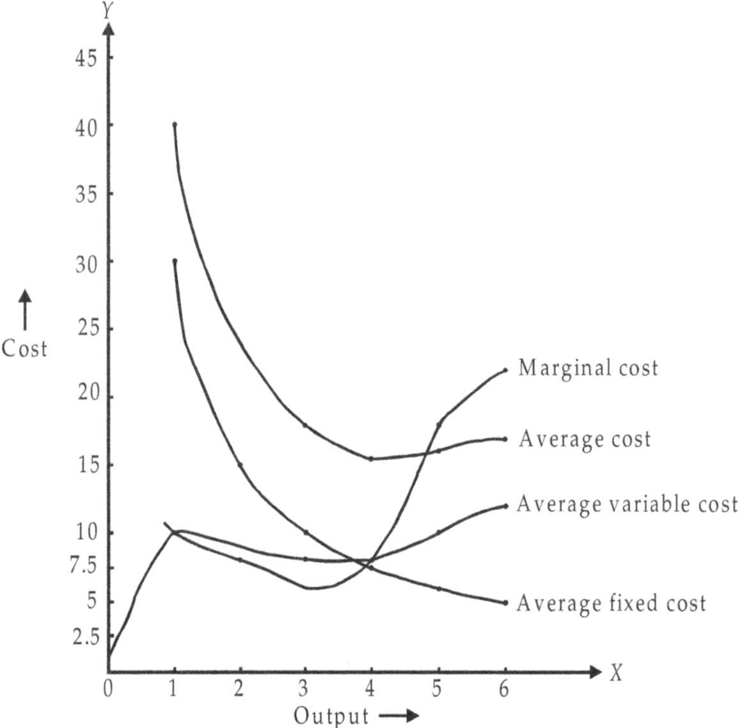

As period of supply is increasing or slope of supply curve is decreasing, the affect of change in demand on equilibrium price is decreasing, (shorter the period affect will be more).

Different Market Forms

I^st^ Classification

1. Perfect market
2. Imperfect market

1. **Perfect Market :** A market is said to be perfect when all the potential buyers and sellers are promptly awayer (aware) of the prices at which transactions take place, and all the offers made by all other sellors and buyers and when any buyer can purchase from any sellor and conversely *i.e.* any sellor can sell to any buyer. Under such conditions the price of a particular commodity will tend to be the same all over or throughout the market.

(Perfect Competitive Market) Competition Characteristics of Perfect Market

1. Large no. of buyers and sellors.
2. Homogeneous product (same product).
3. Free entry and exist.
4. Perfect knowledge of prices and quantities sold and purchased.
5. No transportation cost (absent).
6. Perfect mobility of factors of productions.

Imperfect Market

Monopoly Market : In this situation only one sellor exists in the market for that commodity.

Monopsony market : Only one buyer exists.

Bi-lateral Monopoly Market : One buyer and only one sellor.

For completely or perfectly competitive market.

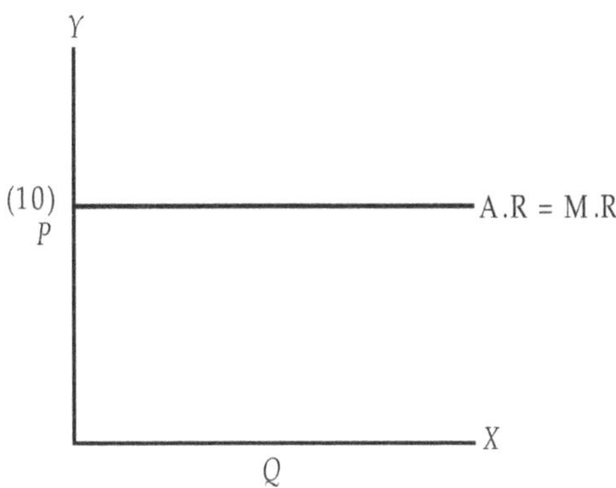

100 units are sold or purchased. Price = 10 Rs./unit

$$\text{Total return} = 1000 \text{ Rs.}$$

$$\text{Average Return} = \frac{1000}{100} = 10 \text{ Rs.}$$

$$\text{Marginal return} = 10 \text{ Rs.}$$

(after selling one additional unit)

A.R. curve is also called demand curve faced by each firm.

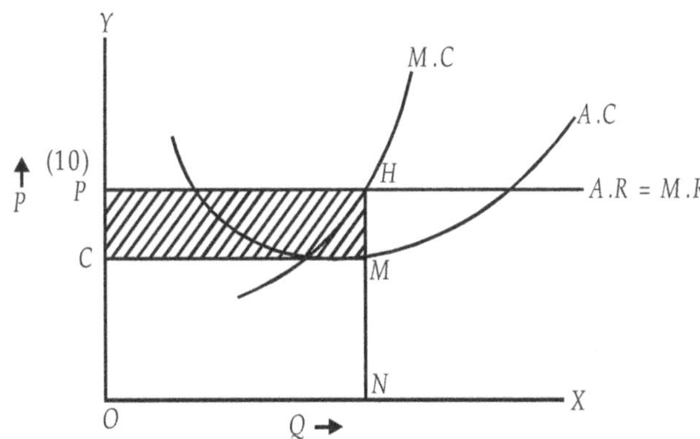

To maximize profit or minimize cost of production conditions are :

1. Marginal return = Marginal cost

   M.R. = M.C.

   cost for producing one additional unit must atleast equal to M.R.
2. M.C. should cut M.R. from below.

ON gives the optimum quantity to be produced at price P (10). By producing ON quantity what is the total return?

A.R. = OP = 10 Rs.

T.R. = OP × ON = (10 × 100) = 1000 Rs.

Average cost = OC

(return per unit = OP) when producing ON quantity output.

Total cost = OC × ON (cost per unit = OC) (say 8 × 100) = 800

Profit = T.R. – T.C.

1000 – 800 = 200 Rs.

(Graphically) ⇒ PC = (profit per unit)

Total profit = PC × ON = PC × CM

= Area of rect. PCMH

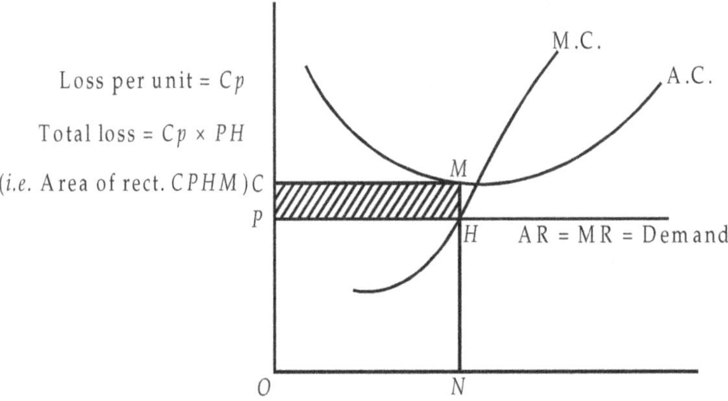

(only one seller)

Monopoly Market Situation : 10 units, at price 5 Rs. to sell more he will decrease price. Demand or A.R. is not 11 to x-axis (it is downward sloping).

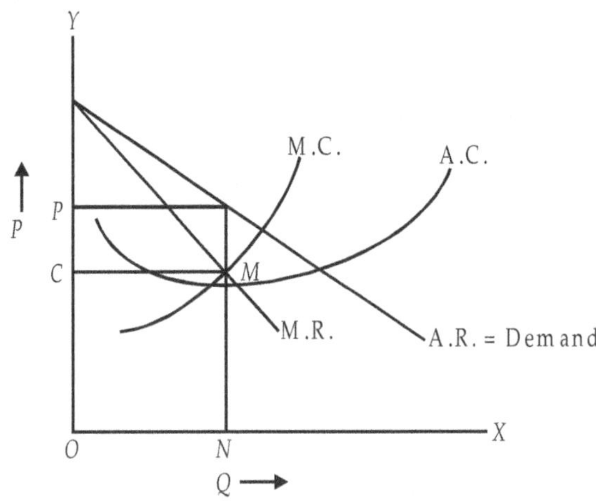

Here the MR curve is below A.R. curve.

Conditions

1. M.R. = M.C.
2. –

OP price of product (unit return)

Total return = OP × ON

Average cost = OC

Per unit profit = PC

Total profit = PC × CM

(*i.e.* area of rectangle PCMH)

Monopolistic : (Not monopoly) Characteristics :

1. Many nos. of buyer.
2. Products are close substituent of each other. (Similar but not identical) *e.g.* cibaca and colgate tooth paste, same purpose served but brand name, taste colour etc. may be different, (curve has less slope then monopoly), as demand increases slope decrease and finally becomes 11 to X-axis.

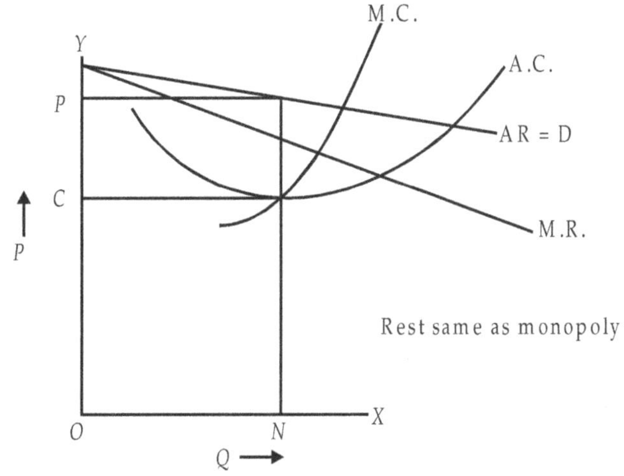

Perfectly Competitive Market : (1) Profit (2) Loss

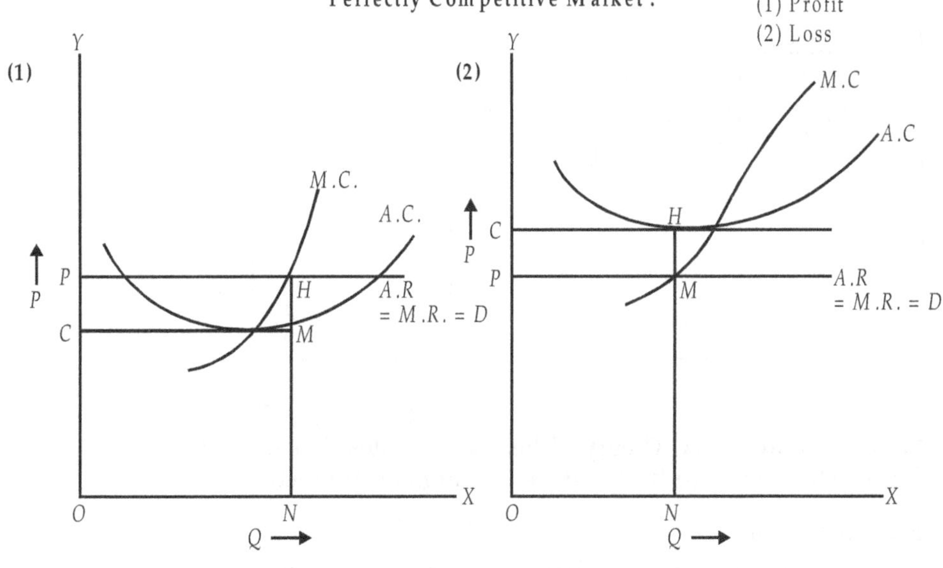

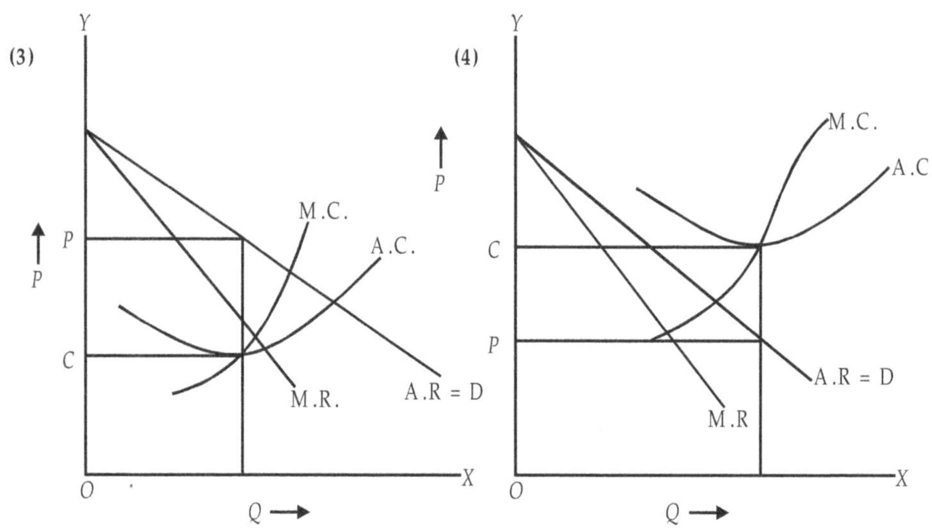

Distribution : Sharing of wealth among the different owners of factors of production. They get their share according as :

Land owner	Rent
Labour	Wages
Capitalist	Interest
Entreprenuer	Profit

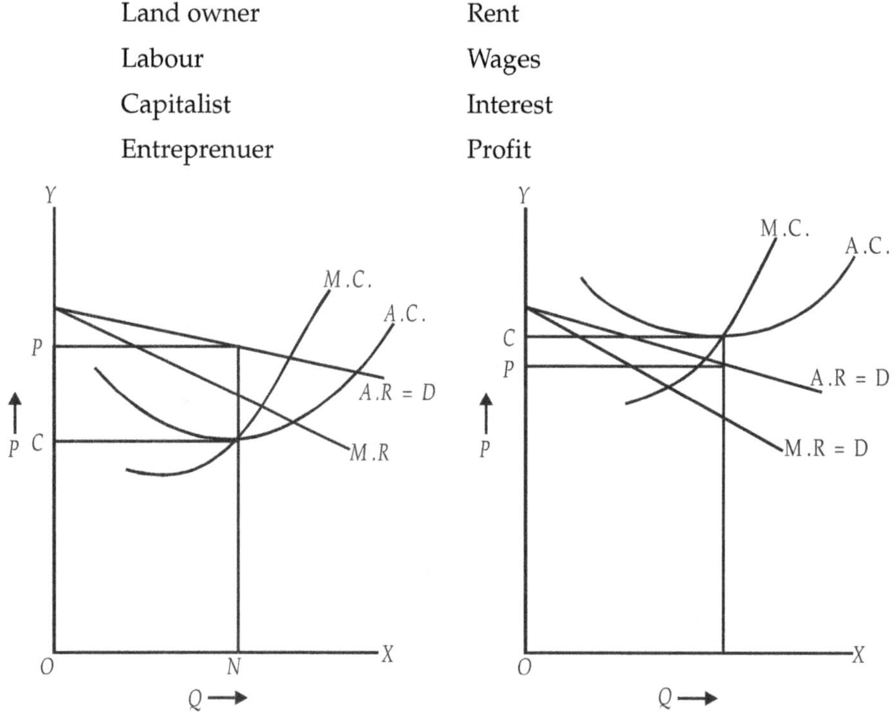

1. **Marginal Productivity Theory :** This theory states that reward or share of each factor of production tends to be equal to its marginal productivity.

Limitation of this theory

(*i*) Because of this theory tells us that at a given level of suppose wages how many labours should be allowed. It does not tells us how wages are determined.

(*ii*) It approaches from demand side only it ignores the supply side completely.

2. **Modern Theory of Distribution or Demand or Supply Theory :** It tells that price of a factor of production is determined by the intersection of demand and supply of that particular item or commodity.

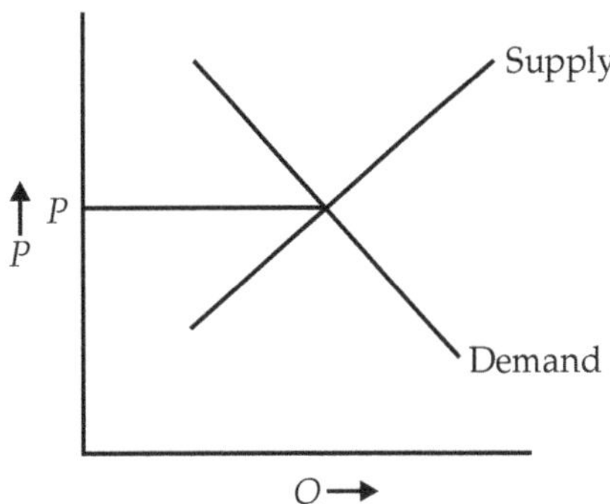

Demand of a particular commodity is not the direct of demand but the indirect or derived demand and it depends upon the demanded price of the finished commodity and the availability of co-operative factors.

Example : If you don't have capital we will not be able to demand labour due to our liability to not pay wages.

Supply Side : Supply of labour depends up on population and nature and ratio of population (child labour, women labour etc.) ratio includes sex ratio.

Expected Income : If labour are expecting high income supply of labour will be high.

Wages (Labour) : It mean payment made for the use of labour.

1. Nominal wages or money wages.
2. Real wages.

1. **Nominal Wages or Money Wages :** Wages interms of rupees or money (payed or received). It does not gives a real picture.
2. **Real Money :** While comparing the standard of living we incorporate all the different factors.

Factors

1. Purchasing power (it is inversely proportional to price level).

$$P.P. \propto \frac{1}{PL}$$

2. **Subsidiary Earning :** Apart from wages a person may be getting other facilities like living quarter, clothes and food.
3. **Extrawork without Extra Payment** : Asking him to work over time without giving him more or increased wages (money).

How in a particular firm given a level of wages for maximum profit how many labour should be demanded? (All industries in a firm pay the same wages) level of wages is not changing.

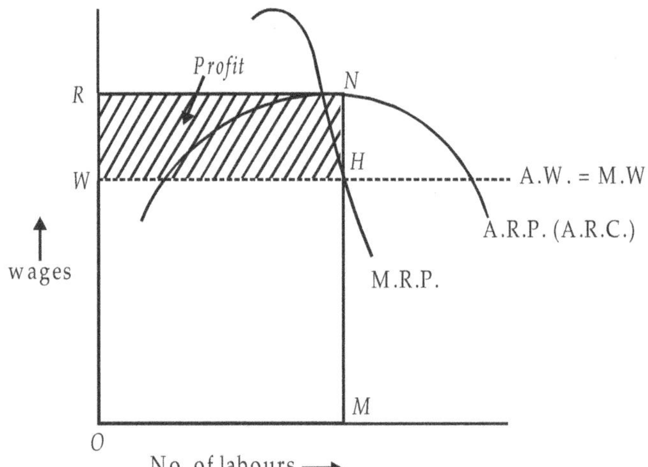

If wages are of same level through out the market (firm) then

Average wages = marginal wages = wages.

A.W. = M.W. = W

Average return = Contribution of each labour.

Average return product (A.R.P.) A.V. curve (AVC).

Marginal Return or Revenue Product : Conditions of equilibrium for a firm for labour market. (Perfect maximizing profit).

1. M.R.P. = MW
   Marginal Revenue product should be equal marginal wages.
2. M.R.P. should cut ARP from above.

OW = wage of one labour *i.e.* cost of one labour.

OW × OM = total cost of all labour.

OR – return per labour

OR × OM = total return

Profit = RW

Per labour

Total profit = RW × OM

= Area of Rect. RWNH for loss

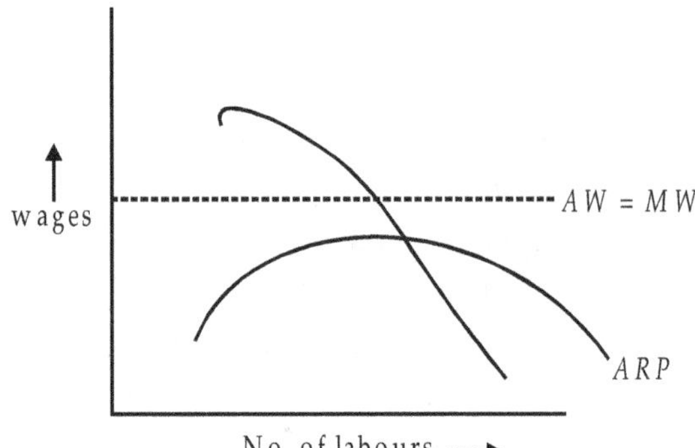

Rent : (Hiring Charges)

Economic Rent : It refers to that part of payment by a tenant which is made only for the use of land (*i.e.* free gift of nature). (Rent may also include other facilities provided by owner such as building, water and electricity.)

This total payment is called contractrant *i.e.* total payment given to landlord by the tenant apart from using the land some charge is for facilities provided by the landlord along with land.

Rent and Transfer Earning

Transfer Earning : (Next best alternative) The minimum price of a factor which should force that factor to be in market or in other words the price which is necessary to retain a given unit of a factor in a certain industry if however the factor is earning over and above its transfer earning then the surplus or excess earning is called economic rent.

Different economic theory on why and how rent arise are given.

(Ricardo) Ricardian Theory of Rent : Rent is that portion of the (if a factor is earning over and above its transfer earning the surplus is called rent) produce of the earth which is paid to the landlord for the use of the original and industructable powers of the soil.

Economic rent according to Ricardo is the true surplus left after the expenses of cultivation as represented by payments to labour, capital and enterprise have been met.

Modern View of Rent of Land : According to modern view rent is determined by the demand of and supply of land.

Modern Theory of Rent : It is the surplus payment in excess of transfer earning of the factor. The economic rent or surplus over transfer earning will arise only when the supply of the factor unit. Land has no supply price.

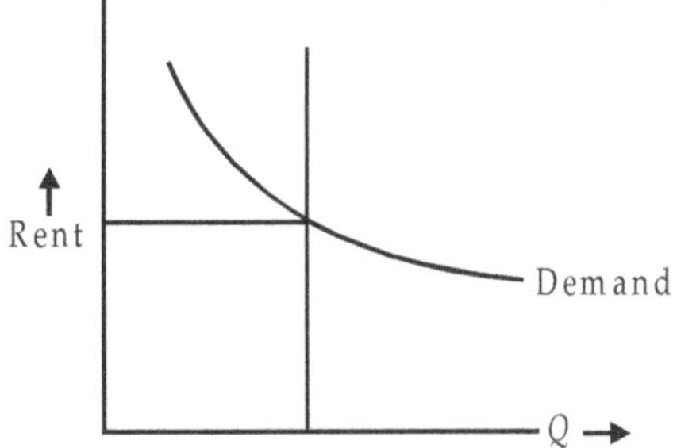

Q–Quantity of land demand of land is determined by extent of use and necessity is less than perfectly elastic.

Why, because, Rent = Present E – Transfer E

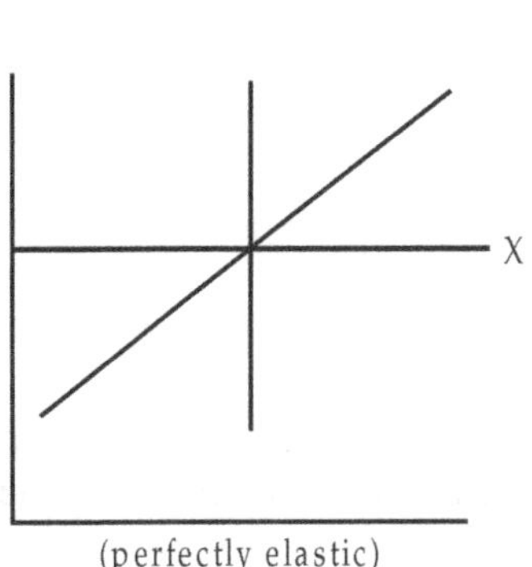

(perfectly elastic)

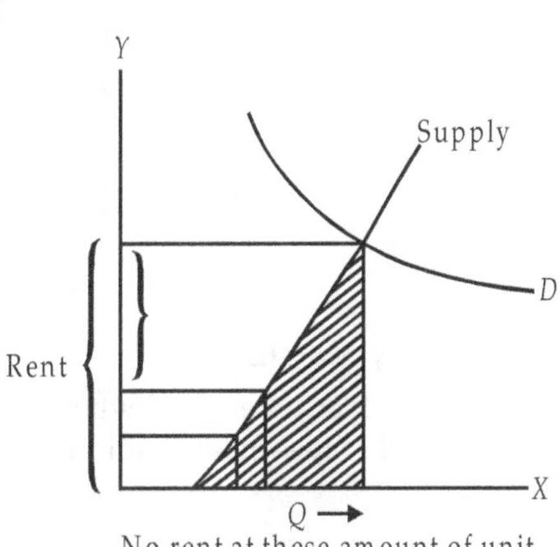

No rent at these amount of unit

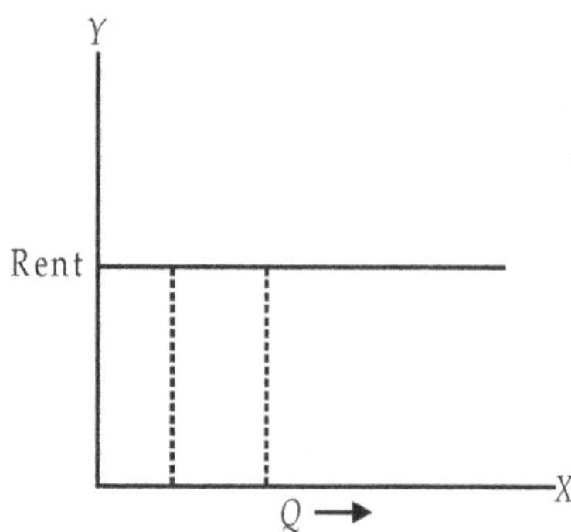

Each and every unit present earning and transfer earning are surplus is nil (infinite) are available at the same price.

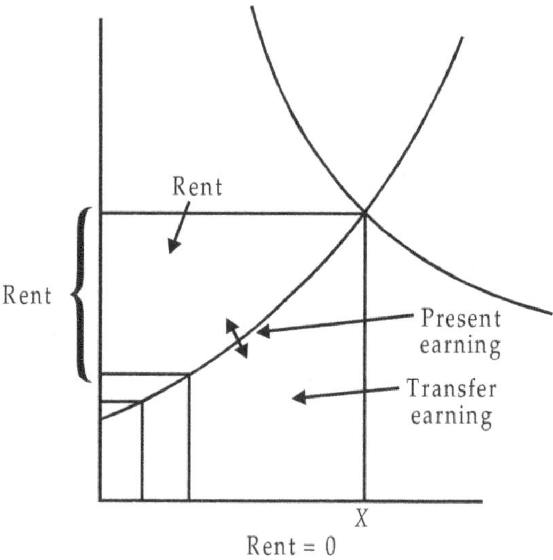

Rent = P.E. – T.E.

In inelastic condition (perfect) T.E. = O

∴ Rent = P.E.

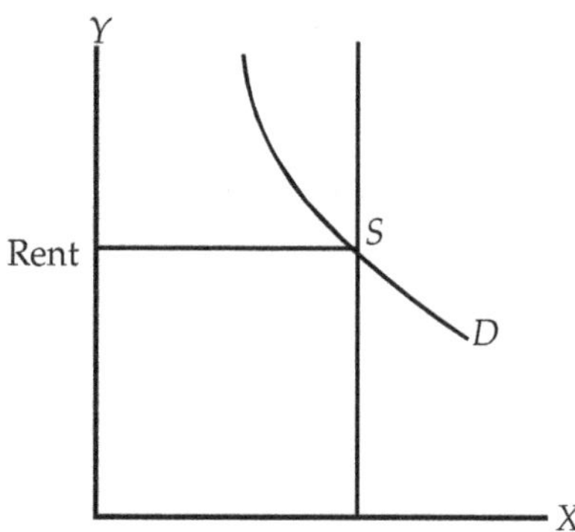

Concept given by Marshal

Quasi Rent : (For other good which are not free gift of nature *i.e.* are man-made). It is a short run phenomenon because in long run supply can be adjusted. Surplus earned by the instruments of production other than land, this term is applied to the income derived from appliances and machines and other instruments which are a produce of human effort, it arises when demand suddenly increase and in demand supply cannot be increased so it a short-run phenomenon.

Interest : Payment made for the use of capital. Or it can be defined as the reward for capitalist or for capital for its services in production.

Gross Interest and Net Interest : The whole of the income received by the lender of the capital from the borrower is not the net or pure interest it is called gross interest.

Net interest is only a part of gross interest. Net interest is the payment for the V capital as such use of the lender renders or provides a number of services and he expects some rewards or fee for those services. So the constituent of gross interest decides net interest or as follows :

1. Insuring against risk.
2. Net interest.
3. Reward for money management.
4. Payment for the inconvenience (inconvenience of not having the money.)

Different theories determining the level of rate of interest are :

I^st Theory

1. **Classical Theory :** Level of rate of interest is determined by supply (saving) and demand (investment) rate curves of interest.

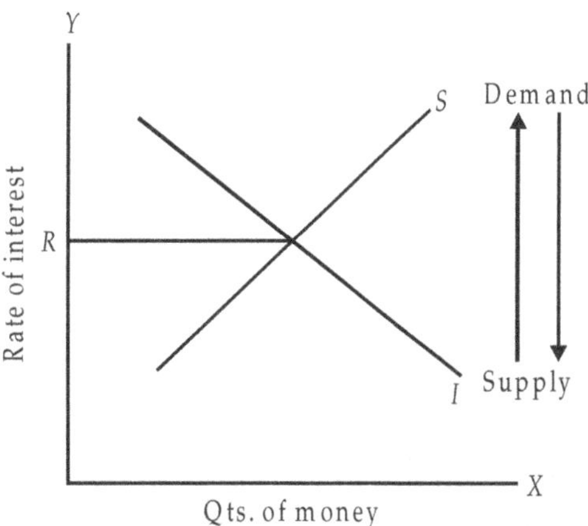

Supply – Saving

Demand – investment

2. **Loanable Fund Theory :** Supply and demand have different constituents. Supply (saving) different constituent of supply side :
   1. Saving out of current income plus bank credit.
      $CI + BI$
   2. Dishording
   3. Disinvestment (not investing money when needed) (*ex.* suppose a thing is to be replaced after 5 years, but after 5 years you still continue using it *i.e.* even depreciating if even after its lite is over.)

Investment Constituents

1. Inhording and dissaving (money saved earlier.)

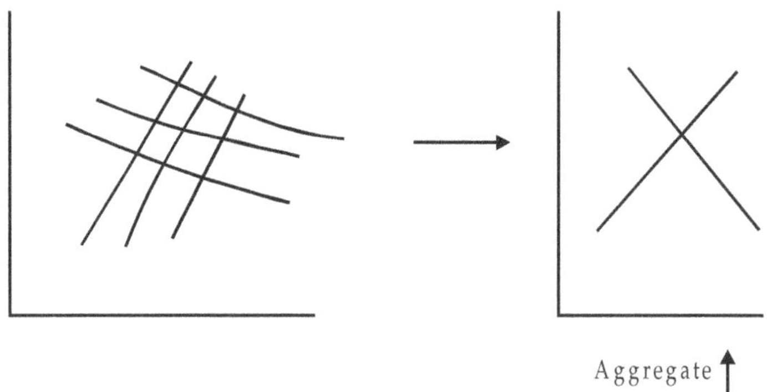

2. **Liquidity Preference Theory :** This theory is given by 'keynes' according to keynes interest is the reward for parting with liquidity for a specific period.

Liquidity : Readily available money *i.e.* cash.

A thing which can be converted easily or readily to money is more liquid. *e.g.,* I year fixed deposit is more liquid than 5 year fixed deposit.

This theory says that desire of the people or demand of people to hold cash, this desire arise due to three motives :

1. Transaction motive (exchange of goods).
2. Precautionary (for unseen accidents *e.g.* silkness).
3. Speculative motive (to gain from the market movements) *i.e.* prices of securities (shares).

Rate of interest in the market will affect speculative motive and not the other two.

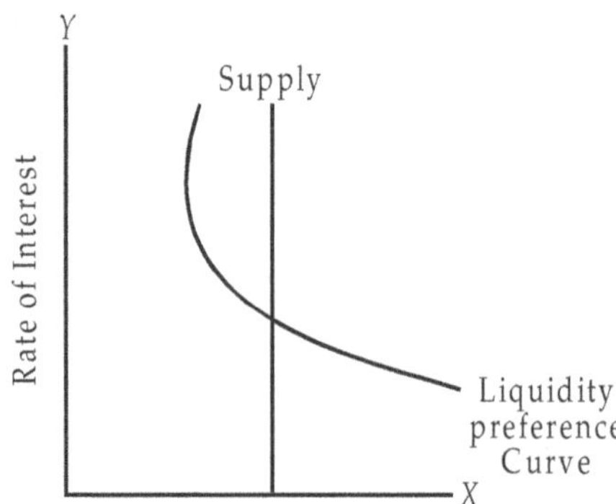

Supply and demand of money higher the rate lower will be demand for money and vice versa.

Profit : Reward of entreprenuer. Net profit and gross profit components of gross profit.

1. Interest of entrepreneur own capital.
2. Rent of the land owned by the entrepreneur.

3. Wages of management (own employer).
4. Reward of the entrepreneur as risk taker.
5. Gains as superior bargainer.
6. Monopoly gains (it he is a single seller in the market.)
7. Chance profit (by chance) *e.g.* for some unforeseen reason prices rises to large extend *e.g.*, war.

Net Profit : Net profit or pure profit is the reward for taking the risk exclusively (risk taker.)

NATIONAL INCOME
(*Income of a Country*)

Definitions

1. **By Marshal :** Labour and capital of a country acting on its natural resources, produce annually a certain net aggregate of commodities materials and immaterials including services of all kinds is called national income.
2. **By Prof. Pigou :** National income is that part of objective income of the community including of course income derived from abroad, which can be measured in money.
3. **By Fisher :** National income refers "solely to services received by ultimate consumers, wheather from there material or human involvement."
4. **Complete Definition :** (Summarised definition) National income may be defined as the aggregate factor income which arises from current production of goods and services by the nations economy. The nations economy refers to the factors of production supplied by the normal residence of the national territory.

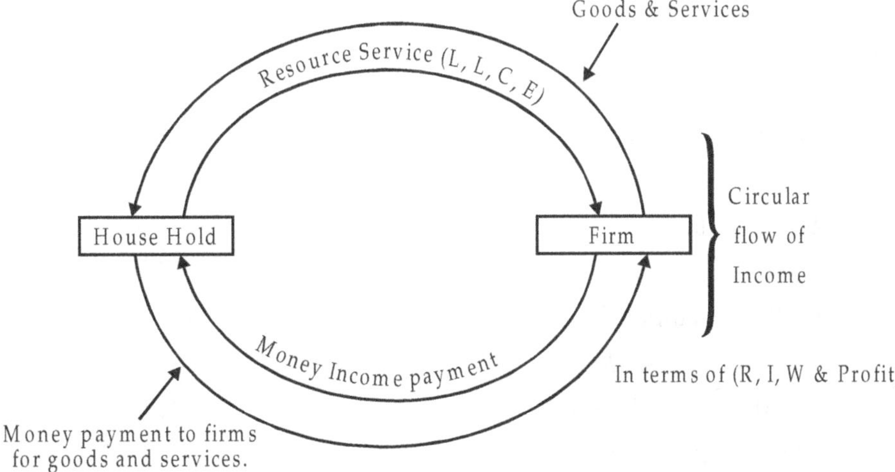

H.H. supplier of factors of production.

Different ways to measure or estimate or calculate national income :

1. As the sum of all the income in cash or in kind accrue to factors of production in a given time period *i.e.* the total income flowing to house hold sectors this is called income method for calculating national income.
2. As the sum of net output arising in several sectors of the nation, called as output or production method.
3. As the sum of expenditure made by all the individuals, as well as government, called expenditure methods.

Different Concepts of National Income

There are 5 concepts of national income :

1. **1st Concept :** It is called as G.N.P. (Gross national product). G.N.P. is defined as the total market value of all final goods and services produced in a year. (Final goods - ready for consumption)
2. **2nd Concept :** Called as N.N.P. (National Net Product) N.N.P. is equal to G.N.P. minus depreciation (*i.e.* NNP = GNP – Depreciation)
3. **3rd Concept :** National income at factor cost (also known as national income) N.I.

   It is equal to NNP minus indirect taxes plus subsidies.

   NI = NNP – indirect taxes + subsidies

Example

1. Shoes its price is 400 Rs. and includes a tax of Rs. 50 ∴ actual cost is 350 Rs.
2. Note book price is Rs. 5 but govt. subsidies it by Rs. 1 ∴ cost = 5+1 = 6 Rs. (part of cost is paid by govt.).

4. **4th Concept :** Personal income (P.I.). The income which is actually received by the individuals. (There are some income which we are earning but not receiving).

   Example : Premium on money or deposits.

   Example : Pension (transfer earning)

   P.I. = NI–SSC–corporate IT–undistributed C.P. + transfer payment. Personal income is equal to national income minus social security contribution (premium of insured money or insuration) minus corporate income taxes, minus undistributed corporate profits plus transfer payment.
5. **5th Concept :** Disposable income (D.I.)

   DI = PI personal taxes

   DI = consumption + saving

transfer payment × pension, unemployment.

28,000 No tax

50,000 20%

12,000 is exempted

28000 + 1200 = 40,000

50,000–40,000 = 10,000

20% of 10,000 = 2,000

50,000 – 1,00,000 – 30%

Disposable income is personal income minus the personal taxes.

Different Methods of Estimating National Income

1. Production or output method.
2. Income method.
3. Expenditure method.

(Different countries choose different method depending upon their needs.)

1. **Production or Output Method :** This method approaches National Income from output side or production side. The economy is divided in to different sectors like agriculture, mining, manufacturing, transport employment fee etc. nation, commerce.

   [Budget : (account of income and expenditure) and other services, then the gross product is found out by adding up net values of all the production that has taken place in different sectors during a given year, it we add in it net income from abroad (net income = export – import) net income from abroad is export – import, we will get G.N.P.

2. **Income Method :** This method approaches N.I. from distribution side. *i.e.* This method measure N.I. after it has been distributed and appears as income earned or received by the individuals of the country. It has ten (10) components :

   (*i*) wages
   (*ii*) interest
   (*iii*) rent
   (*iv*) profit and dividend
   (*v*) income of unincorporated business (self employed persons)
   (*vi*) corporate income taxes
   (*vii*) social security contribution
   (*viii*) undistributed corporate profit
   (*ix*) indirect business taxes
   (*x*) depreciation.

1st April	31st March	Financial year
1st July	30th June	Ag. year

3. **Expenditure Method :** This methods arrives at N.I. by adding up all the expenditure made on all the goods and services in a year. Different components of expenditure methods are :

   (*i*) Personal consumption expenditure *i.e.* what private individuals spends on consumer goods and services.
   (*ii*) Domestic or gross domestic private investment.
   What private businesses spend on replacement, renewal and new investment.
   (*iii*) Net foreign investment.

   NFI = Export – import

   (*iv*) *Government expenditure* : What govt. spends upon purchase of good and services. Indirect business taxes.

Significance or Importance of National Income

1. National income estimates or reveals or tells the overall production performance of a country as it seeks to measure the level of production in a given year.
2. The per capita income (P.C.I.) gives an idea of standard of living of the people in a country.

$$\text{P.C.I.} = \frac{\text{National Income}}{\text{Population}}$$

3. Comparing the national income over a period of time we can know weather the economy is growing, stagnant or declining.

4. National income estimate shows the contribution made by the different sectors of economy.
5. National income estimate throws light on the distribution of national income among different income groups.
6. National income estimates also contains the figures of total consumption, saving and investment in a given year.
7. Guide to economic policy.

Chapter 17

Supply

Supply : Quantity of a product that would be made available to buyers in a particular market at a specific price and time.

i.e., quantity offered for sale in a particular market at a specific price and time. Supply is examined through the view point of a supplier not buyer *e.g.* if the price increases the supplier will increase supply to get more money to over come loss.

Law of Supply : The quantity of a commodity that would be offered for sale varies directly with price under caterus peribus conditions *i.e.* other things remaining the same.

Supply Schedule : It the price of a commodity and the corresponding supplied is arranged in a tabular form then it is known as supply schedule. *Example :*

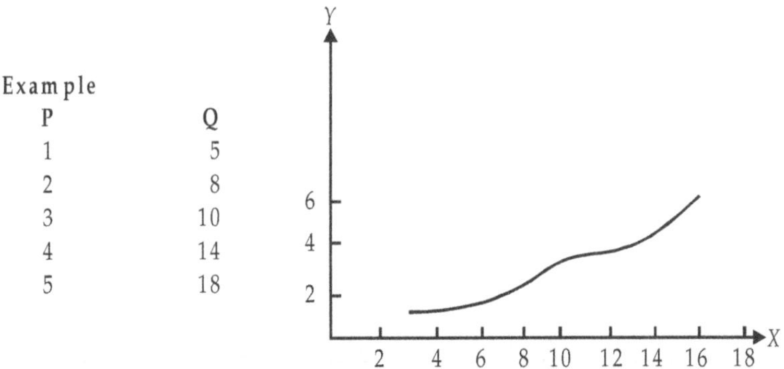

Example

P	Q
1	5
2	8
3	10
4	14
5	18

Supply Curve : Graphical representation of a supply schedule or the locus of different points indicating prices of a commodity and correspond supply. It is a positively sloped curve.

Factor Affecting Supply

1. Price and availability of related goods or substituents. If the price of substituent commodity increase the supply of original good increases. Similarly it the price of substituent decreases the supply of original commodity will decrease.
2. Prices of resources and their availability. The resources here are the factors of production *e.g.* land, capital, labour etc. If the resources are costly or scare they will cause a rise in the price of that commodity.
3. **Change in Technology :** Introduction of better technology increases productivity and thus helps in the lowering of prices.
4. Expectation or fluctuation of future prices. If a supplier that the price of a certain commodity is going to increase than at that time supply will decrease.

5. Weather Conditions : It is specific for agricultural goods, *e.g. rainfalls.* If rainfall is good then in that year production is higher thus prices are low, thus supply is low.
6. Social and institutional influences on a particular day or festival supply suddenly increases.
7. Taxing and subsidising of certain commodities govt. bears a part of the cost of production taxing supplier or seller has to pay a part of profit to govt. that he sells per unit.

If govt. wants to reduce supply to introduce taxes.

These 7 factors are supply factors *i.e.* they shifts the curve to either left or right.

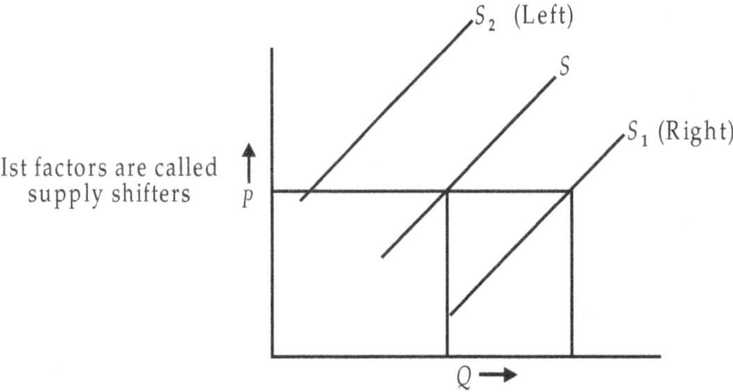

If only the price is changing then there is movement along the curve.

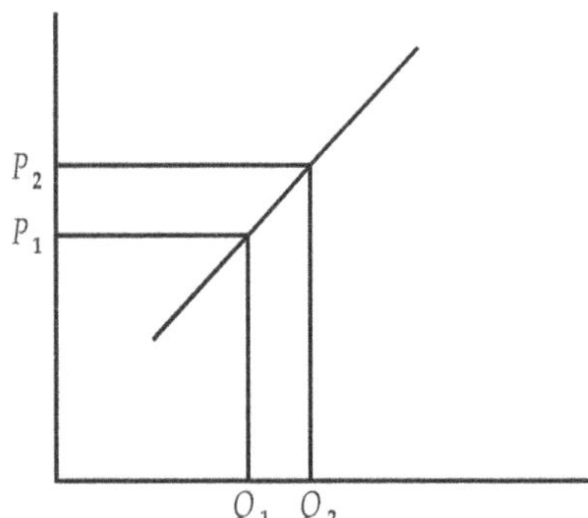

If in demand curve only price changes (movement along demand curve)

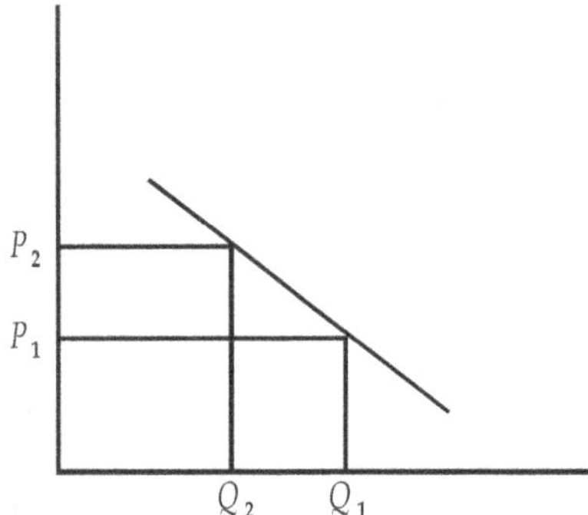

Shift in demand curve (other factors change)

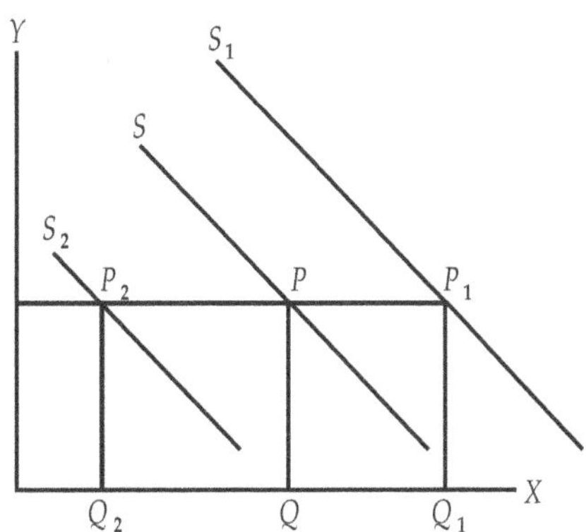

Elasticity of Supply

$$e_s = \frac{\text{Relative change in quantity supplied}}{\text{Relative change in price}}$$

$$e_s = \frac{\Delta Qs / Qs}{\Delta P / P} = \frac{\Delta Qs}{\Delta P} \times \frac{P}{Qs}$$

LAW OF RETURNS

Law of Variable Proportion

It has three different stages namely :

1. Law of increasing returns.
2. Law of constant returns.
3. Law of decreasing returns.

These are depending upon the marginal return rise, remain same or fall respectively. Definition of "Law of variable proportion." It describes the production function with one variable factor, while the quantities of other factors of production are fixed *i.e.* if describes the input and output relationship. The output is increased by the increasing quantity of one factors while others are kept constant.

Output : Return is the function of *i.e.* depends upon labour, land, seed, fertilizer, irrigation.

Y	=	F (L L S F I)
Dependent Variable		Independent variables

I^{st} Stage

Law of Increasing Returns : An industry of firms is subjected to the law of increasing returns if extra investment in the industry is followed by more proportionate returns *i.e.* if marginal product increases. *Example*

Input	*Return*
100	150
110	175

- Industry is a collection of different firms.
- While firm is a individual technical unit. (Producing a brand.)
  e.g. soap industry (industry firm) Lux cod-liver oil etc.

i.e. $\frac{\Delta\ O.P}{\Delta\ I.P.} = \frac{15}{10}$ = Marginal product

O.P – Output

I.P. – Input

2nd Stage

Law of Constant Returns : An Industry is subjected to the law of constant return when increased investment results in proportionate increase in output *i.e.* marginal product is constant.

3rd Stage

Law of Decreasing Returns : An industry of firm is subjected to the law of decreasing returns if extra investment in the industry is followed by less proportionate returns *i.e.* if marginal product decreases.

Example : 1 dose = 10 kg of fertilizer.

	No. of fertilizer doses	*Total production*	*Marginal product*	*Average production*
	1.	80 kg.	–	80/1 = 80
increasing	2.	170 kg.	90	170/2 = 85
	3.	270 kg.	100	270/3 = 90
	4.	370 368 kg.	98 100	368/4 = 92
	5.	430 kg.	62	constant 930/5 = 86
	6.	480 kg.	50	480/6 = 80
Decreasing	7.	504 kg.	29	504/7 = 72
	8.	504 kg.	0	504/8 = 63
	9.	495 kg.	–9	495/9 = 55
	10.	440 kg.	–55	440/10 = 99

$$M.P. = \frac{\Delta\ O.P}{\Delta\ I.P.}$$

$$A.P. = \frac{Y}{X}$$

(One unit input gives how many units of output).

1. **M.P. :** M.P. increases upto *E* and meets A.P. at *F*. Reaches its maximum at *E* decreases from *E* to *M*, zero at *M*, does negative beyond *M* in third stage.

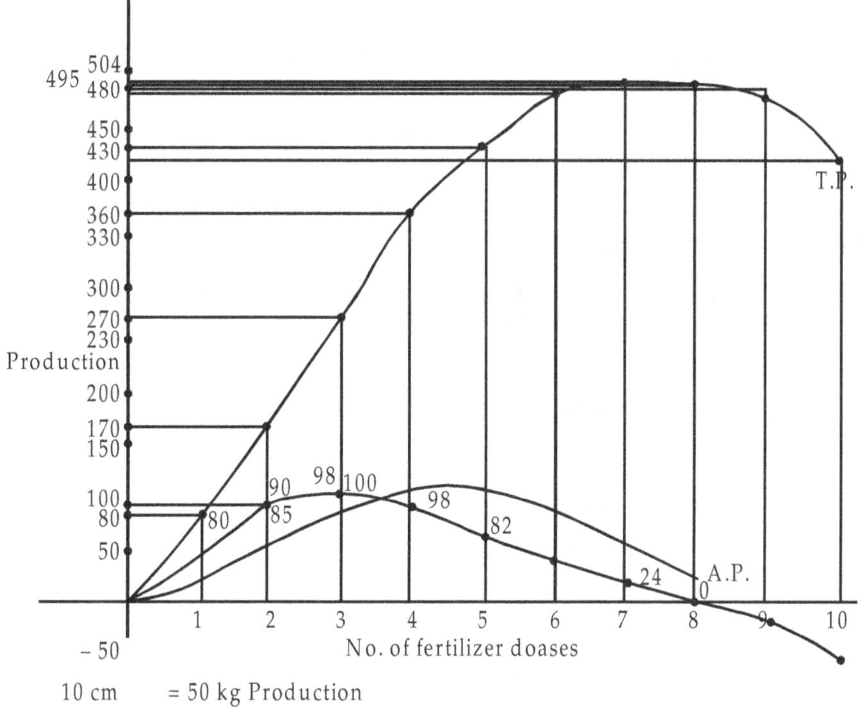

Graph

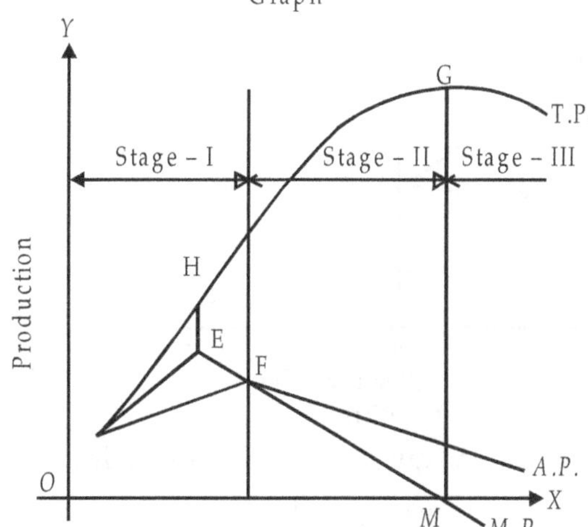

Units of fertilizers

2. A.P. increases up to *F*, meets M.P. at *F* point where it is max. after point *F* it declines but never become zero if at all a single unit is produced.

$$A.P. = \frac{Y}{X}$$

3. T.P. upto *H* point, T.P. increases at increasing rate. (If MP is increasing) it means that T.P. is increasing at an increasing rate).

Point *H* is called inflection point, where the rate of T.P. is switching over. Rate of increasing in T.P. is decreasing.

When T.P. is maximum, M.P. will be zero. Beyond 'G' T.P. decreases that is M.P. is negative.

Stage III : No farmer will use fertilizer beyond 2nd stage because an increase in fertilizer dose causes lower of T.P.

Stage I : Upto *E* point there is no question of stopping because M.P. is increasing. (A.P. is still increasing.)

Stage II : Farmer has to decide this stage were to stop. If input is free of cost farmer goes on putting in the input till upto *M* where M.P. is not yet zero.

If the input is being bought by the farmer then the contribution of additional unit of input is at least the price of that unit otherwise we should useless.

M.C. = Marginal cost.

M.C. = Marginal value product.

M.C. = M.U.P.

where this hits in stage-II we use that much unit of test.

Stage II is called the rational stage of production.

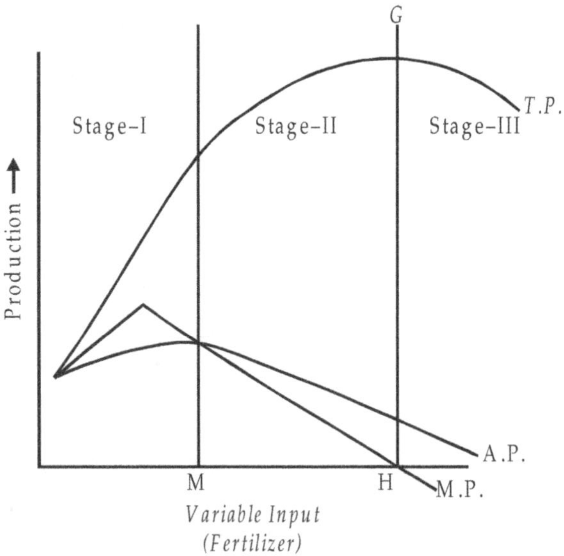

Stage III : M.P. is negative, so no question of going into third stage.

Stage I upto point 'E' A.P. is increasing and T.P, is also increasing *i.e.* efficiency of variable input is increasing as A.P. is increasing T.P. is increasing, *i.e.* efficiency of other constant factors are also increasing. Thus, there is no question of shopping before 'E'.

Stage II Efficiency of variable input is decreasing, A.P. is declining. But at the same time T.P. is also increasing, though the rate of increase is decreasing *i.e.* efficiency of fixed factor is still increasing.

Beyond 'H' efficiency of fixed factors are also decreasing. So, rational stage is in between 'M' and 'H'. Maximum efficiency of A.P. at 'E'. Maximum efficiency of fixed factors at 'G'.

Exact level of fertilizer use will depend upon the price of fertilizer and the contribution of each additional unit of fertilizer to the T.P.

Marginal cost = cost of acquiring each 1 unit.

Marginal return = contribution or return of each unit.

Marginal cost = marginal return.

Return to Scale : It is different from law of returns. In law of returns one input is made variable and it's changed amount is used to examine its affect on production. While in returns to scale QH factors of production are changed by a common proportion and then its affect is seen on the production. *e.g.* all factors increased two times.

Chapter 18

Analysis of Fish Supply and Demand

Introduction

Fisheries in the country since the launching of the First Five-Year Plan in 1951, has witnessed an impressive growth from a highly traditional activity to a well developed and diversified enterprise. The fishery sector during the recent past has played an important role in the Indian economy through employment generation, enhanced income, and earning valuable foreign exchange (Government of India 1996). This sector contributes an estimated 1.37 per cent to the country's GDP, and 5.18 per cent to the agricultural output at current prices (1998-99). The value of export of marine products contributed Rs 5116 crores to the country's exchequer in 1998-99.

Slightly more than half (55 per cent) of Indians are non-vegetarians. The annual per capita consumption of fish is 8 kg per person as against the global average of 12 kg (Government of India 1996). So, scope exists to reach the world average of 12 kg/capita/annum. Empirical studies reveal that a structural shift is taking place in human food consumption towards animal products (Hung & Howarth 1996). Typically, economists have explained such changes in food consumption pattern primarily as resulting from increases in disposable income and changes in food prices. The studies by Kumar (1996), Kumar and Mathur (1996), Kumar (1998) and Bhalla and Hazell (1997) have clearly shown that the composition of food demand across commodities is changing because of change in food habits of the people, change of life style, urbanisation etc., besides change in household income and food prices. The major point emphasised in these studies is that on the whole, direct per capita consumption of cereal as food has declined, while dairy, meat and fish consumption has increased substantially. Integrating fish, livestock, and crop production is an Indian practice from time immemorial. Integrated Fish Farming is attractive to small farmers who are under pressure to produce higher-value commodities, as well as to communities seeking to augment food production and income. Trade liberalization may further add to this pressure. This emerging scenario will have considerable bearing on future demand and supply pattern of fish. The present study was conducted some of these concerns in view with the following specific objectives.

1. To project the demand and supply of fish for 2020.
2. To estimate the supply - demand gap of fish and discuss policy imperatives.

METHODOLOGY

The Data

The demand analysis is based on the data available in National Sample Survey Organization (NSSO) publication of *Consumption of some Important Commodities in India,* NSS 50th round, 1993-94. The data consist on cross-sectional figures on aggregate quantity consumption and values of different food and animal products per person per 30 days for different states by rural and urban categories for the period 1993-94.

The supply analysis is based on time data on quantity of fish production, fish prices, fish seed (million fry) production, and the status of production technologies of fish for the period 1970 to 1998. The important sources of data are *Basic Animal Husbandry Statistics, Agricultural Prices in India, Handbook on Fisheries Statistics.*

THE MODELS

Supply

The quantity produced of a fish like many other foods is hypothesised to be a function of its own prices, prices of inputs used in the production, the existing state of production technology and government policy variables such as supply of credit. It is, however observed that there is a lukewarm response to changes in prices. Such response in assumed to be the result of biological and technical factors.

In this study, we consider a polynomial distributed large to determine the lagged response of the fish production to changes in the fish prices. This model is reported to be quite suitable. Polynomial distributed lag model was originally suggested by Almon (1965) and then modified by Bischoff (1966), Modigliani & Sutch (1966) and Cooper (1972).

The models employed in the study are:

Linear Regression Model

$$Y = b_0 + b_1x_1 + b_2x_2 + b_3x_3 + m$$

where, Y = Quantity of fish production (1970-98)

X_1 = Own price of fish

X_2 = Fish seed (in million fry)

X_3 = Time, which is a proxy for technological change.

Almon Polynomial Price Lag Model

$$Y_t = b_0x_t + b_1x_{t-1} + b_2x_{t-2} + b_3f + b_4T + m_t$$

The transformed model

$$Y_t = a_0w_0 + a_1w_1 + a_2w_2 + b_3f + b_4T + m_t$$

where, w_0 = Price of current variable

w_1 = One lag price

w_2 = Two lag price

f = Fish seed

T = Time (proxy for technological change)

The W's are linear combinations of all the x values (current and lagged)

The weights used in the first constructed variable (w_0) are all equal to unity.

The weights of w_1 will be the simple increasing series of integers.

The weights of third constructed variable w_2 will be squares of the weights of w_1

The estimated response functions incorporate price lags of 1 to 2 years.

It may be noted that here means average fish production without reference to marine/inland, species and types. Similarly, the price refers to average price of fish. There are data limitations to be reckoned with.

Demand

The actual model is specified as

$$Y_i = b_0 \cdot P_{x1}^{b1} \cdot P_{x2}^{b2} \cdot P_{x3}^{b3} \cdot P_{x4}^{b4} \cdot P_{x5}^{b5} \cdot P_{x6}^{b6} \cdot P_{x7}^{b7} \cdot P_{x8}^{b8} \cdot I^{b9} \cdot d_1 D$$

where, Y_i = Quantity of fish consumption per capita over 30 days

P_{x1} = Price of milk

P_{x2} = Mutton and goat meat prices

P_{x3} = Beef and buffalo meat prices

P_{x4} = Chicken prices

P_{x5} = Egg prices

P_{x6} = Fish prices

P_{x7} = Other food prices

P_{x8} = Non-food prices

I = Expenditure (proxy for income)

bi and *di* are the coefficients for the structural and dummy variables, respectively. Dummy variables for regions (north, east, west, south and hills) were specified in the analysis. Both dependent and independent variables are taken in log form in the functional analysis.

When the above demand model was estimated, the coefficient relating to expenditure came out to be negative (–1.47) for pooled analysis and –1.17 for rural and –0.92 for urban (Appendix I). Since these coefficients are not convincing, we examined the data set again and tried to recast the model. Since the consumer behaviour with respect to fish would be normal and stable in fish eating states, we wanted to go for demand analysis with respect to fish eating states only. Fish eating state was decided on the basis of average fish consumption (kg) per person for 30 days estimated for that state nearer to the national average. Thus, 12 urban and 9 rural states/union territories (UTs) were selected and 21 in pooled analysis (Appendix II). Since the number of observations have drastically come down by doing this, we decided to specify the demand model in terms of own price and expenditure unlike demand system estimated earlier. Thus, the final demand model used in the study for rural/urban/pooled sample is specified as :

Recasted Demand Model

$$Y_i = b_0 \cdot P_f^{b1} \cdot I^{b2}$$

where Y_i = Quantity of fish consumption per capita over 30 days.

P_f = Fish prices

I = Income/Expenditure

bi are the coefficients for the structural variables. Both dependent and independent variables are taken in log form in the functional analysis.

ESTIMATION PROCEDURE

Supply

The data for each commodity consists of 29 years observation sets. The estimates of price coefficient generally assume expected positive signs and exhibit a high degree of precision. Linear and Polynomial regression models are used for estimation of regression coefficients. The equations are estimated using the standard *OLS* method. The lagged model is finite and includes only exogenous lagged variables. The estimated response functions incorporate price lags of 1 to 2 years. Elasticities are estimated by using the formulae, where E_p = Supply elasticity, b = coefficient (regional productions), p = the average production, y = average quantity.

The value of R^2 (adjusted) is fairly satisfactory in the supply response functions of fish. This suggests that the relative prices, fish seed and technological and biological developments (proxies for time trend) have played a significant role in enhancing the production of fish in India.

Demand

The data for estimation of fish demand consists of 21 observation sets representing rural and urban populations across 21 states. We have estimated the demand response function with double log specification using Ordinary Least Square (OLS) procedure. The estimated coefficients provided elasticities. The variables included in the model explained 73 per cent of the variability in rural, 45 per cent in urban and 52 per cent in pooled analysis (Table 4).

Projections through 2020 were made by using simple growth rate model based on estimated expenditure elasticities, population and per capita income growth rates and urbanisation.

Results

In the 1960s, India made headlines with its Green Revolution, using high yielding varieties (HYVs) seeds and improved technology to more than double its output of wheat between 1965 and 1972. Today, India is marching ahead with Bule Revolution, by rapidly increasing fish production in small ponds and water bodies, benefiting small farmers, as well as contributing to nutritional food security and national income.

Changes in Structure of Fish Production in India

The fish production scenario during 1950 to 2000 has been shown in Table 1. Fish production in India has increased steadily from 7.5 lakh tonnes in 1950-51 to 56.6 lakh tonnes in 1999-00. Marine fisheries remained the major contributor till 1990-91. Its contribution to total fish production by 1960-61 was over 75 per cent, but it declined drastically to 61.93 per cent in 1970-71. Since then, it remained almost constant till 1990-91. In the nineties, fish production structure underwent substantial changes. The share of inland fisheries increased drastically reaching to 50 per cent in 1990-00. These changes were due to deceleration in growth of marine fish production and a policy shift in favour of inland fisheries, particularly aquaculture.

Trend in Fish Production in India

The trend in fish production during 1950-2000 has been shown in Table 2. Since 1950-51 fish production has been increasing at a rate of 4.12 per cent a year. The inland sector contributed increasingly to the observed growth; inland fish production grew at an annual rate of 5.24 per cent. A desegregated view of pattern of growth shows acceleration in growth of inland fish production during the nineties. On the other hand, growth in marine fish production decelerated to 2.09 per cent during 1990-99 from 4.02 per cent during 1980-90.

Table 1 : Changes in structure of fish production in India

(in lakh tonne)

Year	*Marine*	*Inland*	*Total*
1950-51	5.3 (70.67)	2.2(29.33)	7.5(100)
1960-61	8.8(75.86)	2.8(24.14)	11.6(100)
1970-71	10.9(61.93)	6.7(38.07)	17.6(100)
1980-81	15.5(63.52)	8.9(36.48)	24.4(100)
1990-91	23.0(59.89)	15.4(40.10)	38.4(100)
1999-00	28.3(50.00)	28.3(50.00)	56.6(100)

Figures in parentheses represent percentage to the total.
Source : Economic Survey, 2000-2001.

Table 2 : Growth Trend (%) in Fish Production in India

Period	*Marine*	*Inland*	*Total*
1950-51 to 1959-60	5.20	2.44	4.45
1960-61 to 1969-70	2.16	9.11	4.26
1970-71 to 1979-80	3.58	2.88	3.32
1980-81 to 1989-90	4.02	5.64	4.64
1990-91 to 1999-00	2.09	6.99	4.4
1950-51 to 1999-00	3.41	5.24	4.12

Source : Economic Survey, 2000-2001.

Fish Supply Response

Estimates of fish supply response through linear and polynomial price lag models (transformed) are presented in Table 3. The supply response equations are shown in Appendix III.

Table 3 : Estimates of the fish supply

Equations/Variables	*Linear regression*	*Polynomial regression (transformed model)*
Constant	–246.98	574.23
	(.0-719)	(1.117)
Price/Price W_0	0.1976*	0.8467*
	(2.697)	(2.396)
Price W_1	–	-1.1673
		(-1.004)
Price W_2	–	0.4979
		(0.56)
Fish seed	0.2978*	0.3968*
	(5.818)	(5.933)
Time	0.133	-0.282
	(0.762)	(-1.085)
R^2	0.972	0.978
R^{-2}	0.968	0.968

Figures in parentheses represent to values
*1 per cent level of significance.

Estimates of Linear Regression Model

In conformity with theory, fish price coefficients are positive. Highly significant price coefficients for fish were noticed, implying that higher prices stimulate fish production. Fish seed is used as a variable and it is found to be significant indicating that the availability of fish seed would enhance fish production. Time variable, which represents technological and other structural changes in the fish sector, as expected, is positive.

Estimates of Polynomial Price Lag Model

Fish price coefficients are significant at 1 per cent level, implying that higher prices stimulate the production of fish. It indicates scope of favourable price policy to enable fish farmers to increase investments to improve production of fish. However, time coefficient was negative and non-significant. Fish seed coefficient is positive and significant indicating its availability would increase fish production.

The estimated supply responds functions are robut in terms of explaining variability in fish production. The price impact in the first period is positive and significant, indicating the influence of immediate previous lag price on production of these products. It is interesting to note that the dynamic price impact (as depicted by the delayed price coefficients) increases first with lag, then decreases and finally increases indicating cobbweb type situation leading to rise and fall of production with response to price changes.

Fish Demand

The expenditure and price elasticities of recasted model are shown in Table 4. The demand response equations are shown in Appendix IV. These results clearly show that the expenditure elasticity for fish in general, is elastic. It indicates that if consumer's income increases he would spend more on fish particularly in rural areas. Own price elasticity in rural areas is negative, inelastic; elastic in urban and pooled sample. Thus, both income and price changes affect the demand for fish. It is to be noted that our estimates are based on conventional determinants of demand and the structural ones like change of tastes and lifestyle of people, income distribution of consumers, market availability etc. are left out for want of data. Similarly, the price effects of substitute commodities are not examined. Future work in this are should consider these gaps, to estimate the demand system framework to address fast changing structural transformation in society.

Table 4 : Estimates of the Fish Demand

Elasticity	*Rural*	*Urban*	*Pooled*
Intercept	-5.815	-0.186	-1.273
	(-2.246)	(-0.090)	(-0.918)
Expenditure	2.689*	0.600	1.046**
	(2.866)	(0.80)	(2.006)
Own price elasticity	-0.702***	-1.040**	-1.004
	(-1.755)	(-2.587)	(-0.651)
R^2	0.735	0.46	0.525
R^{-2}	0.646	0.34	0.473

Figures in parentheses represent values.

* 1 per cent level of significance ** 5 per cent level of significance

*** 10 per cent level of significance

Demand – Supply Gap

Projected production (supply) and consumption (demand) figures for fish during 2020 are shown in Table 5. The baseline scenario revealed that the actual production level for fish closely follows its consumption. It may be noted that Kumar's study (1998) under the assumption of 5 per cent GDP growth rate, estimates fish production of 5.7 million tonnes in 2000 and 11.8 million tonnes in 2020 with a growth (1995-2020) rate of 3.75 per cent. However, in 2020, substantial surpluses are expected in fish of about 4.48 million tonnes. The results clearly illustrate the potential of the fish sector and needed strategies to harness it domestically and through exports.

The expected production growth rates of fish, exceeded the corresponding consumption demand by more than 1 per cent. Comparison of projected fish production and consumption in 2020 shows a somewhat different story. The surplus production of 4.48 million tonnes of fish need to be planned for exports or promoting fish eating in non-fish eating states. It is generally felt that post-harvest infrastructure is grossly inadequate in fisheries sector.

Table 5 : Projections of Fish Production and Domestic Consumption

Supply/Demand	*Year 2000*	*Year 2020*	*Growth Rate*
Production*	5.66	13.0	4.4
Consumption**	4.45	8.52	3.3
Surplus	1.21	4.48	

* The production growth rate is assumed to grow at 4.4% p.a. as it was during 1990-2000 (Table 2).

** Consumption figures are weighted average (weighted by rural and urban population divided by total population)

Marketing, transportation, storage, processing and packaging, will be helpful to handle the expected surplus. Also major initiatives are needed for the development of the domestic market (Government of India 1996). It is reported that fish is sold on the roadside and there is no organised effort in marketing of fish. Studies are needed in the areas of price margin, marketable surplus, marketed surplus and price. Studies are also needed in the areas of utilisation by catch and diversification in both harvest and post harvest activities including pharmaceuticals, industrial, chemical and medicinal fields. Providing quality fish to markets away from production centres will be a major challenge in the future. Setting up of inland fish marketing units and development of retail markets in non-fish eating states/places should receive priority attention.

India's share in the booming world trade of fish is less than 2 per cent, which is very low considering the huge export potential for exports. The development of transportation facilities such as availability of steamers to different countries in the world and transportation of refrigerated containers have to be vigorously persuade to clear to clear the surplus. Since several imporing countries are stipulating stringent quality control for marine products, modernisation of the processing facilities to meet international standards assumes significance. Setting standards for intermediary inputs like feed, seed etc., are also critical. Thus, quality control, exports promotion and marketing strategies need to be persued more aggressively, keeping in view the dynamic nature of the export markets.

Conclusions and Policy Implications

India is marching ahead with blue revolution. Fish production in India has increased steadily from 7.5 lakh tonnes in 1950-51 to 56.6 lakh tonnes in 1990-00. Marine fisheries remained the major contributors till 1990-91. The share of inland fisheries increased drastically reaching to 50 per cent in 1990-00. These changes were due to deceleration nine growth of marine fish production and a policy shift in favour of inland fisheries, particularly aquaculture.

The lagged price impact in the first period is positive and significant indicating the influence of immediate previous lag price on production. In is interesting to note that the dynamic price impact (as depicted by the delayed price coefficients) increases first with lag, then decreases and finally increases indicating alternative year rise and fall of production with response to price changes.

APPENDIX I

Estimates of the Fish Demand Response Model

Elasticity	*Rural*	*Urban*	*Pooled*
Intercept	-0.473 (-0.0600)	-9.0742 (-0.9724)	-2.4256 (-0.5600)
Expenditure	-1.1661 (-1.2397)	-0.9227 (-0.9363)	-1.4702* (-2.4735)
Own price elasticity			
Fish	-3.4816* (-4.2787)	-2.1048** (-2.0933)	-3.4203* (-6.4348)
Cross price elasticity			
Milk	6.0601* (4.5147)	6.1637* (2.9403)	5.9048* (6.1283)
Mutton and goat meat	1.7589* (2.1954)	4.6880* (2.9350)	2.9749* 3.6757)
Beef and Buffalo meat	1.1268*** (1.4741)	0.0075 (0.0108)	0.9696* (2.3726)
Chicken	-0.50008 (-0.4621)	-0.9742** (-2.2502)	-0.8096* (-2.4132)
Egg	-3.0541** (-2.1573)	1.0795 (0.6450)	-0.9055 (-0.9066)
Other foods	2.0459 (0.7402)	-2.9341 (-0.8471)	0.4042 (0.2832)
Non food	-2.3181** (-1.7722)	3.78666** (2.0395)	-0.1869 (-0.2290)
Regional Dummies			
North	-0.8713 (-1.2397)	-0.8441** (-1.7878)	-0.8869* (-2.7520)
South	-1.3564* (-2.5330)	-1.5086* (-3.3235)	-1.3257* (-4.9548)
East	-0.7251*** (-1.4608)	-0.7437*** (-1.3561)	-0.8309* (-2.5233)
West	-0.3088 (-0.7156)	-0.2603 (-0.6774)	-0.1144 (-0.4353)
R^2	0.92	0.91	0.88
R^2	0.87	0.84	0.85

Figures in parentheses represent *t* values.

* 1 per cent level of significance,

** 5 per cent level of significance,

*** 10 per cent level of significance.

APPENDIX II

Fish Per Capita Consumption Per Person Per 30 Days

Sl.No.	*State*	*Rural (Kg)*	*Urban (Kg)*
1.	Arunachal Pradesh		0.2.9
2.	Assam	0.43	0.43
3.	Goa	1.36	1.36
4.	Kerala	1.35	1.35
5.	Manipur		0.34
6.	Meghalaya		0.32
7.	Tripura	0.89	0.89
8.	West Bengal	0.54	0.54
9.	A & N Islands	1.40	1.40
10.	Daman & Diu	4.12	4.12
11.	Lakshawdeep	3.61	3.61
12.	Pondicherry	0.69	0.69

APPENDIX III

Supply Equations of Linear and Polynomial Price Lag Model

Linear Model

S_{pf} = – 246.98 + 0.1976 fp + 0.2978 fs + 0.133t R^{-2} 0.968
(0-719)   (2.697)   (5.818)   (0.762)

Polynomial lag model

S_{pf} = 574.23 + 0.8467 f_{w0} –1.1673 f_{w1} –0.4979 f_{w2} –0.3968 fs –0.282t R^{-2} 0.968
(1.117)   (2.396)   (–1.004)   (0.56)   (5.933)   (–1.085)

Numbers in parentheses are *t* values.

The recasted demand model clearly shows that both income and price changes affect the demand for fish. In confirmity with theory, supply price coefficients are positive and highly significant. It clearly shows that production elasticities of fish is highly price elastic. It needs reorientation of price policy to create the environment in which fish farmers will increase investment to further improve production of fish. Since fish seed availability would increase fish production, attention to supply quality fish seed should receive greater attention.

The results relating to supply-demand gap clearly indicated that in 2020, India would be having 4.48 million tonnes surplus in fish produce. The surplus production of this magnitude would need to be either exported or to be domestically consumed mostly by the people in non-fish eating states. This requires substantial investment in post-harvest management, storage, transportation, processing, packaging and marketing. For promoting exports, quality control of both inputs and output, export promotion and marketing strategies need to be persuade more aggressively, keeping in view the dynamic nature of the export markets. So there is need to develop domestic market and also to formulate sound export policy for fish.

Finally, future studies in this area should consider the demand system as a whole and estimate the elasticities to gain better insights for effective policy analysis. However, such studies require detailed and accurate information and data on fisheries production, sale (trade) which are at the moment fragmented and rather inadequate.

APPENDIX IV

Recasted Demand Equations of Rural, Urban and Pooled

Rural

$C_r f = -5.815 \; -0.702P_f + 2.689I \quad R^{-2} = 0.65$

(2.589) (0.400) (0.938)

Urban

$C_r f = -0.186 \; -1.004P_f + 0.600I \quad R^{-2} = 0.34$

(2.067) (0.388) (0.749)

Pooled (Rural + Urban)

$C_r f = -1.273 \; -1.040P_f + 1.046I \quad R^{-2} = 0.47$

(1.385) (0.259) (0.522)

Numbers in parentheses are Standard Errors.

REFERENCES

Almon, Shirley. 1965. The Distributed Lag between Capital Appropriations and Expenditures. *Econometrica,* 33:178-196.

Bhalla, G.S; and P. Hazell. 1997. Foodgrains Demand in India to 2020.

A Preliminary Exercise, *Economic and Political Weekly,* Dec 27, 32; No. 52; A150-A154.

Bischoff, C.W. 1996. Elasticities of substitutions, Capital Malleability, and Distributed Lag Investment Function. Paper presented at the Econometric Society Meetings, SanFransisco.

Cooper, J. Phillip. 1972. Two Approaches to Polynomial Distributed Large Estimation: An Explanatory Note and Comment. *The American Statistician,* 26:1; 32-35.

Economic Survey, 2000-2001.

Government of India 1996. *Report of Working Group on Fisheries for the formulation of Ninth Five Year Plan.* Department of Agriculture & Cooperation, Ministry of Agriculture.

Government of India 1996. Population Projections for India and States.

Report of Technical Group on Population Projections. Planning Commission.

Government of India 2000-2001. Economic Survey, Ministry of Finance, Government of India.

Huang, J. and Howarth E. Bouis. 1996. *Structural Changes in the Demand for Food in India.* Food, Agriculture, and the Environment Discussion Paper 2, IFPRI, Washington, D.C., USA.

Modigliani, F. and R. Sutch. 1996. Innovations in Interest Rate Policy. *American Economic Review* (Papers and proceedings) 56: 178-179.

NCAP 2003. A Profile of People,Technologies and Policies in Fisheries Sector in India (eds. Anjani Kumar, Pradeep K. Katiha and P.K. Joshi).

Kumar, Praduman. 1996 Structural Changes in consumption and small Farm Diversification' in T Haque (Ed), Small Farm Diversification: Problems and Prospects. National Centre for Agricultural Economics and Policy Research, New Delhi.

Kumar, Praduman. 1998. Food Demand and Supply Projections for India. Agricultural Economics Policy Paper 98-01, IARI, New Delhi.

Kumar, Praduman and Mathur, V. C. 1996. Agriculture in Future: Demand-Supply Perspective for the Ninth Five Year Plan. *Economic and Political Weekly,* Sept 28, A131 - A149.

Chapter 19

Marketing

In earlier days the term marketing of fish meant 'buying and selling of fish' at the landing centre. After the Second World War, the concept and functions of fish marketing have taken a new role in business activities. The fisheries have now become highly industrialized in all advanced fishing nations. The new marketing techniques have been adopted so as to sell more fish. The modern fish marketing system lays emphasis in meeting the existing demands for fish, besides tapping the potential demand in the important markets.

In many advanced countries the improved methods of fish marketing are being adopted with the advancement of fisheries development. A progressive fish marketing system will also provide remunerative price to the primary producer though the interest of the consumers is also protected. The fish marketing needs modernization in all the developing countries including India. In Singapore and Philippines, the fish marketing is completely in the hands of private merchants. In Sri Lanka, fish marketing is in the private hands though Government has extended some support to co-operative enterprises. There is much exploitation of fish producers by fish merchants in Malaysia, Indonesia and Philippines. The whispering auctions and cloak auctions are the usual features in Manila (Philippines). Similarly in Hong Kong, bidding is usually done by forming ring amongst buyers. In India, the bidders show some gestures with fingers for their consent for buying at Kanpur and Agra. In Bombay fish markets for the auction of certain species of fish under a vernacular word 'Kodi'. In Malabar area of Kerala State, 'Kalli', a local measure is used for the disposal of sandines and mackarels. In fish markets of Haryana, the fish is sold by 'Kachha maund' and 'Pucca maund' of fish. The marketing of any produce, mainly depends upon the availability, consumption and demand.

Availability, Consumption and Demand of Fish

The demand and consumption of fish depends on the availability, as well as the fish eating population in India. As per Census of 1951, 82.7 per cent of the total population inhabited in rural areas and out of these 46.7 per cent were fish eating. Similarly out of 17.3 per cent inhabitants in urban areas, 43.5 per cent were fish eating. Percentage of fish eating population in rural area is more as compared to the urban area. The percentage of fish eating population increased with the increase of population in the country. As per Census of 1961, 1971 and 1981, the fish eating population in rural area has increased to 50.5, 53.6 and 57.1 per cent respectively. The fish eating population in urban area has also increased to 46.0, 48.7 and 51.8 per cent as per Census of 1961, 1971 and 1981 respectively. In both rural as well as urban areas the percentage of fish eating population has increased. Out of 6,840 lakh population in India, as per 1981 Census, 3,820 lakh were fish eater *i.e.* 55.8 per cent. The fish eater consumes 93 per cent fish by cooking households while only 7 per cent by other way. The fish is cooked at home by 50.5 per cent fish eating population. We want to increase but 2001 it moves around 52 per cent.

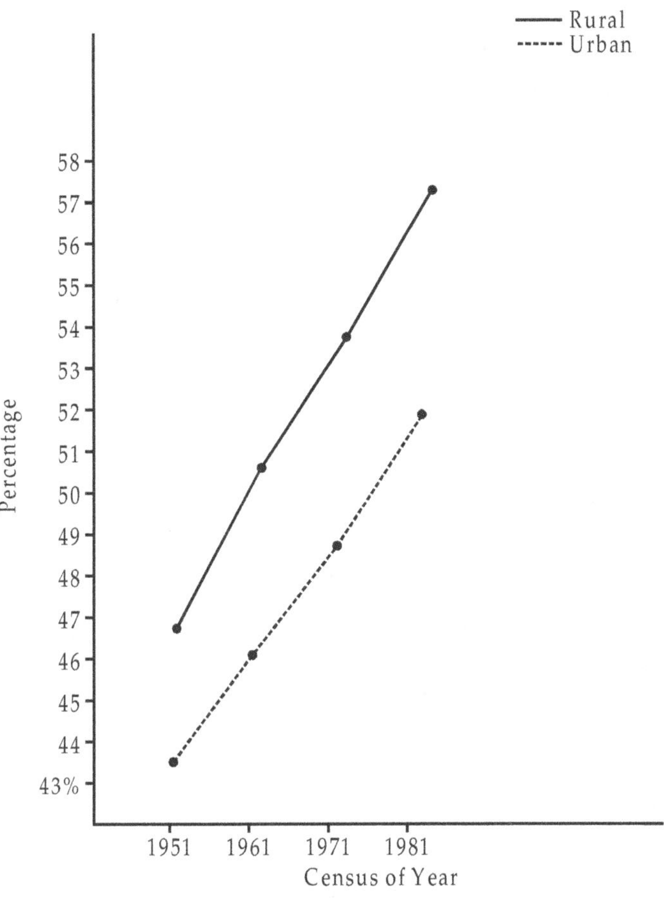

Fig. 1 : Fish Eating Population in India

The fish eating population in Haryana is 15 per cent which is less, as compared to neighbouring States like Punjab 19.15 per cent, Rajasthan 18.7 per cent, Uttar Pradesh 44.4 per cent and Delhi Union Territory 65.7 per cent. As per fish statistics, out of total fish production in the country, 69 per cent fish is consumed as domestic consumption (65.8 per cent in houses and 3.2 per cent outside houses), 14.8 per cent used by feed manufactures, 2.2 per cent by other uses and 14.4 per cent is exported to other countries. The per capita fish consumption in Metropolitan cities is 24.8 grams per day while in urban and rural areas, it is 30 grams per day and 23 grams per day respectively. The average fish consumption in urban areas is more as compared to metropolitan and rural areas. The per capita consumption of fish varies from State to State and within each State from region to region. In general, fish constitutes an important item of food in the diet for the people living in the coastal regions and in the Eastern States of the country comprising. West Bengal, Orissa, Assam, parts of Bihar and Union Territories of North-Eastern India. In other parts of the country, fish is only a supplemental food. The per capita consumption of fish in India at present is 3.2 kilograms per year as compared to World 12.3 kilograms per year. In view of the importance of the fish as a food component being consumed by more than 60 per cent and looking at the facts, we can now calculate tentatively the consumption of fish in the future. Data about the future consumption of fish as food component which has been shown in Table 1.

The requirement of fish forms as one of the main component of animal protein in diet computed 11 kilograms per year per person. It is revealed from the above that the per capita availability of fish is 3.05 kilogram per year in 1961 which increased to 3.20 kilogram per year in 1981 and is projected to be 6.44 kilogram per year by 2001 which is only 41.5 per cent of the requirement. Keeping the view the average 5 per cent growth rate of fish production the projected fish production in 1991 and 2001 is 3666 and 5499 thousand tonne respectively. The fish consumption in India has been shown in Table 1.

Table 1 : Fish Consumption in India

Year	*Population (lakh)*	*Fish eating population (lakh)*	*Fish production (tonne)*	*Fish for human consumption (tonne)*	*Per capita availability of fish kilogram/year)*
1961	4390	2180	961000	663000	3.05
1971	5480	3890	1851600	1277600	3.10
1981	6840	4220	2444000	1686000	3.20
1991*	8065	4758	3666000	2530000	5.31
2001*	9354	5893	5499000	3795000	6.44

* The projection made on the basis of present trend of growth rate.

In many developing countries traditional system of fish marketing is adopted. The methods and practices in trade dealings are based on some customs. These practices have remained unchanged and unimproved over decades. The fish marketing is normally done at the collection centres which are mainly situated in the area of fish landing. The Indian Institute of Management, Ahmadabad has studied the fish marketing practices in Uttar Pradesh, Bihar, Maharashtra, Andhra Pradesh and Karnataka. The Institute recommended that all the markets require modernization.

Definition of Fish Markets

Fish has peculiar feature of its own and gives a big strain and stress on the method of its marketing. The fish landing centres, particularly marine fish, are invariably isolated from good markets, which mean supplies during peak season in those areas without adequate demand. The improvement in fish marketing and distribution can very well eliminate some of the depressed products of malnutrition where people live on subsistence level, by supplying fish at reasonable prices. Fish may not be consistently demanded by few customers due to its peculiar smell. It is an act of fishery technologists to prepare the odourless fish product for wider consumption. Thus the fish marketing should not have the object only catching and selling of fish but the fish marketing should have the wide scope for exploitation, production distribution, preservation and transportation of fish in addition to actual sale of fish by reducing middlemen.

Fish Markets in Haryana

Haryana is a land-locked state having only 15 per cent population of fish eater. Hence, there is no well organized fish marketing system in the State. For promoting intra-State marketing, the Department of Fisheries, Haryana has imposed a condition while auctioning the fishing rights to the fish contractors that every fish contractor shall establish a fish shop in the area of his jurisdiction for supplying fresh fish to the local people. In spite of efforts made by the Department, 85 per cent of the produce is despatched to the other fish markets like Saharanpur, Delhi and Howrah for sale. However, four fish markets were selected for the present studies. The pattern of marketing of fish in the State *i.e.* Ambala, Karnal, Panipat and Sonepat, keeping in view the resources and importance of the cities.

Preparation of Fish Markets

The fishermen visit the fishing grounds and fortunate to strike a catch tends to bring the produce to the nearby market for sale as soon as possible. In some cases the fish catch may be of the good size and variety acceptable to the customers while in some cases it may be poor. There are no effective means for regulating the catch or forecasting the varieties of the fish that would be landed. The quantity of catch is also uncertain, therefore, creates gults and shortages which affect the fish marketing and pricing of fish. The fishermen who actually catch fish play only an insignificant role in the disposal of catches. Their role is only to hand over the fish catch to fish merchants at the landing centres for sale. The final distribution and marketing of catch is done by commission agents who step in at this stage.

Fish Marketing Procedures

(*a*) ***Sale Proceed at Markets :*** At the landing centres, the fish is sold in many ways. Although, it is not possible to draw an exact time of demarcation between different methods of marketing in India. At the landing centres, the fish is assembled and sorted out by the agents or wholesales or fishermen's group leaders. At Chilka lake area, Orissa State, the fish is collected by the leader of fishermen party who is known as 'Bahanias'. Similarly, in the Kakinada area, Andhra Pradesh State, the fish is mostly collected by the "Pettamdars", whereas in Kerala "Thruvilarya". In Gujarat and Maharashtra the primary collector are known as "Tindels". The fresh water fish in most of the cases is sorted out species-wise and also size-wise. Then fish is packed in ice and kept in bamboo baskets or wooden boxes for despatch to the distant markets particularly in Calcutta. These activities are done by the agents on behalf of the wholesalers or on behalf of the commission agents. In some cases, the primary co-operative fish marketing societies directly consigns the fish to the private traders. It is observed in Haryana/UP/Uttranchal that no packing or sorting is done at landing centre. The fish is filled in empty gunny bags and transported to the nearest fish market for disposal. The fish before reaching to the market is handled by 5-6 middlemen in case of fresh water whereas 3-4 persons in case of marine. In Haryana/UP/Uttranchal only 2-3 persons handle fish because the owner of water is generally a commission agent in case of notified water. Of course, in case of private owned waters or ponds, the number of middlemen increases. By and large the prevalent methods in fish marketing are given below :

(*i*) In the primary fish markets, the fishermen or producer directly sell fish to the buyers without having intermediaries. In many places the buyers may be wholesalers or fish merchants or middlemen. In the Southern States particularly in Maharashtra, Andhra Pradesh and Tamil Nadu, the women who belong to fishermen's family traditionally sell fish in retail markets unlike the upper eastern or northern India. The prices are usually negotiated between individual fisherman, seller and buyer.

(*ii*) The fishermen send fish to the commission agents as consignments. The commission agents *i.e.* 'aratdars' auction the fish and the gross sale proceeds are remitted to fishermen after deducting various marketing charges.

(*iii*) Selling of fish is also done by contact method. The prices are fixed before the fishing season starts. The traders make some agreement for the delivery of catches by fishermen at stipulated prices. All the fish caught is disposed of at the fixed price. In Gujarat, the Mandela Committees are formed where the fishermen's representatives, Government representatives and traders assemble and fix the price based on the previous year's prices and also to some extent on the Bombay fish market prices. In this case generally, prices do not fluctuate much.

It is observed that the fish being perishable commodities are brought nearest fish markets and mostly cleared daily. In all markets the fish is put for sale without gutting and cleaning. However, the sorting is done size-wise and group-wise. There are two main groups 'Chilkar' and 'Nanggi'. Different heaps are made before sale by auction. Live fish 'Sol' and 'Singhi Mangur' are sold in live condition in water. It is usual practice in almost all the fish markets that the entire catch assembled at the wholesale fish markets for disposal by auction to the merchants and the retailers between fixed time from 9.00 A.M. to 11.00 A.M. The fish is generally auctioned in small heaps. The vendors or retailers assess the weight of fish and give the bid as per his estimate only. The fish put to sale is neve weighted at fish markets somewhere bid is given per kilogram of fish. The bid is released to the highest bidder for the particular heap and weighed after the auction. The spoilage starts in fish no sooner it is taken out of water. Hence, the care is taken at every stage to maintain the quality of fish. It is observed that unpreserved fish become unsaleable within eight hours. Due to lack of well organized fish market in Northern States, almost entire catch is being sold in fresh conditions.

During the period of study the number of fish markets not increased in Northern States. There are several variables in fish marketing practices in Northern States. These includes sale, preservation, pricing method, partnership, selling price by variety, payment and borrowing by producers from fish merchants. There are many channels of flow for sale in Northern States.

(*a*) ***Sale Proceed at Production Centres :*** The fish producers sell their produce at the ponds site to the rural masses at the time of harvesting. It is not popular practice and only a fraction of total production is sold.

(*b*) ***Sale at District Headquarters :*** The producers bring their produce to the district headquarters and sell fish door to door as vender. It is also not a popular practice and only small quantity is sold in this manner.

(*c*) ***Intra-Districts Marketing :*** Although there is a Government contractor shop at every headquarters but fish is transported from one district to the popular fish market of other district with the idea to have good price. The fish produced in area is transported either to other as fish markets.

(*d*) ***Sale of Fish at the Fish Markets, Outside the State :*** It has been observed that good portion of fish is sent to Delhi or Howrah fish markets for sale. It is observed that more than 85 per cent of the fish farmers of Haryana sent their produce to fish markets outside the State.

Fish Marketing Intermediaries

There are many fish intermediaries market in northern states of India from fish producers to consumers forms a complicated network which is given below :

Fish Market Intermediaries	*Notation*
Fish farmers	FF
Fishing workers	FW
Fish farmers-cum-contractors	FFC
Fish farmers-cum-contractor-cum-wholesellers	FFCW
Fish farmers-cum-wholesellers	FFW
Fish farmers-cum-retailers	FFR
Fish farmers-cum-fishing workers	FFFW
Commission agents	CA
Commission agent-cum-wholeseller	CAW
Whole-seller	W
Whole-seller-cum-retailers	WR
Fishing worker-cum-retailers	FWR

Contd....

Fish Market Intermediaries	*Notation*
Fish farmers-cum-Vendor	FFV
Fishing worker-cum-Vendor	FWV
Fish contractor	FC
Retailers	R
Vendor	V
Consumer	CUM

(*a*) ***Fish Market Network :*** There are various networks in fish marketing system which are not understandable and in systematic manner. A model has been designed keeping in various persons engaged in fish marketing. The fish producers or fishermen have various options to adopt the channel to minimize the middle-men. The present model of fish marketing has been shown in Figure 2. The options with fish producers or fishermen for the sale of fish are as under :

1. Sold fish directly to consumers.
2. Sold fish through commission agents.
3. Auction the fish to fish contractors.
4. Sale the fish to fishing worker adopting royalty system.

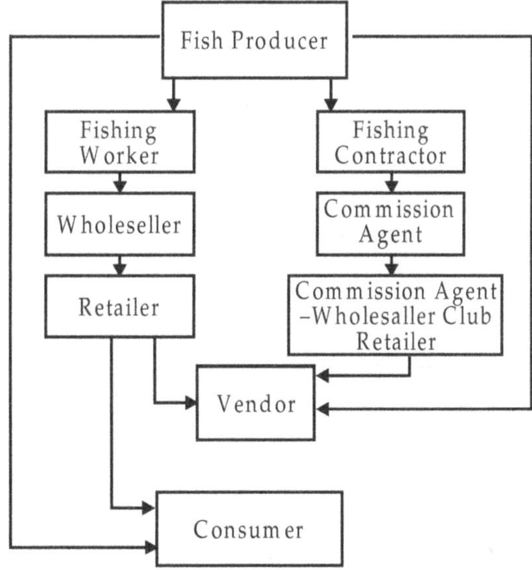

Fig. 2 : Model of Fish-Marketing

The first person in the channel is fish farmer/fish producer and last is consumer. Lesser the gap between two will produce more profit to both. The market intermediaries take the profit at various stages and the cost of fish increases. The merchandise charges are exorbitant and prevalent practice is very cumbersome.

(*b*) ***Merchandise Charges :*** The commission agents recover the following charges for clearing and selling the consignments brought by fish producers.

(*i*) *Clearance and Cartage Charges :* These charges includes the cost incurred by the commission agents for assembling the produce at the various landing centres and further transporting them to the fish markets.

(*ii*) *Packing the Ice Charges :* The cost of basket, gunny bags and ice are charged from the producer for landing to the fish market. It is observed that in very few cases the ice is used from landing centre to fish market, however, storage in ice is done at shop.

(*iii*) *Octroi and Terminal Taxes etc. :* These expenses are charged from the producers to pay such taxes to municipal committee etc.

(*iv*) *Charity :* The prevalent practice is to take out a fish weighing approximately 2 kilograms out of a lot of 40 kilograms, on the pretext that its sale proceeds shall be utilized for donation to religious institutions. This fish is sold at the end of the auction and no record of the same is ever kept.

(*v*) *Commission :* During marketing proceedings 7 per cent commission is charged on the gross sale value of fish from the purchasers. At Panipat fish market 3 per cent extra commission is also charged from the producers.

(*vi*) *Miscellaneous Charges :* The cost of the supervision by the commission agents and others unforeseen charges are received under this head like postage, telephone charges, munshiana etc.

(*c*) ***Details of Market Expenses :*** During the course of study of Karnal Fish Market during the year 1985-86 and 1986-87, various chargeable expenses on the sale of fish are given in Table 2.

Table 2 : Chargeable Expenses on Sale of Fish

Item of Expenditure	*Amount (Rupees)*	*Percentage to total charges*
Price of fish at the rate of Rs. 70 per kilogram for 40 kilogram		
Transport charges		
Empty gunny bags/boxes baskets and packing charges per 40 kilogram		
Ice and storage charge		
Municipal Committee charges		
Overhead expenses/charity per 40 kilogram of fish		
Commission at the rate of Rs.		
Total Marketing Expenses		
(*i*) Total expenses including original cost of fish		
(*ii*) Percentage of market expenses on 40 kilograms of fish		
(*iii*) Average retail price at the rate of Rs. 11 per kilogram for the sale of 40 kilograms of fish		
(*iv*) Margin of profit over the purchase price of 40 kilograms of fish		
(*v*) Percentage of profit to retailers		
(*vi*) Percentage of marketing charges and profit of retailer on the initial cost of 40 kilograms of fish		

(*d*) ***Margin of Profit at Various Levels :*** A detailed study was conducted during the course of research to work out the margins of profits earned by marketing intermediaries at various levels.

(*i*) *Primary Fish Producer :* The primary fish producer is a fish farmer who culture fish in his own pond or leased pond. It is estimated that there are 4,038 fish farmers.

It is observed that 44.6 per cent of the total fish production comes from pond culture. Hence, fish farmer plays an important role in fish marketing process. A case study was conducted of four fish farmers regarding the assessment of production cost, exploitation cost and marketing procedures. The abstract of the information regarding the cost of production of fish and other expenses on sale of fish are given in Table 3.

Table 3 : Cost of Fish Production and Expenses on Sale

Name of the fish farmer	*Expenditure on fish culture (Rs.)*	*Fish production (kg)*	*Cost of production per kilogram of fish (Rs.)*	*Expenditure on exploitation and transportation of fish per kilogram (Rs.)*

(*ii*) *Fish Contractor :* As per policy of Government, the fishing rights of notified waters are leased out to the fish contractor for catching fish for one year under state rule Haryana Fisheries Rules, 1966 and Punjab Fisheries Act, 1914 in the open auction. The fish contractors have to pay the amount of auction in three instalments. The fish production and income from the notified waters which are auctioned to fish contractors during the last five year have been given in Table 4.

Table 4. : Cost of Fish Production from Notified Waters

Year	*Fish production from notified waters (kg)*	*Income from notified waters (Rupees)*	*Cost of one kilogram of fish (Rupees)*
1	2	3	4

The amount paid by fish contractor to Haryana Government in lieu of auction money for one kilogram of fish. Expenditure for one kilogram of fish on exploitation and transportation. Total expenditure on one kilogram of fish. Average sale of fish for one kilogram at wholesale rate. Net profit to fish contractor on one kilogram of fish. Percentage of profit to fish contractor on one kilogram of fish.

(*iii*) ***Profit to Fishing Worker :*** Although there is no fisherman community, yet the persons engaged in fish capturing activities are known as fishermen. The number of identified fishermen. There are various methods for the exploitation of fish :

(*a*) Primary fish producer auction the fish pond to the fishermen parties.

(*b*) The fish pond is auctioned for exploiting for one kilogram of fish.

(*c*) The fishermen parties exploit the fish on 50:50 basis.

In southern state there are regular and identified fishermen.

Table 5 : Expenditure and Income to Fishermen Parties

Depreciation of nets' cost (One drag net, two phaslas, one castnet, 5 handnets, 6 khallas)
Depreciation cost of crafts etc. (simple boat)
Depreciation cost of temporary hutment etc.
Recurring cost over ropes, threads and other materials
Total
The average cost of fisherman per day for exploitation
Average expenditure in exploitation of one kilogram of fish
Average charge of exploitation per kilogram of fish
Net profit to fisherman per kilogram of fish
Percentage of profit to fisherman

(*iv*) ***Profit to Fish Commission Agent :*** The fish commission agent who is a primarily a fish wholesellar, comes in the picture when the produce arrives at the fish market. Fish contractor does the work of commission agent as well as of wholeseller. The 'Pucca Arhait' system of fish sale is not prevalent in so many fish markets of various states.

Table 6 : Expenditure and Income to Fish Commission Agents

Rent of the shop
Salary to a *munshi*
Salary to a supervisor
Salary to two labourers
Electric and water charges
Overhead and unforeseen charges
Total
Average expenditure on one kilogram of fish as 274.3 tonne fish arrived during 1985-86
Income towards commission on the sale at the rate of 7 per cent for per kilogram fish
Miscellaneous charges at the rate of 5 per cent on sale of one kilogram of fish
Net income on one kilogram of fish
Net profit on one kilogram of fish to fish commission agent
Percentage of profit to commission agent

(*v*) ***Profit to Retailers :*** The retail sale of fish is done at the retail shop which are five in number at Karnal. The retail marketing of fish is in the private hands. In Panipat and Ambala, the retail fish shops are allotted by the Municipal Committees. Thus the retailers have to abide the rules and regulations of local Municipal Committee. The retailer decides the retail price of fish on the basis of the market arrival and its demand by the consumers. Moreover, he includes the amount of expenditure incurred by him *viz.*

transportation from wholesale market to the sale point and other unforeseen charges to be spent during the day and also by covering the risk of suspected spoilage. During the course of study of Karnal retail market during 1985-86, the average profit earned by the retailer is given in Table 7.

Table 7 : Profit to Retailers

Average wholesale price of one kilogram of fish at the rate of Rs. 40 kilograms
Average retail price of one kilogram fish
Profit of retailer on one kilogram of fish
Percentage of profit on one kilogram of fish by retailer

In case of capture fishery the production cost for one kilogram of fish is Re. 0.87 and consumers purchase at the rate of Rs. 11.00 per kilogram, hence the margin is Rs. 10.13 per kilogram. While in the case of culture fishery, the production cost of one kilogram of fish is Rs. 2.30 and sale price is Rs. 11.00, thus it creates the margin of Rs. 8.70 per kilogram of fish. The profit earned at various levels from primary produce to consumer is given below :

Fish farmers	44.33 per cent
Fish contractor	104.70 per cent
Fishermen/Fishing workers	425.00 per cent
Fish commission agent	500.00 per cent
Fish retailer	27.20 per cent

It is observed that maximum profit is earned by the fish contractors, fishing parties and fish commission agents. The profit earned by fish farmer's and fish retailers are moderate. In any way fish consumer is exploited. There are four possible channels of flow of fish from producers to consumers. Shorter the chain more the profit at particular stage.

Fish Marketing Channels

The earning in selling fish by fish producer is different in adopting different channels of fish sale.

(*a*) *Channel I :* If fish producer sells fish directly to the consumer at the pond site, the total expenditure is incurred by the fish farmer on raising one kilogram fish and exploitation would be Rs. 4.45. If fish farmer sell the fish at the rate of Rs. 11.00 per kilogram then the net profit gained by him is Rs. 6.55 over every kilogram of fish. Thus fish farmer earns 147.2 per cent net profit.

(*b*) *Channel II :* If fish producer sells fish through retailer then cost of production, exploitation, packing and transportation on one kilogram of fish comes out to be Rs. 4.85. The retailer purchases fish at the rate of Rs. 8.65 per kilogram and then profit to the producer is Rs. 3.80 per kilogram of fish *i.e.* 78.4 per cent.

(*c*) *Channel III :* If producer sells the producer to wholesaler at the rate of Rs. 7 per kilogram of fish then he earns Rs. 2.15 on one kilogram of fish *i.e.* 44.3 per cent.

(*d*) *Channel IV :* If fish producer sells his produce through commission agent where 3 per cent commission is also charged from the seller then net profit to the fish producer comes out to Rs. 1.94 per kilogram of fish *i.e.* 40 per cent.

Thus by adopting different channels of fish sale, the producer earns minimum profit *i.e.* 40 per cent and maximum profit *i.e.* 147.2 per cent depending upon the channel adopted by him.

Price and Price Determinator of Fish

The fluctuation in fish price is very prominent. The changes are so frequent to predict any trend. There may be one price in the morning while another in the evening in the same market. Sometimes prices change at short intervals of time *i.e.* from minute. The prices no only due to the sudden supply and demand of particular variety of fish but also due to prices of other varieties in the market. The perishability tendency of fish, have definitely a role to play in determining the price of fish at the markets. The fish price often vary from spot to spot in one fish market where many auctions are taken at a single time. When fresh fish reaches early in the morning a high price is quoted. Generally, the fish price fall as the day advances. The uncertainty of supply and demand plays a great role in price determination. In the fish markets at the time of auction, the pace of selling fish is so fast that it cannot permit news of low prices. There is a great price risk because of high perishable tendency of fresh fish. The fish markets are having certain timings. Generally, the auctioneers start the bidding price based on the experience. It is also observed that some buyers in some instances purchased one type of fish at a higher price from one particular agent while some other brought the same fish at a lower price from a neighbouring agent, which indicated the big variation in prices. In all the fish markets of Haryana prices do not go down according to the theory of demand and supply during glut season. This may be due to the ignorance of the seller and buyers about the fish arrivals in the markets. It is also observed that the theoretical equilibrium price level mechanism may not strictly apply atleast in short run period for whole market in its practical connotation. The large quantity of fish used to come to Bombay Fish Market early in the morning, generally disposed off at a higher price. This was mainly due to the anxiety of buyers who wanted to take the fish to retail market as early as possible for the fear that there would not be any sale after 9.00 A.M. to 10.00 A.M. in retail markets. The demand of fish is usually high in the morning in Haryana, therefore, the retain buyers are prepared to pay high price for without bothering about the increase of supply and lowering of rates in later part of the day. Time factor plays a prominent role in almost all the fish markets. It is also observed that increase in market arrival cannot create a study demand of fish. Therefore, there is a wide range of fluctuation in prices which indicates the instability in selling and purchasing of fish.

In all the fish markets, one shop belongs to Government fish contractor who also acts as a commission agent or wholesaller. He receives whole consignment from notified waters which are taken by him on lease for fishing. Thus, he receives about 55 per cent of total arrival. The catch from natural waters comprises most of cat fish which are liked by the local meat eater in the State. Thus, there is a monopoly of the fish contractor in fixing the prices of fish. It is also observed that some of the retailers always chinch to a specific agent even if they would get fish at some cheaper rates from other merchants due to the fact that they get some credit facility from the fish contractors.

The demand of fish in Haryana is steady as only 15 per cent of the population is fish eater. It is, therefore, there is hardly any increase in demand of fish during glut period moreover, during storages, the local people select other alternative of fish such as meat or fowl instead for purchasing fish at higher rates.

For any extremely perishable commodity like fish, the method of auction is most feasible because it reflects the tendency of consumer's demand in the wholesale market through the retailers. The fish producers and fishermen hardly gets any remunerative price for the produce in wholesale fish markets where the monopoly of fish contractors exist. Normally, the producer share to sell his produce at the lower rates foregoing his profit with the fear of blockade of money and spoilage of fish. The wholesale distributive machinery in Haryana is not sufficient due to less number of merchants in this trade. It is seldom possible either for producer or seller to impose a higher price at the wholesale markets. The short-term changes in demand and supply are responsible for frequent price variation to some extent. The long term demand and supply are the important factors in bringing down the cost of production for the whole industry.

The aggregate supply of fish in the short period, theoretically speaking, will remain constant over a momentary time and space; and even if there is some change in actual supply and demand of fish will remain constant for finding out the equilibrium market price. Under such circumstances, if there is a high demand, it will give a higher price and *vice versa.* Huge supply of fish normally means that, if there is unchanged demand, buyers will ask for fish at a very low price. The low price can be increased if any public agency or Government purchases the excess supply of fish by giving a higher support price. So as to regulate the price of fish, the report on fishery policies in Western Europe and North America have rightly observed, "Measures aiming at stabilizing first hand prices, it may be justified, *inter alia,* on account of the heavy fluctuation in catches in many countries. With the aim of contributing the overall economic stability in the industry, for which it is necessary that the fishermen receive an adequate return for catches, the Government or the industry itself of member countries have established more or less elaborate measures. In Norway, for instance, a minimum price policy was introduced by the sales organizations, which are co-operatives in character, in consultation with the government. When fishermen/fish farmers in the past were virtually unprotected against heavy fluctuations in the prices caused by frequent changes in supplies, price support measures were introduced. All the fish produced in Norway cannot be sold lower than the minimum price *i.e.* first hand wholesale prices, stipulated under the Raw Fish Act, 1951. The White Fish Authority and Herring Industry Board of the United Kingdom now called 'Sea Industry Authority' and the Agriculture Marketing Board of Sweden have also adopted such price measures in regulating price of the fish. The need for a similar minimum price system is highly required in Haryana to give a boost to fishery trade.

Table 8 : Average Wholesale and Retail Fish Prices in Haryana

	Year		*Year*	
	Wholesale (Rs./kg)	*Retail (Rs./kg)*	*Wholesale (Rs./kg)*	*Retail (Rs./kg)*
In all the markets	8.35	11.00	9.50	11.25
Major carp group	8.70	11.40	10.70	12.40
Cat fish group	10.50	12.50	10.80	12.55
Miscellaneous group	5.80	9.10	7.50	9.20

A detailed study was conducted to see the price pattern of fish in Haryana after survey of four main markets situated at Ambala, Karnal, Panipat and Sonepat during the years 1985-86 and 1986-87. It was observed that average retail price during the year 1985-86 was Rs. 11.00 per kilogram while Rs. 11.25 per kilogram in 1986-87. The wholesale price was Rs. 8.35 per kilogram and Rs. 9.50 per kilogram in the year 1995-98 and 1986-87 respectively. The retail as well as wholesale prices during the year 1986-87 were higher than the prices of proceeding year. The increase in retail as well as wholesale prices was 2.3 per cent and 13.8 per cent respectively. The retail price was higher by 31.8 per cent than the wholesale price during the year 1985-86. Similarly, the gap between wholesale and retail prices during 1986-87 was 18.5 per cent. The gap between retail and wholesale prices reduced during 1986-87 which is a good sign for the market development. The average wholesale and retail prices of cat fish group were higher than the other groups. The trend of fish prices in different fish markets in Haryana is shown in Annexures II and III. The average prices in Haryana during the study course is given in Table 8.

The increase in wholesale and retail prices was 23 per cent and 9 per cent respectively for major carp group. The increase in the wholesale and retail prices for cat fish group during 1985-86 and 1986-87 was 2.8 and 0.4 per cent respectively. The wholesale as well as retail prices of cat fish group were high than the other groups during both the years. The increase of wholesale and retail price of miscellaneous group was 29.3 per cent and 1.0 per cent respectively. The pricing trend is different in the different fish markets in Haryana. There are various factors which affect the price of fish.

(*a*) Factors Affective Prices

The following are the main factors influencing the prices of fish in Haryana State.

(*i*) *Elasticity of Demand* : Delhi and Saharanpur fish markets play most dominant role on the price of fish in Haryana. Haryanavis are basically vegetarian, hence most of produce is sent to Delhi or Saharanpur fish markets. Therefore, whenever there is a demand of fish in Delhi and Saharanpur fish markets, the price of fish in Haryana tends to rise. Delhi fish market plays so much role that even in period of no demand of fish at Delhi fish market, the fishing is stopped in Haryana. During the particular days near festivals, the fish consumers in Haryana take no fish due to religious conception such as 'Norate and Sarad'. Similarly to take no fish in the English months where 'R' not fall in the word also plays important role in affecting the price pattern in the month particularly May, June, July and August.

(*ii*) *Type, Sex, Weight and Quality of Fish* : The demand of cat fishes being more in Haryana, therefore, the price is generally more as compared to other groups. Large size and fresh fish also fetches good price in the markets.

(*iii*) *Distance of Procuring Centres to Markets* : The distance of procuring centre to fish markets also plays an important role in the pricing of fish. Low catch is normally transported by buses early in the morning. In number of occasions the bus conductor refused to allow the fish to be placed in the bus. The fishermen tend to hide it from the conductor and pack it in such a way to escape from the eyes of the conductor. Such a case was observed during the survey of Panipat fish market where a fisherman brought the fish from Munak Head to fish market appeared to be a bundle of clothes. With the result of that the quality of fish generally get spoiled. The larger the distance so covered, the greater would be the spoilage which naturally has adverse bearing on the fish price.

(*b*) Study of Fish Price in Various Markets of Haryana

The pattern of prices of fish in different fish markets in Haryana are given below :

(*i*) *Fish Market Ambala* : The average wholesale price in Ambala fish market during the years 1985-86 and 1986-87 was Rs. 10.40 and Rs. 10.27 per kilogram respectively. Similarly, the average retail price of the corresponding years was Rs. 13.10 and Rs. 12.10 per kilogram respectively. The wholesale price of fish during 1986-87 was lower by 1.25 per cent than the wholesale price of 1985-86. Similarly the average retail price during the year 1986-87 was 6.9 per cent low than the year 1985-86. The average wholesale as well as retail prices of cat fish group were higher than other groups during both the years under study. During the year 1985-86 the margin between wholesale and retail price was (+) 26 per cent while the margin in the year 1986-87 was (+) 17 per cent. Thus gap between wholesale and retail price was reduced by (+) 9 per cent during the two years. The average wholesale as well as retail prices of cat fish group showed the increasing trend during the two years, while the retail price of major carp group was higher in 1985-86 than the year 1986-87. The average wholesale as well as the retail price of miscellaneous fish group during the year 1986-

87 was lower than the year 1985-86. The wholesale as well as the retail prices were remained stable from October 1985 to March 1986. There was a sudden rise in price of major carp in the month of June 1985. The price of major carp was again stable from November 1986 to January 1987. It is observed that price of major carp group was higher in winter months as compared to other months in both years. The wholesale price fluctuated between Rs. 9 to Rs. 11 per kilogram during 1985-86 while Rs. 10 to Rs. 13 during 1986-87. Similarly the fluctuation in retail price was from Rs. 12 to Rs. 14 per kg during 1985-86 and Rs. 11 to Rs. 14 during 1986-87.

The cat fish group dominates the other groups of fishes so far as the prices are concerned. The wholesale and retail prices of cat fish group were Rs. 16 and Rs. 18 per kilogram respectively in the month of September 1985. The rates were remained stable from October 1985 to March 1986 in the year 1985-86. During the year 1985-86, cat fish price was never lower than any group of fish. The price of cat fish was low in the months of April and May 1985 as compared to other months. During the glut season *i.e.,* monsoon where 27.7 per cent fish arrived to this market showed no change in price. During the year 1986-87, the price was high in the months of July, August, November and December 1986 as compared to other months. During the months of January and February 1987 the price of cat fish group was lower than major carp group. The prices of major carp as well as the cat fish were same in the month of March 1987.

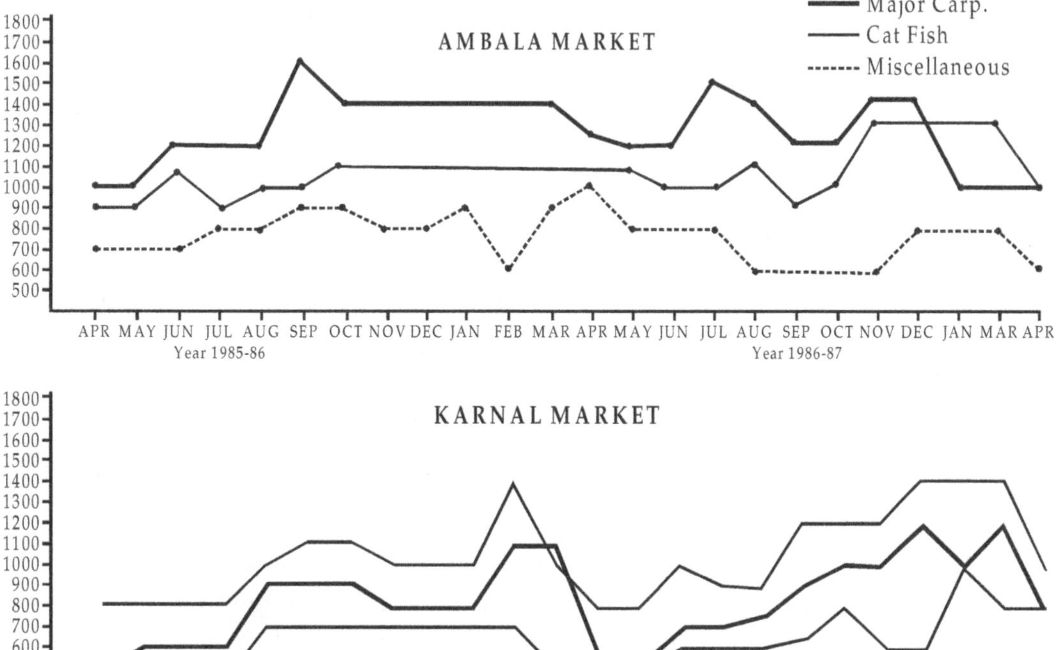

Fig. 3 : Wholesale Fish Price

The price of miscellaneous group of fishes was remained low as compared to other group during both years under study. The price of miscellaneous group was found lower during the year 1986-87 as compared to 1985-86. The highest price of this group was during the months of September, October 1985, January and March 1986. During the months or April, May and June 1985 the price remained low. The fluctuation of prices of this group for wholesale and retail were Rs. 7 to Rs. 9 per kilogram and Rs. 10 to Rs. 12 respectively during the year 1985-86. Similarly the price fluctuated form Rs. 6 to Rs. 10 per kilogram and from Rs. 8 to Rs. 12 respectively for wholesale and retail, during the year 1986-87. The price of miscellaneous group in the month of April 1986 was the highest competing the price of major carp group. The price during the months from August to November 1986 and March, 1987 was lowest as compared to other months. The trend of wholesale and retail fish prices are shown in Figs. 3 and 4.

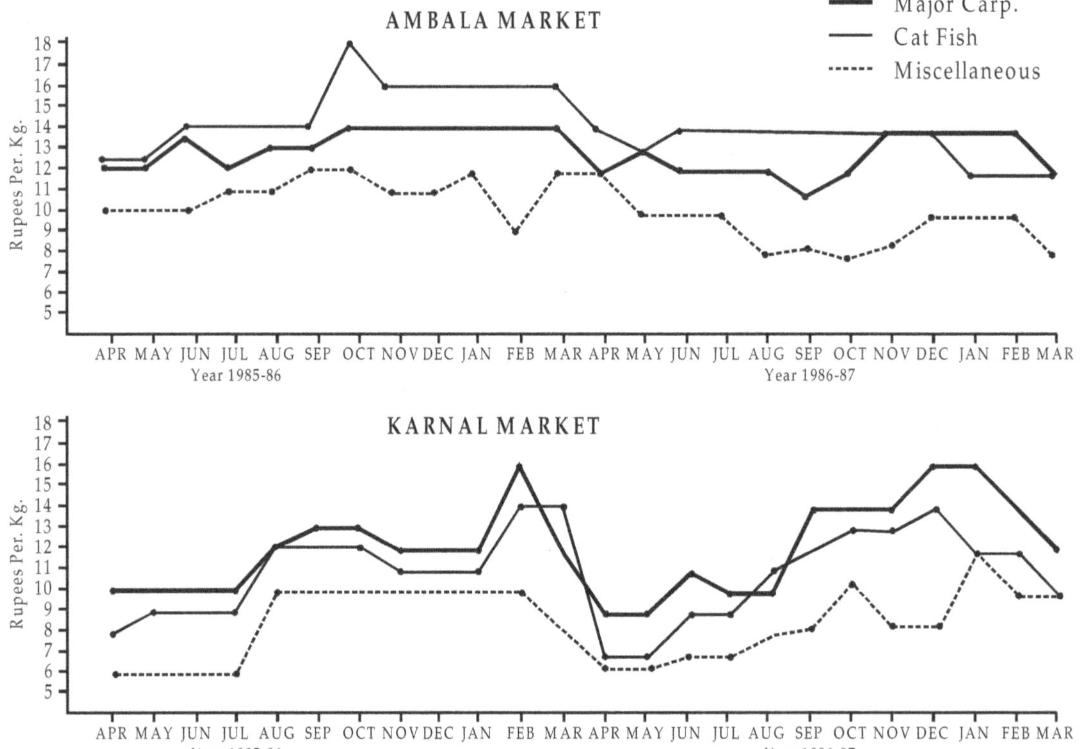

Fig. 4 : Retail Fish Price

(*ii*) *Karnal Fish Market* : The prices in Karnal Fish Market showed the upward trend. The average wholesale and retail prices of fish during the year 1985-86 increased by 14.2 and 1.5 per cent respectively during 1986-87. The margin between wholesale and retail market fish prices during 1985-86 was Rs. 3.05 per kilogram which was reduced to Rs. 2.16 per kilogram during the year 1986-87. The wholesale as well as retail prices of cat fish group were highest during both the years as compared to other groups of fishes.

The prices of wholesale and retail of major carp group fluctuated between Rs. 5 to Rs. 11 per kilogram and Rs. 8 to Rs. 14 per kilogram respectively during 1985-86. Similarly, prices fluctuated between Rs. 5.50 to Rs. 12 per kilogram and Rs. 7 to Rs. 13 per kilogram respectively during 1986-87. The highest price was in the months of February and March, 1986 and lowest was in the month of April 1985. The price remained stable in the months of May, June and July 1985; from August 1985 to December 1985 and January 1986. The price in the month of December 1986 was highest. The price in the months of April and May 1986 was lowest. The wholesale and retail prices were equal in the month of February 1987. The price of cat fish group showed the upward trend. The average wholesale and retail prices during the year 1985-86 rise by (+) 2 per cent and (+) 5 per cent respectively during 1986-87. The wholesale and retail prices during 1985-86 fluctuated between Rs. 8 to Rs. 14 per kilogram and Rs. 10 to Rs. 16 per kilogram respectively during 1985-86. Similarly, during 1986-87, the wholesale and retail prices fluctuated between Rs. 8 to Rs. 14 per kilogram and Rs. 9 to Rs. 14 per kilogram respectively during 1986-87. The price was highest in the months of February and December 1986 and January 1987. The price was lowest in the months of April to July 1985 and April and May 1986. The price remained stable in the months of June to August 1985 and from September to November 1986. The prices of wholesale and retail were equal in the month of February 1987.

The price of miscellaneous group of fish also showed an upward trend. The average whole sale and retail prices of fish during 1985-86 raised by (+) 55.8 per cent and (+) 12 per cent respectively during 1986-87. The margin of whole sale and retail prices were Rs. 4.20 per kilogram during 1985-86 while it was Rs. 1.90 per kilogram during 1986-87. The average wholesale and retail prices during 1985-86 fluctuated between Rs. 3 to Rs. 7 per kilogram and Rs. 6 to Rs. 10 per kilogram respectively during 1985-86. While in the year 1986-87 the wholesale and retail prices fluctuated between Rs. 5 to Rs. 10 per kilogram and Rs. 6.50 to Rs. 12 per kilogram respectively. A stable but higher price was noticed during the months of August 1985 to February 1986 and January 1987. The price was low in the months of April to July 1985 and from April to May 1986. The price of miscellaneous group was the same as that of major carp in the month of January 1987. The trend of wholesale and retail fish prices are given in Figures 3 and 4.

(iii) Panipat Fish Market : The price at Panipat Fish Market showed an upward trend. The average wholesale price was Rs. 8.50 per kilogram and retail price Rs. 11.15 per kilogram during the year 1985-86 which increased by (+) 11.8 per cent and (+) 2.3 per cent during 1986-87. The margin between wholesale and retail prices was Rs. 2.65 per kilogram and Rs. 1.90 per kilogram during the years 1985-86 and 1986-87 respectively. The average price of cat fish group remained higher in the year 1985-86 while price of major carp was higher in 1986-87.

The wholesale and retail prices of major carp fluctuated between Rs. 7 to Rs. 11 per kilogram and Rs. 10 to Rs. 14 per kilogram respectively in the year 1985-86. While in the year 1986-87, the wholesale and retail prices fluctuated between Rs. 7.50 to Rs. 12.50 per kilogram and Rs. 9 to Rs. 14 per kilogram respectively during 1986-87. The highest price of major carp group was in the months of November, December 1985 and February, 1986 and from September to December 1986 and January 1987. The lowest price was from May to September 1985, March 1986 and June 1986.

The price of cat fish group showed the downwards trend. The wholesale and retail prices of the year 1985-86 were down by (–) 14.3 and (–) 6.0 per cent respectively in the year 1986-87. The margin between wholesale and retail prices were Rs. 2 and Rs. 2.75 per kilogram in the years 1985-86 and 1986-87 respectively. The highest price of this group was in the months of November December 1985, November and December, 1986 and January, 1987. The lowest price was in the months from June to August 1985, April and May 1986.

The price of miscellaneous group of fishes showed an upward trend. The wholesale and the retail prices for the year 1985-86 were raised by (+) 40.6 and (+) 9.0 per cent in the year 1986-87. The margin between wholesale and retail process were Rs. 3.00 and Re. 1.25 per kilogram in the years 1985-86 and 1986-87 respectively. The highest price of the fish was in the months of April, October to December 1985, November, December 1986 and January 1987. The lowest price of the fish was in the months of June 1985 to September 1985, March and May 1986. It is very interesting to note the prices of miscellaneous group were same as that of cat fish group in the month of September 1986. The trend of wholesale and retail prices of fish has been shown in Figures 5 and 6.

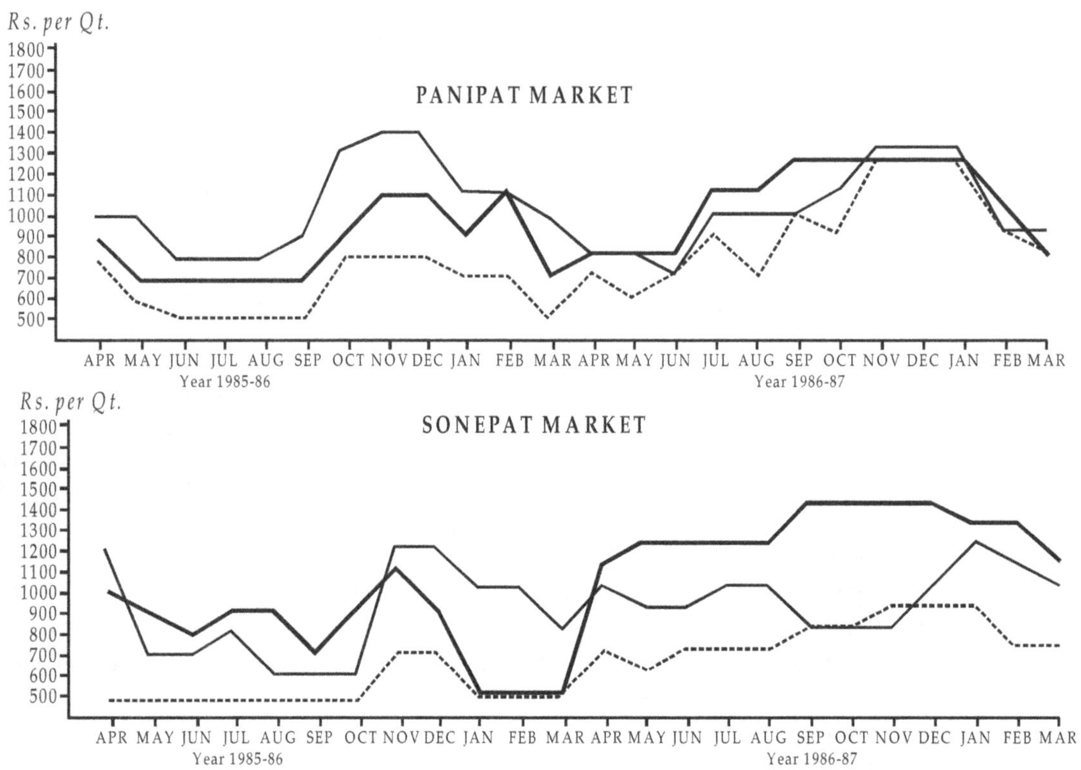

Fig. 5 : Wholesale Fish Price

(*iv*) *Sonepat Fish Market :* The price in Sonepat fish market showed upward trend. The wholesale and retail prices of the year 1985-86 were raised by (+) 37.2 and (+) 17.6 per cent in the year 1986-87. The margin between wholesale and retail prices were Rs. 2.30 and Re. 1.35 per kilogram in the years 1985-86 and 1986-87 respectively. The price of cat fish group was highest during 1985-86 while it was lower than major carp group in the year 1986-87.

The wholesale and retail prices of major carp were higher by (+) 57.5 and (+) 36.0 per cent in the year 1986-87 than the prices of preceding year. The margin between wholesale and retail prices were Rs. 2.00 and Re. 1.00 per kilogram in the years 1985-86 and 1986-87 respectively. The wholesale and retail prices fluctuated between Rs. 5 to Rs. 10 and Rs. 8 to Rs. 13 per kilogram respectively during the year 1985-86. While these fluctuated between Rs. 8.00 to Rs. 12.50 for wholesale and Rs. 13.00 to Rs. 16.00 per kilogram in 1986-87. The highest wholesale price was Rs. 10 per kilogram whereas the highest retail price was Rs. 13.00 per kilogram during 1985-86. The highest prices were in the months from September to November, 1985 *i.e.* Rs. 14 for wholesale and Rs. 16 per kilogram for retail was noticed. The lowest price was noticed in the months of January to March, 1986 where it was Rs. 5 per kilogram for wholesale and Rs. 8 per kilogram for retail during the year 1985-86. It is a abnormal phenomenon. The lowest price *i.e.* Rs. 11 per kilogram for wholesale and Rs. 13 per kilogram for retail were noticed in the months of April 1986 and March 1987.

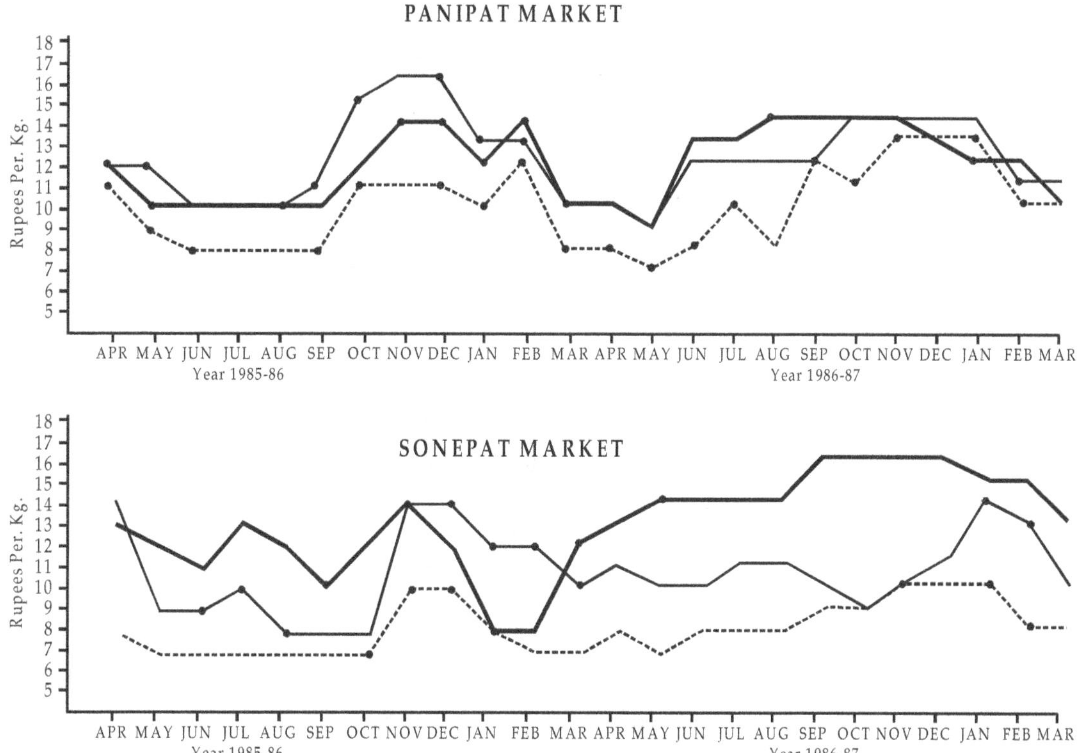

Fig. 6 : Retail Fish Price

The cat fish group also showed an upward trend of marketing. The average wholesale and retail prices during 1986-87 were higher by (+) 10.5 and (+) 3.3 per cent than the preceding year. The margin between the wholesale and the retail prices were Rs. 2 and Re. 1.45 per kilogram in the years 1985-86 and 1986-87 respectively. The wholesale and retail prices fluctuated between Rs. 6 to Rs. 12 per kilogram and Rs. 8 to Rs. 14 per kilogram respectively in 1985-86 while in 1986-87 it fluctuated between Rs. 8 to Rs. 11 per kilogram and Rs. 9.50 to Rs. 14 per kg. The price of cat fish group was higher in the months of April, November and December 1985 and January 1986. The lower price was noticed in the months of August to October 1985 and October 1986.

The miscellaneous group also showed the upward trend. The wholesale and retail prices in the year 1985-86 were increased by (+) 50.0 and (+) 13.2 per cent in the year 1986-87. The margin between the wholesale and retail price was Rs. 3 per kilogram and Rs. 1.70 per kilogram respectively in the year 1985-86 and 1986-87. The wholesale and retail prices fluctuated between Rs. 4 to Rs. 7 per kilogram and Rs. 7 to Rs. 10 per kilogram respectively during 1985-86. While the wholesale and retail prices fluctuated between Rs. 6 to Rs. 9 per kilogram and Rs. 8 to Rs. 10 per kilogram respectively in 1986-87. The highest price was in the months of November 1985, November 1986, December 1986 and January 1987. The lower price was noticed from June to October 1985 and January to March 1986 and May 1986. The trend of wholesale and retail fish prices have been shown in Figures 5 and 6.

(c) *Inferences of Fish Price Trend in Haryana*

After the study of four main fish markets in Haryana State, the following inferences were drawn :

(*i*) The average wholesale price of all kind of fish.

(*ii*) The average retail price of all kind of fish.

(*iii*) There is a price rising trend in the sale of fish. The rate of increase in wholesale and retail price is (+) 13.0 per cent and (+) 2.3 per cent respectively.

(*iv*) In general the margin of profit to retail fish merchant from (+) 31.7 per cent to (+) 18.5 per cent.

(*v*) The wholesale as well as retail fish prices in Ambala fish market were higher as compared to other fish markets.

(*vi*) Ambala fish market showed downward trend of fish prices.

(*vii*) The rates are lower as compared to other market.

(*viii*) On an average the fish price of cat fish group remained high as compared to other groups of fish.

(*ix*) Comparison of prices among different groups/species of fishes.

(*x*) In particular months, the prices of miscellaneous fish group were equal to that of major carp or cat fish group various fish markets.

(*xi*) The price of fish is not affected by the market arrival *i.e.* glut season or shortage season.

(*xii*) The fish price during winter months from November to February is higher as compared to other months of the year.

(*xiii*) The margin of profit to fish parties, fish contractors and fish commission agents are more as compared to other marketing intermediaries.

(*xiv*) There is no support to fish markets by the Government.

(*xv*) Most of fish markets in Haryana are not regulated by any law.

Strategy for Fish Market Development in Haryana

Since all the fish markets in Haryana are not organized in well manner, a thrush is required to reform the markets by modernizing the traditional fish marketing methods by introducing new management techniques. The strategy of fish market management can be created by analyzing the present pattern of marketing, setting the objectives, developing the fish demand, formulation of new plan, marketing operations and market control which has been given in Fig. 7.

There are many environmental opportunities for fish marketing as there are many places where fish is not marketed. The Tourist Complexes established by respective state Tourism Corporation where good fish sale stall can be established. Likewise, in Universities Campus and Sports School such stalls can be set-up. This kind of environmental possibilities exist in many areas and more areas can be find out by conducting survey.

A fishery firm can take up a relevant marketing action in which it is likely to enjoy a differential advantage over the firms. This is because of its experience and technical resources available with the firm. A fish firm can set up ice plant and net fabricating unit to give a boost to the fish marketing process.

A particular firm can have the specific advantage by dealing the particular kind of fish. In Haryana, cat fish group is in great demand particularly in winter season. The need of people can be met by importing cat fish from neighbouring states and exporting the major carp group in lieu of this. It can also explore the possibility of marketing of major carp as well as miscellaneous group side by side the cat fish.

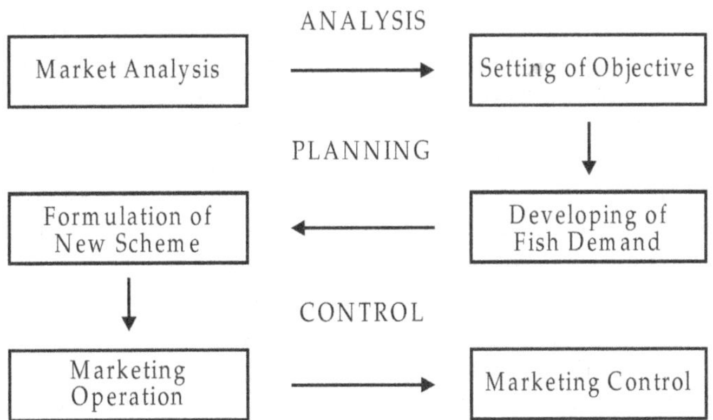

Fig. 7 : Strategy of Fish Market Development in Haryana

A core marketing need to be designed for the market development. This system has three type of classifications :

(*i*) Intensive growth

(*ii*) Integrative growth

(*iii*) Diversification growth

(*i*) ***Intensive Growth :*** Some opportunities are hidden in the present marketing efforts or traditional marketing which have to be explored. The following methods can be adopted to intensify the growth in the market :

(*a*) *Market Penetration :* It includes the more aggressive fish marketing efforts to increase the sale of fish. These efforts include the publicity of the fish by various methods to depict the nutritional value. In general Haryanavis is a vegetarian and to inculcate the habit of fish taking requires more efforts. It has been accepted throughout the world that the fish is rich in protein and phosphorus. Fish is considered as one of the best nerve tonic. Fish can be supplied instead of meat products in schools and colleges messes. In addition to this, a number of people who like to take but do not know the process of cooking and fish recipes. Short term courses can be conducted by the Food Processing Institutes in the State to impart the knowledge of fish cooking. Free distribution of palatable fish products can be distributed to the general public to create taste. Sale can also be promoted by changing the nomenclature of major carp as vegetable fish which can fetch more attention of the people who are basically vegetarian in Haryana.

(*b*) *Market Development :* Efforts of present fish sales form those markets can be diverted to new markets where new sale centres can be opened in different areas. Identification of the regions, according to consumer's taste is necessary so that new segmented market may be approached for future marketing of fish or fish products. The possible areas can be railway stations, bus stands, tourist complexes etc. so these can be supplied easily.

(*c*) *Products Development :* The main hindrance in the sale of fish is its peculiar ordour. New fish products can be evolved which have no smell and can give a good taste to the consumer. Improved fish products can be created like Fish Bournvita, Fish Biscuits, Fish Fillets, Fish Fingers, Fish Wafers etc.

(*ii*) ***Integrative Growth :*** The basic fish market may be combined with another concern unit by making backward or forward or horizontal integration.

(a) *Backward Integration* : A firm selling fish may seek the ownership in other firm who is supplying fish or fish products. A wholesaler can combine with the commission agent or can do the work of fish exploitation. In Haryana 425 per cent profit is gained by the fishing parties for the exploitation of fish. As the fishermen population in Haryana is negligible and the fishermen parties who belong to neighbouring States have the monopoly. Similarly the fish commission agent earns 500 per cent profit. Thus by clubbing the wholesale with the commission agent as well as fishing workers the number of intermediaries will be reduced and would provide more profit to producers to wholesaler and consumers.

(b) *Forward Integration* : The fish production and distribution are the two main components of fish marketing. The fish marketing machinery if combined with the distribution authority then market development would be easier. If the wholesaler combines with retailer and fish hawker then the inter-competition between channel who tends to earn more profit can be reduced. Thus consumer would get good fish at reasonable price which would promote the marketing in future.

(c) *Horizontal Integration* : A fish firm can join hands with other parallel firms who are producing ice, baskets, packing material etc. so that the competition can be reduced to stabilize the market.

(iii) ***Diversification of Growth :*** The fish markets should also take up the act of diversification or products for promotion of the market.

(a) *Concentric Diversification* : Introducing the products which are technically and market synergies with the present fish commodities can promote the fish marketing. The introduction of new products like 'fish ham' and 'fish sausages' may attract more customers. This type of change will also create interest to the new customers who intends to take fish in Haryana. This will help in inculcating habit of fish taking among Haryanavis.

(b) *Horizontal Diversification* : If a new product like frozen fish kheema or fillets are supplied alongwith the chilled fish that would create more customers. If the dressed fish is also sold along with the raw fish then more customers would be attracted as the dressing of fish for non-traditional fish eater is tedious task.

(c) *Conglomerate Diversification* : It comprises to add new product to the new customer because of deficiency of old one. In Haryana, the heavy demand of cat fish can be substituted by live fish or Rohu fish by common carp.

In addition of above, the following factors must be taken into account for the development of fish markets in Haryana :

(i) *Setting up of Firm Object* : The objectives of a firm should be clear-cut with the firm's target to obtain the goal. The strategy should be designed in such a manner that firm objectives can be achieved easily.

(ii) *Segmentation* : Market segment is the basic factor that every market should have different needs, tastes, styles and links of the buyers. No single type of fish and fish product would satisfy all buyers. The market segments like, geographical situation, use of the products, types of buyers etc. must be take into account.

(iii) *Marketing Positioning* : A particular pattern of market concentration can give the maximum results to achieve the main goal. The firm cannot go to all places even if there are opportunities. It should go after viable positions. The main features are as under :

(a) Market segment is of sufficient current size.

(b) Segmented market is not over occupied by present competition.

(*c*) It should have the potential of growth.

(*d*) Segmented market has some relative unsatisfied needs in which a particular firm can serve well.

It is observed that one type of marketing of product is not economically advantageous, but in combination with two or three products, marketing becomes economical. If a retail fish merchants sales aquarium fish and other meat products like eggs and meat etc. then market of all the products would increase.

(*iv*) *Market Entry Strategy* : It is seen that many fish merchants leave the trade when they are enable to have the notified waters on lease from the Government. The Government should render support to them who have the thorough knowledge to the trade. The established firm should have technical staff for conducting survey.

(*v*) *Marketing Mix Strategy* : This is one of the main factor to see the response of the buyers toward the product. Advertisement for sales promotion and feedback method should be adopted to make the people familiar with the fish and firm. The people of Haryana are unaware about number of fish products which can give a good market in future. These products should be tried in public at reasonable rates.

(*vi*) *Timing Strategy* : Proper timing is the important factor for promotion of fish markets. If a regular fish customer takes a particular variety of fish then he may be asked to try the fish of other kind also, which might create a good taste for him and thus the market would expand.

(*vii*) *Formulating Plans* : A firm and sound planning is necessary to obtain the objectives of the firm. The commitment of targets should be achieved with full efforts. A sale target based on the past experiences, areas, regions etc. be fixed. The market budget should be based on the targets. The Fisheries Department, while auctioning the fishing rights of notified water should also fix the sale target of exploited fish. Government should render the help to the contractors in achieving the targets.

The management of fish markets are not up to the mark. The pattern of marketing is traditional. All the existing market in most of states improvements in working on the basis of modern management techniques as already mentioned. All the components of the marketing such as production, exploitation, transportation, preservation, distribution and even technique of using fish as food requires improvement for the promotion of fish marketing system in most of State. The Fish Farmers' Marketing Co-operative Societies should be formed for performing the combined fish marketing activities.

SHRIMP MARKET REPORT

US shrimp imports continue to increase; however, growth has slowed compared to previous periods. This matches the behavior of shrimp imports for previous years which by the end of the year reduce volume compared to previous months, as can be seen in the graph below. Despite this, in October 2006, a record level for monthly imports was reached at 67 230 MT. In the period between January and October 2006, shrimp imports grew both in volume and in value, reaching 468 526 MT (+ 13%) valued at US$ 3 245.5 million (+ 14%).

As for the origin of imports, Thailand remains the main supplier for the US market, with a 32.6% share of the imported volume. The second exporter is now China which accounts for 11.3% of imports, followed by Indonesia (10.9%) and Ecuador (10.5%). These four countries concentrate 65.3% of total shrimp imports, which shows a continued trend to increased origin concentration given that in the same period in 2005, the same countries accounted for 59% of imported volumes. The growth of Chinese imports is noteworthy as volumes and values increased by 62% and 73% respectively. Other countries, with lower import share but with strong growth were Guatemala, New Caledonia and Brunei. On the other hand, countries, such as Costa Rica, Brazil and Saudi Arabia showed strong reductions in the volume sold to this market.

Looking at the unit value of imports from different countries, uneven trends are evident, with countries with a strong growth in unit value, such as Vietnam (+11.8%), Bangladesh (+10.6%), Guyana (+14.2%), Brazil (+32.5%) and Costa Rica (+49.8%) contrasting with countries such as Mexico and Guatemala (–21.1% and –24.4% respectively).

Anti-Dumping

Currently, the WTO is investigating the pertinence of the anti-dumping measures taken by the US against Ecuador and Thailand in relation to international commerce rules. On another subject, India announced that it will resort to the WTO regarding the bonds policy applied by the US that enforces Indian shrimp exporters to take out a customs bond that equals almost the annual value of duties, as they consider it is an arbitrary and discriminatory measure. The US have also applied anti-dumping measures to Brazil, China and Vietnam.

Main Imported Products

Except for some sizes of headless shell-on frozen shrimp, all other imported products volumes were maintained or increased. The most significant growth was for the "other frozen preparations, category which had a 40.2% increase. The "others" category grew both in volume and value, +37.8% and +42.9% each. This category includes canned products that had a 37% increase in imported volume and an 80.7% increase in value, which reflect a 31.9% increase in unit value. The other products included in this category increased in volume, but this was in part natively compensated for by a decrease in unit value. Frozen breaded shrimp, one of the value-added products, grew 4.8%, keeping unit values almost unchanged.

Headless and Shell-On : This remains the biggest import category, accounting for 43.2% of the total imported volume. This represents a reduction compared to the same period of 2005, when this category's participation in total imports was 47.7%. The total increase for this category was modest at 2.1%, reaching 202 257 MT, valued at US$ 1 407.6 million. The most imported size was 31/40, however, volumes declined compared to last year, from 38 712 MT to 37 389 MT (–3.4%). Imports of 15/20 also had a reduction (–21%) with a 22.6% reduction in the imported value. A similar result characterized the 21/25 size, with a reduction in both volume and value (–2.7% and –6.1% each). Import shares for this category remain highly concentrated. For the 41/50, 51/60, 61/70 and >70, the main suppliers are Ecuador and Thailand; for 26/30 and 31/40 the main providers are Thailand and Indonesia; for 15/20, India, Bangladesh and Mexico; and for 15 the main providers are Vietnam, Bangladesh, Mexico and India.

US Shrimp Imports by Product : January-October, 2005-2006

	2005		2006		Change		Unit Value		
	MT	*1000US$*	*MT*	*1000US$*	*Vol*	*Val.*	*2005*	*2006*	*Change*
Breaded frozen	34100	163027	40859	194672	19.8	19.4	4.8	4.8	0.3
Frozen other preparations	63329	442080	88777	62965	40.2	42.4	7.0	7.1	1.6
Other preparations	968	7704	975	6381	1.0	17.4	8.0	6.5	18.2
Peeled frozen	115424	813250	131372	970460	13.8	19.3	7.0	7.4	4.8
All sizes	198190	1384932	202257	1407619	2.1	1.6	7.0	7.0	0.4
<15	18903	248037	19056	251401	0.8	1.4	13.1	13.2	0.5
15/20	18323	182280	14472	141168	21.0	22.6	9.9	9.8	1.9

Contd...

1	2	3	4	5	6	7	8	9	10	11
	21/25	17801	164188	17328	154105	2.7	6.1	9.2	8.9	3.6
Headles and	26/30	26003	189015	28244	208327	8.8	10.2	7.3	7.4	1.5
Shell-on frozen	31/40	38712	230774	37389	229082	3.4	0.7	6.0	6.1	2.8
	41/50	26564	137700	28667	154338	7.9	12.1	5.2	5.4	3.9
	51/60	23392	109944	26975	133868	15.3	21.8	4.7	5.0	5.6
	61/70	15171	67297	16539	75529	9.0	12.2	4.4	4.6	2.9
	>70	13320	55698	13589	59801	2.0	7.4	4.2	4.4	5.2
Others		3103	25682	4277	36709	37.8	42.9	8.3	8.6	3.7
Total General		**415112**	**2836675**	**488516**	**3245498**	**12.9**	**14.4**	**6.8**	**6.9**	**1.4**

Peeled Frozen : Imports of these products grew both in volume and value, with total volumes between January and October at 131 732 MT valued at US$ 970.5 million. These represent increases of 13.8% and 19.3% in volume and value respectively. This also means a 4.8% increase in the unit value. The main exporters to the US market for these products in the period under review were Thailand, Indonesia and Vietnam, with a 32%, 18% and 11% share each.

Breaded Frozen : This line of products had a 19.8% increase in imported volume, and its unit value remained stable. China is the origin of 80% of the trade of frozen breaded shrimp, followed by Thailand with a participation of 13%.

The item "other frozen preparations" increased in the period its share of total shrimp imports in its different varieties.

Domestic Supply

Towards the end of the year, domestic shrimp prices showed a slight increase, which represents a recovery after the decreasing trend of recent months. However, the slight recovery does not compensate for the total drop in the level of prices. This was due mainly to the abundance of shrimp coming from the Gulf, as the industry recovered following hurricane Katrina. This created an over-supply situation which had a negative impact on prices, with considerable discounting by suppliers.

Recent Trends

The holiday market has been inundated with Asian value-added products, such as shrimp rings and other cooked and breaded products. This implies continued difficulties for domestic and Latin American shrimp producers which sell mainly raw headless and shell-on blocks.

Main Headless and Shell-on Shrimp Exports to the US Market, 2006 (by volume and Sizes)

<15	Vietnam	19.9%
	Bangladesh	15.6%
	Mexico	14.1%
	India	13.6%
15/20	India	24.9%
	Bangladesh	23.5%
	Mexico	22.3%
21/25	Mexico	37.8%
	Thailand	14.3%

Contd...

	India	12.8%
26/30	Thailand	38.8%
	Indonesia	15.9%
	Mexico	11.6%
31/40	Thailand	30.8%
	Indonesia	17.9%
	Ecuador	15.6%
41/50	Thailand	29.6%
	Ecuador	27.4%
	Indonesia	8.7%
	Venezuela	8.5%
51/60	Ecuador	33.7%
	Thailand	33.1%
	Venezuela	6.3%
61/70	Ecuador	41.5%
	Thailand	30.4%
>70	Ecuador	44.3%
	Thailand	18.2%
	Malaysia	6.3%
	Venezuela	5.8%

MINISTRY OF AGRICULTURE FISHERIES BOARD, INNOVATION PROJECT LAUNCHED

Agriculture received special attention at the highest levels in the government. Two meetings of Agriculture Coordination Committee were held under the Chairmanship of the Prime Minister. A meeting of Minister of Agriculture and Allied Sectors of all States was called on December 22, 2006 to discuss the recommendations of the National commission of Farmers. The Commission has proposed a National Policy for Farmers.

A National Bamboo Mission was set up to promote development of bamboo sector. Rs. 568.23 core scheme will work for increasing area under bamboo, promoting marketing of bamboo products and generating employment in the bamboo sector.

Funds for the National Horticulture Mission, launched last year, have been increased by over 30% to Rs. 1000 core. Significant progress has been achieved by the Mission in one year of its operation towards envigorating horticulture sector in the country. A Central Institute of Horticulture has been established in Nagaland.

Considering the importance of rainfed areas in India's agriculture, a National Rainfed Area Authority was created in November 2006. The Authority will support upgradation and management of dryland and rainfed agriculture. It will also converge the various schemes being operated by different Ministries in the area of watershed development.

A micro-irrigation scheme was launched this year to promote water efficiency in farming activities.

Agriculture Ministry has taken initiatives to promote modern terminal markets in important cities for fruits, vegetables and other perishables. These markets would provide modern facilities for electronic auction, cold chains and necessary logistics to farmers to sell their produce.

An Agri-marketing Summit was held in September in collaboration with CII, aimed at creating awareness on the agri-marketing reforms and catalyzing private sector investment in agrimarketing business. The North-East Agri Expo 2006 was organized at Dimapur, Nagaland involving all north-eastern States and Sikkim.

Agricultural credit from institutions had already crossed Rs. 1.14 lakh core by October 2006 (*i.e.*, in the first seven months of 2006-07) as compared to Rs. 1.68 lakh core in 2005-06). The target of Rs. 1.75 lakh core set for this financial year is likely to be fully achieved by March 2007.

As against the target of bringing in 50 lakh farmers every year into the institutional credit fold, over 48 lakh new farmers had availed of loans from banks and cooperatives by October itself.

Starting Kharif 2006-07 the rate of interest on crop loans up to Rs. 3 lakh has been reduced to 7%. Government has also finalized a package for revival of the short-term rural cooperatives credit structure. The Rs. 13596 core package will usher in reforms in the cooperatives sector and help the cooperatives to provide easy credit for farming and allied activities.

The scope of Kisan Credit Card (KCC) has been enlarged to include term loans and also to meet consumption needs. By September 2006, more than 6.25 core KCCs have been issued to farmers.

An Agriculture Summit 2006 was organized jointly with FICCI for promoting public-private partnership in various areas of agriculture. Besides industry and trade representatives, a large number of farmers from across the country participated in the Summit.

A constitution amendment bill has been introduced in the Lok Sabha in May 2006 for empowering cooperatives. The bill is presently under consideration of the parliamentary Standing Committee on Agriculture.

The Prime Minister and Agriculture Minister visited Vidarbha in Maharashtrain June-July to have a first-hand look at the distress of farmers in the region. The Prime Minister announced a special package on 1st July for six districts of Vidarbha region. Subsequently the Government approved special rehabilitation package for affected districts of other States. Taken together, the package covers 31 districts of Maharashtra, Andhra Pradesh, Karnataka and Kerala and involves a sum of Rs. 16978.69 core. This comprises subsidy/grants of Rs. 10579.43 core and loans to the tune of Rs. 6399.26 core.

Animal Husbandry, Dairying and Fisheries

The National Fisheries Development Board was set up in September 2006. The Board has started promoting, supervising and coordinating developmental activities in the fisheries sector. It activities cover the entire range of fisheries including aquaculture, brackish water fisheries, deep-sea fishing, seaweed cultivation, and processing and marketing

The Department has allocated over Rs. 800 core to States for promoting livestock, dairy and fisheries sectors. The focus of diary sector has been on rapid genetic upgration of cattle and buffaloes and provision of health cover and fodder.

A Dairy/Poultry Venture Capital Fund has been started. Assistance from the Fund is provided for dairy/poultry projects and is supported further by NABARD.

Livestock Insurance has been started with the aim of providing relief to farmers in case of death of their cattle and buffaloes. The scheme to this effect is being implemented initially in 100 selected districts across the country.

Over 53 thousand fishermen have been extended financial assistance for construction of houses under the welfare programme for fishermen till October 2006. Under the scheme over 12 lakh fishermen are provided insurance cover every year.

From 15 December, 2006 Government has started on-line registration for obtaining Sanitary Import Permit for livestock products.

A number of steps were initiated during the year to control avian influenza (bird flue). A contingency plan has been prepared and a Joint Monitoring Committee regularly monitors the ground situation. Following outbreak of bird flu in some parts of Maharashtra, Gujarat and Madhya Pradesh in February this year, strict control and containment operations were initiated. This also involved culling over 10 lakh birds and destruction of 8546 tonnes of feed and feed ingredients. A package was also announced to provide relief to the poultry industry affected by bird flu. The package included credit relief and provision of maize at concessional rates.

In 2006, Kndia was declared a rinderpest free country.

BELGIUM-SMALL COUNTRY BUT BIG OPPORTUNITIES FOR SEAFOOD BUSINESS

Belgium is a small country situated in the North-Western region of Europe. It has 66 km (41 miles) of coastline along the North Sea. Belgium has an area of 30, 522 sw. km (11, 785 sq. miles), and a population of about 11,000,000. It has very cold winters with a high percentage of rainfall. The summer is hot and often humid. Four countries surround Belgium, one of them is the world's smallest country, the Grand Duchy of Luxembourg. The other three countries are France, Germany and the netherlands. Ravaged, ransacked and plundered by wars, foreign occupation [under the dictatorial regimes of Italy, France, Austria and Spain] and the Vikings, Belgium has a long and interesting history of survival and establishing its own identity. At last, rising like a Phoenix from the ashes, the Kingdom of Belgium was formed in 1831.

Belgium is a ready consumer of seafood, considered to have mature tastes. The Belgian market represents a real live test platform, all the more because of its dynamic distribution networks. It is the second most Michelin-starred country in Europe after France [Michelin-world renowned type manufacturer based in France but it is also famous for its Red and Green travel guides that include seafood restaurants].

With only 10 million consumers, Belgium could be considered a small country, but in terms of commercial opportunities, it is in fact large. More than 60% of products consumed in Belgium are imported and 60% of production is exported. In some sectors where there is little local production of seafood, supply is based mainly on importation.

A lot of Belgium's food is French. Beer is the most popular drink in Belgium, and oysters are a popular seafood dish.

Belgium is an excellent distribution center for much of Europe's seafood industry. The per capita consumption of seafood in Belgium has reached a high of over 30 kilos compared to an average of 12-15 kilos for the rest of Europe. There is a steady increase in the seafood consumption pattern in Belgium. Although the traditional fish business remains strong, there is also a trend towards high profile, high quality value-added products. The most promising sub-sectors for these products are lobsters, fresh/frozen salmon and frozen value added seafood.

The Ostend Fish Auction in Ostend in Belgium is one of the pre-eminent fish auctions in the world, just as this country is one of the most important seafood markets in the world. Approximately, 50 per cent of the fresh fish sold in Belgium is sold at this auction. Therefore, the auction provides opportunities for the export of bulk fish seafood products into the European market. There is a growing deficit for European Union (EU) domestic seafood supply and producers of seafood from Asian countries like India can make preliminary forays into the Ostend auction.

The European seafood market is clearly a very lucrative market on a global scale and it is extremely important for the province's seafood producers to have opportunities to break into this marketplace. There are certainly challenges and the government is focusing on solutions such as tariff relief and transportation options.

The EU is made up of 25 countries, including approximately 450 million people. The EU marketplace has a strong preference for seafood and there is an inadequate supply at present.

The western European seafood market is one of the largest in the world. Europe is facing a 30-40 million pound deficit in mussel centre in Yerseke, the Netherlands. The mussel centre is an excellent example of government and industry working together to create opportunities for industry and the economy as a whole.

Developing innovative and effective ways to develop and market our province's seafood products is an important part of government's plan for growing the province's seafood industry. It is critical that entrepreneurs effectively compete in the global marketplace and further establish the province's seafood products as world class in terms of mussel supply, and there are many opportunities for producers of this item to break into this lucrative seafood market. There is much one can learn from the production side of the western European seafood industry.

Since 1993, Brussels is host to the annual European Seafood Exposition [ESE] which is held in April. This international event is known in seafood circles as Europe's premier seafood fair and attracts exhibitors and visitors from all around the world. About eighty percent of them are non-Belgians. The most important message that the show conveys is that the country's seafood industries capable of effectively competing in the global marketplace. ESE is the most important seafood exhibition in Europe, and even in the world. Here all participants/exhibitors get great opportunities to meet their trading partners from all over the world and strike mutually beneficial, long-term business deals within a period of three days.

The European seafood industry is currently undergoing a huge metamorphosis. Total allowable catches [TAC] are being imposed on member states in order to build severely depleted stocks and protect those in sound biological state. At the same time, a long term strategic plan has been developed to maintain the economic activities of the fleets concerned. The entrepreneurs in the seafood industry are being enouraged to develop their business by adopting new production techniques and finding new resources. The Belgian economy has great depth and diversity. A highly developed market economy, it is heavily reliant on international trade. The country's Gross Domestic Product (GDP) is dominated by a very large service sector (70 per cent of GDP), followed by manufacturing (25 per cent) and agriculture (2 per cent). Exports account for more than 70 per cent of Belgium's GDP, making it one of the highest per capita exporters in the world. In addition to its own exports, Belgium functions as a transit and distribution center for many other countries to the rest of the European market. Consequently, almost 75 per cent of Belgium's trade is with other European Union countries. This highlights the country's importance as a commercial axis in Western Europe.

Belgium has a number of factors that contribute to its attractiveness for trade and investment. Its capital, Brussels, is an urbane city and home to the headquarters of the European Union and NATO, as well as hundreds of international institutions, associations and multinational corporations. Antwerp is the second-largest port in Europe and Belgium's second-largest city. An outstanding network of roads, rails and inland waterways enable goods shipped into Antwerp to be moved quickly and cheaply to European manufacturing and distributions centers. Geographically, Belgium is within a 600-kilometer radius of 70 per cent of the EU market.

Belgium has an excellent transportation network of ports, railroads and highways, including Europe's second-largest port. Belgium has an excellent network of distributors who are often regarded by the French and Germans as a neutral source of goods. In addition, Belgium adheres to EU laws and directives, and Belgian business continues to benefit significantly from a single European market.

In addition to the Commercial Service, there are numerous banks, professional organizations, service companies, financial organizations, which are prepared to advise and assist parties.

COMMISSION SEEKS HELP IN DEALING WITH EU TRADE FEUDS

Trade Commissioner Mandelson has launched are view of the EU's controversial system for protecting its producers against unfairly traded and subsidised imports from third countries.

Back Ground

The EU, like most other importing economies, operates a system of trade defence instruments (TDIs). These instruments anti-dumping, antisubsidy and safeguard measures-allow the EU to defend its producers against the following kinds of distortions in competition that are harmful to the European economy :

- "Dumping" where third-country companies export goods at below production-cose prices, as the EU claimed was the case regarding imports of shoes from China and Vietnam (EurActive 24/03/06);
- "Subsidisation" where non-EU exporters benefit from internationally illegal subsidies allowing them to produce a good excessively cheaply; as in the EU case against South Korea for unfair subsidisation of the semi-conductor producer Hynix, and;
- "Large and sudden surges of imports of goods into the EU" putting European industries at risk; as was the case when WTO limits on imports of textiles from China to the EU were removed, causing a sudden flood of Chinese clothes to enter the EU (EurActive 18/05/05).

During the ten years that the EU has operated its current trade-defence system, the global economy has changed significantly, with business and workers' interests increasingly linked to production outside the EU.

The rise of China as an export power has underlined a split within the EU between those that are reaping the benefits of cheap imports and those that are under pressure from heightened competition.

These conflicting interests and divisions among EU countries have made it increasingly difficult to define what constitutes "Community interest."

Trade Commissioner Peter Mandelson has, already twice this year in disputes over imports of Chinese textiles and leather shoes, found himself stuck between free marketers such as Britain, Germany and Sweden-which said that imposing TDI in these cases was protectionist and would raise prices for consumers-and manufacturing countries such as Italy, France, Spain and Poland, which claimed that imports of under-priced Asian goods were putting their industries at risk and threatening thousands of jobs (EurActive 04/10/06).

In order to avoid such situations in the future, the Commission, on 6 December 2006, launched as full-scale review of its trade-defence system, with a Green Paper that asks questions such as whether the criteria for using TDIs need to be toned down, whether other measures could be used, and whether the interests of importers and consumers should be given more consideration in investigations.

Issues

Predictably, European improters and consumers tend to urge for restraint in the use of trade-defence instruments (TDIs), whereas EU producers see them as an effective tool in the fight against "unfair trade practices" and oppose any loosening of the current rules.

Supporters of TDIs often view them as the only means currently available to the EU for addressing predatory trade practices by non-EU companies. They point out that in certain countries, particularly in China-the world's biggest target for anti-dumping measures-state interference makes it necessary to maintain TDIs.

Opponents of the EU's trade defence system say that, although TDIs could be legitimate if they were really used to remedy abusive behaviour, generally the sole aim is to protect outdated and uncompetitive industries. They say that imposing TDIs goes against the interests of consumers and those industries that have succeeded in modernising and remaining competitive.

MARKET REVIEW FOR FISH AND FISHERY PRODUCTS

Fishery Highlights : 2003-2005

The fish consumption has increased world wide. In the global supply Wild catches has dwindled but global supplies has increased steadily supported by aquaculture sustainability has become a major concern in aquaculture practices.

Aquaculture has generally contributed to higher supplies of fishery products. However, sustainability has become a major issue for both capture fisheries and aquaculture nowadays.

Higher demand for fishery products in developed and developing countries as also generated to the increased trade flow in the domestic and international markets. This trend is further supported by increased demand for quality fishery products, semi-processed products and convenience seafood particularly in developed nations and fast growing developing, countries and territories.

However, market accessibility particularly in the traditional developed countries is becoming more stringent and challenging for the export industries in developing countries. Qualitative issues (such as antibiotic, dioxin, mercury, histamine, malachite green, bird/Avian flu and other health related issues), tariff barriers such as anti-dumping duties, quotas and other non-tariff issues such as tractability, sustainability, eco-labeling, organic certification etc. continue to make processing costs higher, profit margin lower for seafood processors/exporters. The latest malaise to the fishing industry is the global hike in fuel prices making many fishing operations uneconomical.

On the other hand international market prices did not recover at all, particularly for conventional block frozen fishery products, to compensate the increasing fishing, farming and processing and transportation costs.

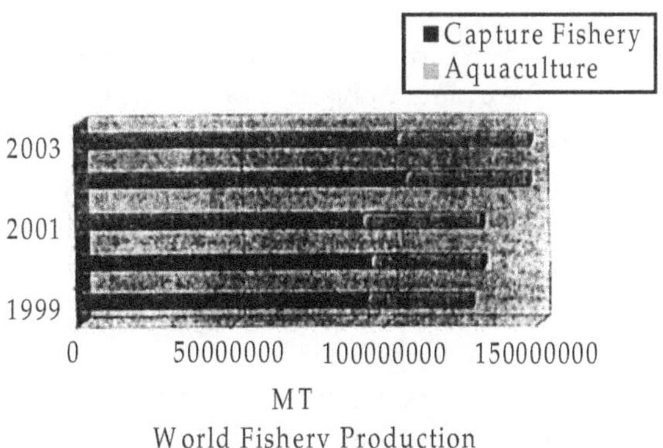

World Fishery Production

Despite several odds, the sector maintained its growth rates over the years through diversification, adaptation, innovation and value-addition in various stages of fisheries including fishing, aquaculture, harvesting, handling, processing, product development and marketing.

World Production

Global harvest of fishery products has totalled about 146.30 million tons in 2003. Despite a 6.6% growth in the aquaculture sector, total production increased only by 0.35% against 2002 attributed to the negative growth in the capture fisheries.

China, was the leading supplier for farmed and wild-caught fishery products, accounting for 38% of the global fishery harvest followed by Japan–the largest fishing nation, Peru, the USA and Chile.

Aquaculture

The total supply of aquaculture products increased to 54.78 million tons with a value of US$ 67.31 billion compared to 51.38 million tons in 2002. Over 77% of these was made of tin fish, crustaceans and molluscs amounting at 42.30 million tons. The balance represented aquatic plants and other organism such as seaweeds, sea sponges etc.

Over the years, China had nearly 70% share in global aquaculture production including fin fish, crustaceans and molluscs. The trend continued in 2003.

NATION-WIDE FISH MARKETING NETWORK

There have been suggestions for the past over 30 years to develop a Domestic Fish Marketing Network in the country but these have not yet moved towards the desired result. The main reason for the failure, as assessed, has to be attributed to the weak initiatives in this regard.

The suggestions are that the network of the domestic marketing system has to progressively cover all the States of the country, linking centres of production to consuming points. For establishing the network, State Governments have to play a major role. For them to do this, guidelines for setting up a Statewide network and linking it to national system would have to be circulated by the Centre among the State Governments for follow-up, in relation to an overall plan of action envisaging the development of a national network of the system, to be monitored by the Centre eventually. In other words, this national programme would have to entail the preparation of a master plan for its implementation in a co-ordinated manner. Considering the massive work involved, a separate organisational set up both at Central and State levels with a functional integration mechanism at the Centre, to link the effort at State levels with the national grid becomes imperative. The needed funds have to be earmarked and a well briefed set of officials under a dedicated leadership has to be placed in position for the project to gain the needed shape on the ground and acquire strength to make the system a vibrant force to keep the people supplied with fish on an enduring basis in a hygienic condition. As organised supply of wholesome fish to the consumers, which is the long desired and cherished final link of the ultimate objective of fisheries development, has to be achieved and the network is needed in this context.

There is an overwhelming consensus among those concerned at various levels dealing with fisheries that a domestic fish marketing network on a national scale is of utmost importance and is a necessity considered from any angle. Such a network will streamline and strengthen the marketing activity which presently functions in the country in patches and in a haphazard manner, with resultant coverage and economic disparities.

Sometime back the Amalgan Group in Cochin, under the leadership of Mr. Abraham Tharakan, announced initiatives in the direction of setting up of a domestic fish marketing system, but the present status of the efforts is not known.

A few decades back, there had been a route formed in Karnataka for the supply of fish to the people, operated under the co-operative sector, from coastal production centres in South Canara

District to consumers in Bangalore city and to Coorg district. There are also certain known routes of fish supplies that developed from heavy production centers both along the coast and from inland areas, mainly to metropolitan cities like Mumbai, Kolkata and Chennai. Another development, of late, is the flow of farmed fish from States like Andhra Pradesh to Orissa, West Bengal, Bihar, Assam and several other northeastern States. A Study of the dimensions of this marketing activity would provide a wealth of information to serve as guiding basis for the development of a nation-wide marketing system. In fact, it should be possible to absorb the present coverage by ongoing routes into the overall planning, to be made by organising additional facilities as needed. Other routes on the same line but with improvements to the ongoing ones could be promoted, incorporating solutions to the problems that the fish transporting vehicles now face while crossing state borders, for repacking fish with fresh ice at identified points etc. Provision of infrastructural facilities at determined points for production and supply of ice, for repacking of fish along the long distance routes of fish transportation can well be achieved with a measure of effort.

The way to proceed further on this vital subject seems to consist of first appointing a National Level Committee to prepare and present to the Union Department of Animal Husbandry, Dairying and Fisheries a detailed plan of action for establishing a nation-wide domestic marketing system. This Committee, supported by a secretariat, could be entrusted with the task of first undertaking a detailed study of the existing domestic fish marketing system in the various States and in the inter-State arena. After a critical appraisal of the structure and working of this system, the Committee could recommend intra-State and inter-State marketing systems for adoption, covering infrastructural, management and other needs. Matters related to inter-State trade, the facilities that are required to be provided to ensure quick passage.

Principles and Strategies

The enhancement of fisheries through stocking of fish seed had been a common practice throughout the world. It has emerged as one of the most widespread management tools for inland fisheries, for the reason that it has often been biologically successful. As in other parts of the world, this management strategy has also gained importance in India, especially with regard to inland fisheries of rivers, lakes, reservoirs, etc. Many of the States are practising stocking activities in their own rivers, reservoirs and lakes in order to enhance production and for the purpose of stock improvement. The reason for this shift in approach is that the more conventional approaches to management by control of fishery have proved incapable of limiting fishing efforts, and for compensating the short fall of recruitment caused by overfishing and environmental damage. Many stocking programmes are carried out in India without properly defined objectives or evaluation of the potentials or actual success of the exercise, despite considerable evidence of improvements in the yield from fisheries all over the would, achieved through application of measures for revival of fisheries based on defined objectives. Unless the stocking activity is maintained taking into account many of the under values that many in impinge on the outcome of the stock enhancement exercise stocking programme will never result in enhancement of the fisheries. This paper outlines the principles and strategies for stocking including main issues and options that should be considered before a stocking programme is initiated and recommends or protocol, to evaluate and improve the effectiveness of stocking as a management tool.

Principles of Stocking

Stocking of fish into its natural habitual is generally in response to degraded wild fish stocks, resulting from habitual alternations due to various reasons. Depending on the problem, stocking can be considered to be either a permanent or a temporary solution. EIFAC (1984) set out the motives for stocking which include the four main principles : (*a*) Compensation, (*b*) Enhancement, (*c*) Maintenance, and (*d*) Conservation.

Compensation : This encompasses stocking of native species to compensate the losses due to the human disturbance to the environment, for example, when there is lack of critical habitat such as spawning ground, or when obstacles of any king have to be overcome. Fishes planned for stocking may be released into unaffected parts of the river catchments or lakes taking into account the impact of stocking on wild stocks in the areas concerned. The degree of dependence of the particular fishery on stocking depends on the extent of modification to the ecosystem. Many of the stocked fisheries are based on the species such as exotic carps, that do not breed to a significant extent in the water body. This activity is particularly important when the stocks of the species are degrading and eventually die out without new individuals being added to the population.

Enhancement : Enhancement stocking is principally used to maintain or improve stocks where production is actually or perceived to be less than what the water body could potentially sustain, and where the reason for the poor stocks cannot be identified. Enhancement efforts include activities carried out to strengthen the quality and quantity of spawning stock, so as to improve the natural reproduction potential. Majority of the stocking activities carried out in the past probably fall in this category. However, the accuracy of the assessment of the state of stock and its potential production has a profound effect on the success or failure of the stocking exercise. Stocking for fishery enhancement is considered to be significant in respect of commercial as well as recreational fisheries.

Maintenance : Most of the fisheries around the world are assessed to be systematically overfished to the point where reproduction of the exploited populations is impeded. In order to support these species and to compensate for recruitment overfishing, stocking exercise is carried out. Maintenance of the degrading species in the principal purpose of this stocking exercise.

Conservation : In many places water bodies are stocked with fishes in order to conserve the stocks of a species that are threatened and face extinction. Stocking may take place in refugia or other areas that are not subjected to the endangering threats. However, of the species have to be maintained in areas where the threats still exist through continuous inputs of new materials from hatcheries.

Stocking Strategies

In any stocking activity, planning and a stepwise approach for the implementation of the plan is the most important aspect to be considered so as to take care to the ecological and practical aspects and to ensure that the activity is a successful one. Guidelines are available in many countries in respect of stocking exercises, which are often species specific or related to a particular type of water body. It is an essential part of stocking strategy to identify its objectives, and mechanisms by which it will be carried out, keeping in view the potential ecological and environmental risks. Appropriate implementation strategies are inevitable for ensuring the success of the stocking programme. In a broad sense, the different strategies to be considered can be divided into: (1) pre-stocking appraisal; (2) Stocking process; and (3) Post-stocking evaluation and management.

Pre-stocking Appraisal

Whenever stocking of fish for enhancement is to be carried out, the fish and foremost step to be considered is pre-stocking appraisal. The suggested strategy for planning a stocking exercise is given in Fig. 1. Three main points to be taken into account in this respect are given hereunder.

MARINE PRODUCTS EXPORT

1. Aquaculture and Growth of Marine Products Export

Of late, sustained growth of marine products export from India is largely dependent on the progress of aquaculture production. Out of 138085 tons of shrimp exported during the financial year 2003-04, 87,066 tons is from aquaculture that constitute 63% by quantity and 83% by value.

MPEDA envisages achieving 15% growth rate in foreign exchange earnings from export of marine products to achieve an export of US$ 5.0 billion by 2010-2011. To achieve this, a four-pronged strategy is being adopted.

(*i*) Increase production from capture fisheries, particularly by resource specific fisheries.
(*ii*) Increase production from culture fisheries and diversified aquaculture.
(*iii*) Introduction of new technology and modernization of processing facilities for improvement of quality and value addition.
(*iv*) Market promotion by re-establishing existing markets and penetrating to new markets.

2. Present Status of Aquaculture

Major portion of foreign exchange earnings in the seafood export presently is from shrimp export. Shrimp has been the mainstay of our marine products export for a long period. Coastal aquaculture has been concentrating mainly on brackish water shrimp farming due to the high unit value realization of shrimp. Freshwater prawn (scampi) farming has also been picking up in certain inland and maritime states.

India is bestowed with different environmental and agro-climatic conditions. A wide range of systems are used with different degrees of intensity. Different Species being cultured in freshwater areas at present include finfishes such as *Cyprinus carpio, Labeo rohila, Cirrhinus mrigala, Catla catla, Ctenopharyngodon idella, Hypophthalmycthys molitrix, Clarias* spp, *Onchorhynchus mykiss* and Crustaceans like *Macrobrachium rosenbergii, M. malcolmsonii.* The salt water species suitable for culture are finfishes like *Lates calcarifer, Chanos chanos, Mugil cephalus* and crustaceans such as *Penaeus monodon, Penaeus indicus* and Molluscs like *Crassostrea madrasensis, Perna viridis* and *Paphia* spp.

3. Aquaculture Activities in the Coastal Region

In the coastal region, aquaculture can be taken up as coastal farming on land and as mari culture on sea. Diversification provides opportunity for crop rotation, utilization of diverse environmental conditions, greater flexibility to suit market requirements and resultant wider market access. Development of technology for breeding, availability of commercial feeds, technical feasibility, ready marketability, economic viability and environmental impacts, sustainability are the factors to be considered for effective diversification programmes. Rajiv Gandhi Centre for Aquaculture, a non profit society under aegis of MPEDA, mandated to take up research and development and field trials and dissemination of aquaculture technology on commercial lines has initiated programmes like breeding and culture of sea bass, fattening of lobsters, breeding of mud crabs etc. Some projects like production of Artemia etc. are also in the pipeline. Considerable effort is required for tapping the resources of other potential species.

4. Present Status of Diversification

Although, several experimental efforts have been made in India to standardize the farming technology of other aquatic organisms, the effort towards diversification had only little impact on commercial aquaculture. MPEDA has already taken up some projects for diversification of coastal aquaculture including culture of fin fishes like sea bass, crustaceans like crabs, lobsters and molluscs such as mussels, clams etc.

4.1 Fin Fishes

Global production of fin fish through culture is to the tune of 3,420,000 tones. Although there are several fin fishes of commercial importance, the breeding technology has been successful only for a few species. Hence, commercial activities will be concentrating mainly on these species. Tilapia, sea bass, grouper, milkfish, mullets etc. are the potential species for which commercial aquaculture activities could be initiated.

4.1.1 Sea Bass

Sea bass commonly known as barramundi in trade circles, is produced mainly in Indonesia, China, Taiwan and Malaysia. In 2003 these countries produced 5,483 tons, 4,811 tons and 4,211 tons respectively. This fish is sold at US $ 2 per kg.

Asian Sea bass, Lates calcarifer is the species suitable for aquaculture in India. It is a eurihaline species, and can be cultured in freshwater, brackish water or seawater. It is a fast growing fish, with considerable market demand in India and abroad. Breeding and seed production technology for sea bass has already been established successfully by the Central Institute of Brackishwater Aquaculture (CIBA), Chennai and the Rajiv Gandhi Centre for Aquaculture (RGCA). The bottleneck in promotion of commercial production is non-availability of commercial feed suitable for sea bass culture. Presently, the farmers depend upon trash fish or farm made pellet feed. Commercial feeds are available in the international market. Import of sea bass feed seems to be expensive therefore alternate sources will have to be explored. In spite of these constraints, sea bass remains the most important candidate for immediate diversification. Culture period for table size is approximately 9 months.

European sea bass, Dicentrarchus labrax, is produced mainly in Greece and Italy. During the year 2003, these two countries produced 24,838 tons and 9,600 tons respectively. Technology for commercial seed production is available. Culture period for table size is 7-8 months. This fish is sold at US $ 1 per kg.

4.1.2 Milkfish

Milkfish, *Chanos chanos,* has demand in the world market, particularly in South East Asian countries. It is a fast growing, herbivorous fish. Hatchery seed production technology is not available for commercial seed production. Commercial aquaculture of milkfish is being practiced in Indonesia. There is good potential for growing of this brackish water species in India. Philippines produced 246504 tons of milk fish followed by Indonesia 226114 tons during 2003. Culture period for table size is 7-9 months. This fish is sold at US $ 1 per kg.

4.1.3 Catfishes

In 2003, 3,200 tons and 4,024 tons respectively, of North African cat fish *Clarias gariepinus* was produced in The Netherlands and Nigeria. India and Indonesia produced 43,340 tones and 70,826 tons of Torpedo shaped catfish *Clarias* spp. Torpedo shaped cat fish, *Clarias* spp. African catfishes, *Heterobranchus bidorsalis* are commonly cultured in Liberia. Commercial seed production technology is established for cat fishes. Culture period for table size is 9 months. This fish is sold at US $ 2 per kg.

4.1.4 Mullets

Grey Mullet, *Mugil cephalus* is an important candidate for aquaculture. However, seed production under controlled hatchery conditions is yet to be developed for boosting the culture activities. In 2003, Egypt topped production with 1,35,609 tons. Indonesia also produced 10,725 tons of other mullets (Mugildae). Culture period for table size is 9 months. This fish is sold at US $ 1 per kg.

4.1.5 Tilapia

Tilapia, commonly known as aquatic chicken is an important candidate for aquaculture in the global perspective. This fish is sold at US $ 1 per kg. *Oreochromis mossambicus,* the species commonly found in India is not important in the trade as it fetches only low value. The exotic species, *O niloticus, Nile Tilapia,* seems to be the ideal species for commercial production. Since seed production technology is yet to be established in India, seeds will have to be imported from elsewhere. Israel is pioneering in seed production. Tilapia culture is fast developing world over. In the year 2003, China produced 8,05,859 tons followed by Egypt with 1,99,557 tons. Culture period for table size is 9 months.

4.1.6 Pearl Spot

The brackish water species, the pearl spot, *Etroplus suratensis* has very good demand in the middle-east region. Organized production of pearl spot could be attempted for boosting its export. Technology for commercial seed production is not available in the country for taking up commercial ventures. Culture period for table size is 8 months. This fish is sold at US $ 2 per kg.

4.1.7 Cobia

Tremendous interest has been taken worldwide to culture this fast growing pelagic species, *Rachycentron canadum.* In India, we are yet to initiate culture of cobia fish. China and Taiwan are main producers producing 20,667 tons in 2003. Technology for commercial seed production has been developed in Queensland, Australia. Culture period for table size is 7 months. This fish is sold at US $ 1 per kg.

4.1.8 Trouts

Trouts are suitable for cold waters of hilly tracts of India. Projects on turn key basis are needed for popularizing Rainbow trout, *Onchorhyncus mykiss,* Chile is main producer with 1,09,578 tons, during the year 2003. Hatchery technology for production of fingerlings for stocking is well developed for trouts in Norway. Culture period for table size is 12 months. Fish of 1 kg and *above* is sold at US $ 3.6 per kg.

4.1.9 Groupers

The commercially important groupers belong to the genus, Epinephelus. In fact, live groupers fetch a premium price in the international market. Although breeding of this fish is yet to be standardized in India, Australia and Japan have already perfected this technology. Therefore, for the time being we may have to import technology for seed and feed production for initiating any major attempt on grouper culture in India. During 2003, China topped in production with a production of 26,790 tons followed by Taiwan (11,564 tons), Indonesia (8,665 tons).

4.2 Crustaceans

Crustaceans other than shrimp and scampi have not undergone any significant development in the field of commercial aquaculture. There is considerable scope to develop commercial projects for crabs, lobsters, crayfish, Artemia etc. considering the global demand. Commercial technology is yet to be standardized for the production of seed and feed required for the culture of some of the species.

4.2.1 Crab

The Indo-Pacific swamp crab (Mud Crab) is produced mainly in Indonesia. 7152 tons was produced in 2003. Mud crab fattening is practiced in some places, as *live* crab has good demand in the export market. *Scylla serrata* and *Scylla tranquebarica* are two important mud crabs available in India. Since commercial crab breeding technology is not available in the country, MPEDA started a project through RGCA to initiate crab-breeding programme. Once the technology is perfected and crab seeds are made available, it is anticipated that crab culture programmes can be promoted on a large scale. Culture period for table size is 5-6 months. Mud crab is sold at US $ 4 per kg.

4.2.2 Lobster

There are *several* species of lobsters, with commercial importance. *Panulirus homarus, P. ornatus, P. polyphagus* etc. are some of the important species, commonly found in Indian coast. There were some attempts to fatten juvenile lobsters under controlled conditions. MPEDA also took up a lobster-fattening programme through RGCA for demonstration of the technology and for finding out suitable feed combination for lobster fattening. Lobster is sold on an *average* at US $ 16 per kg.

4.2.3 Cray fish

Cray fish is an exotic species, which is mostly confined to temperate regions. The most important species is Red Swamp Cray fish, *Procambarus clarkii*. Considering the availability of large extent of freshwater area in the country and demand of crayfish in the global market, we may initiate pilot scale culture of cray fish in the country, by importing the required seed and feed initially.

4.2.4 Artemia

Brine shrimp, (Artemia spp) is mainly produced in USA, China, Russia, Iran, Vietnam, Brazil, Argentina, Columbia, Kazakhstan & Turkmenistan. It is estimated that about 300 shrimp hatcheries have been established in the country to supply quality seed to coastal aquaculture. Artemia is a vital input for the hatchery operations and presently, the entire requirement is met through import. Although we have considerable land under saltpans, where Artemia could be harvested, no systematic approach has been initiated in this line. Production of quality Artemia within the country is an area, with considerable scope for development. This could also provide an opportunity for import substitution. RGCA is implementing a project for Artemia production, under the consultancy of an expert. In 2003-04, the total wet artemia cyst produced in the world was 4,000 tons. The cost of cyst increased from US $ 10 to 55 per kg from 1991 to 2005.

4.3 Molluscs

Bivalve molluscs are widely distributed in the Indian Coast, and they are considered very nutritious and protein rich food. Edible oysters, mussels, clams, pearl oysters are the commercially important species.

4.3.1 Edible Oysters

Edible Oyster is a sedentary animal, attached to a hard substratum. The important species suitable for culture are *Crassostrea madrasensis* and *C. gryphoides*. The oysters are filter feeders and their food consists of organic detritus and phytoplankton. They form subsistence fisheries among the fisher folk. Artificial breeding of oyster has been successful at Queensland, Australia. So far no organized culture effort has been made in India. Isolated attempts cannot ensure a ready market and hence, organized effort is required for proper marketing of cultured oysters. In 2003 China produced 36,68,237 tons of Pacific cupped oyster *C. gigas* and USA produced 65,632 tons of *C. virginica*. Commercial seed production is yet to be introduced in India. Culture period for table size is 12 months.

4.3.2 Mussels

India has potential resources of two important mussels *Perna viridis* (green mussel) and *Perna indica* (brown mussel) distributed along the coastline. In addition to the availability of wild seeds, hatchery technology is also available for seed production. Pilot and experimental efforts to demonstrate mussel culture has been proved successful at Queensland, Australia and large scale culture projects could yield promising results. Technology for commercial seed production is available. Culture period for table size is 12 months.

4.3.3 Clams

Although several clams contribute to Indian fisheries, the species suitable for culture are *Meretrix meretrix, M. casta, Katelysia opima, Paphia malabarica, Anadara granosa, Villorita cyprinoides, Tridacna maxima* etc. The clams mainly serve as subsistence fishery for the fisher folk. Clams have restricted movements and they burrow into the substratum. For farming purpose, clam seeds can be collected from the natural grounds or produced in hatcheries. Culture technology is relatively simple; as it involves transplantation of seed to suitable grow out area and protecting by pens. The pen enclosure also prevents the movement of clams, away from the culture site. China is the major producer of blue cockle *A. granosa*. China produced 317,870 tons in 2003. Culture period for table size is 12 months. Spain is the pioneer in developing hatchery technology. In India, commercial technology is yet to be developed.

4.3.4 Pearl Oyster

Cultured pearls have excellent demand within the country as well as in the export market. *Pinctada fucata* and *P. margaritifera* are two important oysters identified in the Indian coast for pearl culture. Research programme on breeding as well as culture were initiated in the country by the Central Marine Fisheries Research Institute (CMFRI) in Tamil Nadu coast. *P. fucata* is distributed in the South East Coast. *P. margaritifera,* known as the black lip pearl oyster, is found off the coast of Andaman & Nicobar Islands. The most ideal locations for starting pearl culture projects would be Gulf of Mannar and Andaman & Nicobar Islands. It takes about one and a half years for pearl production.

4.4 Sea Plants

India has vast resources of sea plants along its coastline. The important species fall under green, brown and red algae. The red algae, *Gelidiella acerosa, Gracilaria edulis, Gelidiellum* spp. etc. are agar yielding plants. The species brown algae like *Sargassum; Turbinaria* are algin yielding plants. Besides, agar and algins, phytochemicals such as carrageenan, mannitol, iodine, laminarin etc. are also extracted from marine algae. The chemicals extracted from sea plants are widely used as gelling, stabilizing and thickening agents in many industries like food, confectionary, pharmaceutical, dairy, textile, paper, paint, varnish etc. Many protein rich sea plants are utilized for food purpose in the form of jelly, jam, wafers, soups, salads etc. A part of sea plant requirement is presently harvested from the natural sea plant beds. In order to sustain our natural resources, efforts must be made to culture sea plants. Proven technology for sea plant farming is available and could be promoted to enhance our sea plant resources. The bays and creeks along the East and West Coast, especially off the coast of Tamil Nadu and the seas around Andaman & Nicobar Islands and Lakshadweep islands offer excellent ground for sea plant culture. There are reports that corporate bodies have already shown some interest in culture of sea plants. In 2003, Chile produced 69,650 tons of *Gracillaria*, China produced 7,27,530 tons of Nori, (*Porphyra tenera*) and Indonesia produced 2,31,900 tons of Red plants.

VALUE ADDITION – THE PARADIGM SHIFT FOR SUSTAINABILITY OF THE SEAFOOD INDUSTRY

India's seafood export has seen steady growth over the last few years-post liberalization. However recent international trade issues have had an effect and exports had fallen by 11.83% in terms of quantity and 11.47% in terms of value during 2003-04 before picking up in 2004-05. An in-depth analysis reveals that the major chunk of our seafood exports is even now in the frozen form. Our average unit value realization has been hovering around 3 US $ per kg for over a decade. Idle capacity is high and many smaller firms have not been able to withstand the competition. Coupled with these inherent limitations in the industry, has been tariff and non-tariff barriers imposed by buyers. Another phenomenon that can be observed in the consumer market is the gradual disappearance of the conventionally processed products and their re-emergence in new styles and forms. It is thus time for a paradigm shift, to value addition maintaining high quality standards and with diversified products and processes, to ensure sustainability as well as profitability of the industry.

Current Status of Value Addition

Value addition basically involves adding value to the product so that returns can be increased. Value can be added to fish and fishery products according to the requirements of different markets and the products can range from live forms to ready-to-serve products. It is the retail end that is to be targeted. With the increasing number of home-makers taking up jobs, it is imperative that products that are easy to use should be developed and marketed. Value addition is as much a form of marketing as it is developing a product. Packaging is also thus of prime importance in the whole product development process.

The trend in the export of value added products has been rising through the last two decades but the rise has been significant from 2000 onwards. The quantity has risen from 40710.87 tonnes in 1993 to 58522.15 tonnes in 2003. Value-wise it rose from 517.30 crores to 1480.65 crores during the same period. Value added products contributed 14.28% to quantity and 24.59% to the value of the total seafood export from the country. The value added products currently being exported from India includes accelerated freeze-dried shrimp, battered and breaded products, value added frozen products, IQF products, live and ready-to-eat products and by-products. Among the products the contribution of IQF products was the most significant with the percentage contribution hovering around 40% both in terms of quantity and value over the last decade. The contribution of all other products is only marginal. There is immense scope for live and processed items and also by-products. As of now their contribution to value added exports, is only around 3%. Canned products, which are the popular form of sale at the retail supply end, has disappeared from our product mix mainly due to the high production costs. Ready-to-eat products also have a growing market both international as well as domestic, which can be tapped by the industry.

The various products currently being exported are accelerated freeze dried shrimp; battered and breaded products like cutlets, burgers, breaded shrimp, butterfly shrimp; frozen crab products like soft shell crab, pasteurised crab, crab meat, cut crab with claws; other frozen products like cooked salad shrimp, sushi, analogue products; IQF shrimps, crab, baby octopus, octopus tentacles, lobster, cuttle fish and squid; live crabs, aquarium fishes, snail, lobster, shrimp and baigai; fish, prawn and clam pickles; ready to eat products like fish curry, spiced and fried shrimps prawn chutney, fried fish and edible fish powder; byproducts like chitin, chitosan, glucosamine hydrochloride, agar, fish maws, is in glass, shark fin, shark fin rays and squalene.

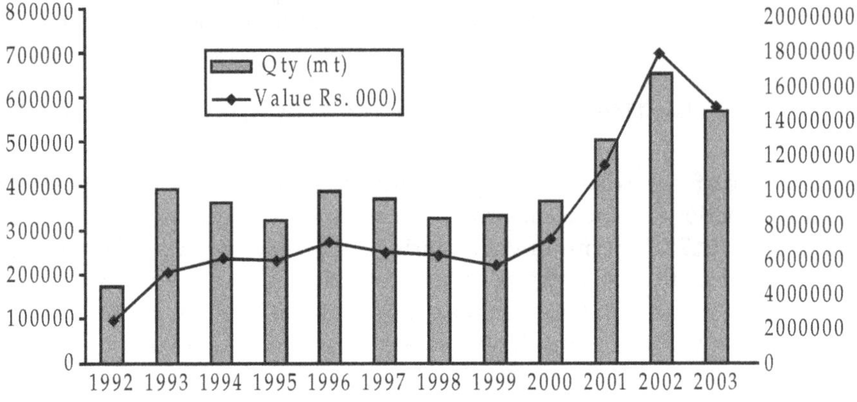

Export of value added products from India

Technologies from CIFT in Value Addition

Value addition has been a major focus of the research activities undertaken by CIFT. The Institute has developed processes and products that can add value to the basic product. Some of the technologies available with the Institute are listed below:

1. Coated Products: They fall in that category of products that are convenient for use and widely valued by consumer. Many products can be coated and immediately frozen, or they may be pre-fried, and then frozen for distribution and sales to consumers and food service establishments. They can be quickly reconstituted by conventional heating methods. Some products are designed for reheating in microwave ovens.

   Battering and breading enhance food product's appearance and organoleptic characteristics in addition to improving its nutritional value. Coating acts as a moisture barrier, minimizing moisture losses during frozen storage and microwave reheating. The most important function of coating is value addition by increasing the bulk of the substrate thereby reducing the cost element of the finished product. Most coated products are now available with a three-way cook option. They can be baked in a conventional oven, prepared under the grill or fried. The hunt is now on for coatings, which are suitable for use in a microwave oven.

   The production of battered and breaded fish products involves several stages. The method varies with the type of products and pickup desired. In most cases it involves seven steps. They are portioning/forming, pre-dusting, battering, breading, pre–frying, freezing and, packaging and cold storage. Some of the products are cutlets, fish balls, burger, nuggets–developed from low value fishes. Products that has good export value include breaded shrimp, breaded squid rings, coated fish fillets, crab claw balls etc.

2. **IQF products :** An important improvement in freezing is the shift from the conventional block frozen to the individually quick frozen products. With the advent and spread of aquaculture of shrimp, individual quick freezing has become very popular. Farmed prawn offers the advantage of harvesting at a predetermined time and hence can be frozen in the freshest possible condition. Because of this, most of the farmed prawn is frozen as whole IQF. Lobster, squid, cuttlefish, different varieties of finfish etc. are also processed in the individually quick frozen style.

   IQF products fetch better price than conventional block frozen products. However, for the production of IQF products raw-materials of very high quality need to be used, as also the processing has to be carried out under strict hygienic conditions. The products have to be packed in attractive moisture-proof containers and stored at –30°C or below without fluctuation in storage temperature. Thermoform moulded trays have become accepted containers for IQF products in western countries. Utmost care is needed during the transportation of IQF products, as rise in temperature may cause surface melting of the individual pieces causing them to stick together forming lumps. Desiccation leading to weight loss and surface dehydration is other serious problem met with during storage of IQF products.

   Some of the IQF products in demand are prawn in different forms such as whole, peeled and de-veined, cooked, headless shell-on, butterfly fan tail and round tail-on, whole cooked lobster, lobster tails, lobster meat, cuttlefish fillets, squid tubes, squid rings, boiled clam meat and skinless and boneless fillets of white lean fish. IQF products can be easily marketed as consumer packs, which is not possible with block frozen products. This is a distinct advantage in marketing.

3. **Extruded Products :** Extruded products like noodles, wafers, flakes, etc. from vegetable sources are well established in the consumer market. But fish based extruded products are yet to gain popularity. Fish based extruded products have got very good marketing

potential. Formulation of appropriate types of products using different fish mince, starches etc., attractive packaging for the products developed, market studies, etc. are needed for the popularisation of such products. Such products can command very high market potential particularly among the urban elites. The technology can be employed for profitable utilization of by catch and low values fish besides providing ample employment opportunities.

4. **Speciality Products from Shrimp :** Shrimp continues to be the favoured seafood the world over. Many speciality products with shrimp as the raw material have been developed by CIFT, which have good international market. Some of them are centre-peel shrimp, cooked centre peel shrimp, easy-peel shrimp, cooked easy-peel shrimp, shrimp skewer, fantail round, coated fantail round, butterfly shrimp, coated butterfly shrimp, butterfly "sushi" shrimp, stretched shrimp (nobashi), breaded "nobashi", shrimp single kebab (barbecue), shrimp vegetable kebab.
5. **Fish Mince and Mince Based Products :** Minced meat is the meat separated from fish free of bones, skin etc. In principle, meat separation process can be applied to any species of fish, but when it is applied to low cost fishes significant value addition will accrue. Flesh can be separated from filleting waste also. Minced meat can be used as a base material for the preparation of a number of products of good demand, Minced fish can be used for the preparation of a number of products like fish sausage, cakes, cutlets, patties, balls, pastes, surimi and surimi based kneaded products like kamaboko, chikuwa, hampen, fish ham, texturised product etc.
6. **Canned Products :** Canning is making a come back with cheaper raw material for cans becoming available like TFS cans. Technology for canning already exists for shrimp, crab, fish, mussel, calm and squid.
7. **Retort :** Pouch Technologies for preparation of region specific ready-to-eat products. This can be used to diversify into the domestic markets.
6. Byproducts like chitin, chitosan, glucosamine hydrochloride, agar, isin glass, shark fin rays, gelatin, squalene, fortified fermented fish meal, fish silage, squid meal etc.

   Value addition and diversification of products, processes and markets is the only way out for long term sustainability of the seafood industry. Along with the international market the vast, emerging, dynamic, urban domestic market with a demand for convenience products needs to be tapped. Technologies available at National Institutes like CIFT can be made use of effectively for achieving this goal. A sustained institute-industry interaction is also essential.

REFERENCES

1. R.A. Lawson, "Report on Credit for Artisnal Fisherman in South-East Asia". *Report of Food and Agriculture Organisation* (Rome, 1972), No. 122.
2. P.S. Rao, *Fishery Economics and Management in India* (Bombay : Pioneer Publishers and Distributors, 1983), p. 198.
3. Kodi means 22 pieces of fish.
4. Kalli means 508 maunds approximately.
5. Kachha maund means 44 kilograms while Pacca maund is of 40 kilograms.
6. T.A. Mammen, *Long Term Perspectives of Fish Marketing in India* (Ahmadabad : Indian institute of Management, 1982).
7. Government of India, *Report of National Commission on Agriculture* (New Delhi : Ministry of Agriculture, 1976), Part VIII, p. 239.
8. Government of India, *Report of National Commission on Agriculture* (New Delhi : Ministry of Agriculture, 1976), Part VIII, p. 239.

9. Government of India, *Handbook of Fisheries Statistics, 1986* (New Delhi : Ministry of Agriculture, 1987).
10. Information personally collected from the Ministry of Agriculture, Government of India, New Delhi During the course of study.
11. Government of India, *Report of National Commission on Agriculture* (New Delhi : Ministry of Agriculture, 1976), Part VIII, p. 240.
12. Indian Institute of Management, *Inland Fish Marketing in India* (Ahmedabad, 1983), Vol. III, pp. 103-125.
13. S.C. Agarwal, "Capture Fisheries in Karnal Tehsil in Haryana State", in *Journal of Fishing Chime,* 1982, Vol. 2, No. 8.
14. P.S. Rao, *Fishery Economics and Management in India* (Bombay : Pioneer Publishers and Distributors, 1983), p. 205.
15. P. S. Rao *Fishery Economics and Management in India* (Bombay : Pioneer Publishers and Distributors, 1983), pp. 216-217.
16. Chilkar and Nanggi mean fish with scale and without scales respectively.
17. S.C. Agarwal, "Capture Fishery in Karnal Tehsil in Haryana State", in *Journal of Fishing Chime,* 1982 Vol 2, No. 8.
18. P.S. Rao, *Fishery Economics and Management in India* (Bombay : Pioneer Publishers and Distributors, 1983), p. 178.
19. P.S. Rao, *Fishery Economics and Management in India* (Bombay : Pioneer Publishers and Distributors, 1983), pp. 215-240.
20. P.S. Rao, *Fishery Economics and Management in India* (Bombay : Pioneer Publishers and Distributors, 1983), p. 221.
21. Koilu Phillips, *Marketing Management* (New Delhi : Prentice-Hall, 1979), pp. 45-63.

Chapter 20

Food Quality, Safety and Risk Management

Introduction

Food is a big business. Food production is now scientifically based and transportation of food over a long distance to arrive it's destination in the most wholesome condition is now possible. Consumers world-wide have access to a wider variety of high quality foods in greater quantities than *ever* before. The other two developments in the food trade are the dramatic increase in countries engaged in food production and export and the internationalization of food tastes and habits. So it's encouraging to see companies in the fishery investing in seafood quality and safety as part of their business strategies to stay ahead.

It is the responsibility of food safety authorities to meet consumer expectations and to guarantee them a high level of health protection, by adopting the necessary measures. Risk management is one of the essential tools for setting up food safety systems and it seems appropriate to share experiences in this field so that all the countries have access to information which will allow them to adopt the necessary measures to protect the health of consumers.

What is Risk?

"Risk is a function of probability of an adverse effect and the severity of that effect, consequential to a hazard(s) in food." "Hazard(s) is a biological, chemical or physical agent in, or condition of, food with the potential to cause an adverse health effect."

The risk to the world's population from hazards in and on food depends largely on the degree of control exercised by producers, processors and official food control authorities to prevent or minimize the risks to acceptable safe levels. Food safety risk analysis is an emerging discipline, and the methods used for assessing and managing risks associated with food hazards are still being developed. In contrast, risk is the estimated probability and severity of adverse health effects in exposed populations consequential to hazards in food.

Understanding the association between a reduction in hazards that may be associated with a food and the reduction in the risk of adverse health effects to consumers is of particular importance in the development of appropriate food safety controls. Unfortunately, there is no such thing as 'zero risk' for food (or for anything else).

Risk Analysis consists of three elements :

- Risk assessment
- Risk management
- Risk communication

What is Risk Management?

It is primarily one of the three aspects of risk analysis, the others being risk evaluation and risk communication. The Codex Alimentarius has adopted the following definition: risk management is the process of weighing up the various possible policies, taking account of the evaluation of risks and other factors involved in the health protection of consumers and the promotion of fair trade practices, and taking decisions accordingly, *i.e.* choosing and implementing the appropriate prevention and monitoring measures.

The management of food-related risk is therefore a political prerogative which involves balancing the recommendations formulated by the experts commissioned to scientifically evaluate the risks, and the resources of all types that social and commercial groups and manufacturers can set aside for dealing with these risks.

Structured risk assessment and risk management processes should help decision makers develop effective solutions for the reduction of foodborne illness. The public should be involved in the risk management decision making processes. These processes should not be construed as prescriptive and should allow flexibility. It is most important that the decisions made and actions taken are rational, defensible, and arrived at in an open and transparent manner. It may not be necessary to spend extensive time on each task, particularly if a management issue is well characterized.

Functional separation is essential to the conduct of risk analysis activities in order to maintain the scientific integrity of the risk assessment process and to avoid political pressures that would undermine the objectivity and the credibility of the conclusions. Separation of risk management and risk assessment helps to ensure that assessments are not biased by pre-conceived opinions related to management solutions. However, there is a need for frequent interaction between risk managers and risk assessors in order to arrive at effective risk management decisions. Active interaction is necessary to ensure that the assessment will meet the needs and answer the concerns of the risk manager.

The assessors must understand the manager's questions and both parties must acknowledge any constraints, which may impact on the risk assessment. The strengths and limitations of the assessment must be properly communicated so that people using the risk assessments can properly understand the results. Interactions between assessors and managers do not end with the completion of the risk assessment. There will often be exchanges of information and input from assessors during subsequent risk management activities, for example, during the option assessment stage and in communication of results to interested parties.

HOW CAN FOOD SAFETY REGULATORS MANAGE A KNOWN OR FUTURE RISK TO PROTECT THE HEALTH OF CONSUMERS?

1. By basing Policies and Measures Adopted on an Evaluation of the Risks

The WTO Agreement on the Application of Sanitary and Phytosanitary measures (SPS Agreement) states, in fact, that WTO members should base their sanitary and phytosanitary measures on risk evaluation.

It should be noted, in this respect, that risk evaluation is a scientific process consisting of stages of identifying and characterizing the dangers, then evaluating exposure to these dangers in order to characterize the risk (probability that the danger will be expressed in real terms).

2. The Principle of Precautionary Measures, in the Absence of Sufficient Scientific Proof

There is, however, an exception to the obligation to base sanitary and phytosanitary measures on a risk evaluation. This allows governments to adopt sanitary and phytosanitary measures even when the risk evaluation is incomplete and to use precautionary measures to protect their citizens. The SPS agreement states that in cases where relevant scientific proof is insufficient, a WTO member country may provisionally adopt sanitary and phytosanitary measures based on the relevant information available. Under such circumstances, the countries should then strive to obtain the additional information necessary for a more objective evaluation of the risk and should re-examine the sanitary and phytosanitary measure accordingly, within a reasonable time-frame.

Scientific uncertainty cannot, therefore, serve as an excuse for a decision-maker to fail to act in response to a food-related risk. Thus when a potentially dangerous and irreversible situation begins to emerge, but the scientific evidence is lacking for a full scientific evaluation, risk managers are legally and politically justified in adopting precautionary measures without waiting for scientific confirmation.

It is, in fact, the responsibility of decision-makers to adopt the necessary measures to protect consumers. It should be noted once again in this respect that citizens are more demanding today than formerly as regards food safety. They give priority to health safety over other criteria which might have prevailed in the past, in a context in which the food supply is large enough to offer replacements.

3. The "Farm to Table" Approach

To be sure of the safety of foodstuffs, all aspects of the food production chain in continuity must henceforth be considered, from primary production (including animal protection and health aspects) and the production of animal feed, to the distribution of foodstuffs to the end consumer. Each component may have an impact on food safety.

4. Traceability

Traceability is an essential requirement in guaranteeing food safety. When a danger threatens (for example food poisoning), the risk manager should be able to determine the food responsible, rapidly carry out a precise, targeted withdrawal of dangerous products, inform consumers or agents in charge of monitoring foodstuffs, go back along the whole length of the food chain if necessary to identify the source of the problem, and put it right. Traceability studies thus allow risk managers to limit exposure of consumers to the risk and thus the economic impact of the measures by targeting products at risk.

For it to be effective, the traceability system must involve all stages in the pathway, from the live animal or raw material to the product undergoing final processing, from stock-rearing to food sector companies via companies in the animal feed sector.

5. Management of Health Risks in an Emergency and in Emerging Risks

Despite the checks carried out by risk managers, incidents are always possible. To ensure consumer safety, it is important that risk managers are informed as soon as possible of an incident and have access to the most precise possible evaluation of the risk in order to be able to implement the necessary measures and avert the danger. Health surveillance is thus vital, and within this framework the circulation of information is essential.

Furthermore, managing health risks in an emergency or emerging risks requires good cooperation between the monitoring services in charge of food safety and effective procedures for withdrawing suspect products from the market.

6. Taking Account of Socio-Economic Concerns

The implementation of regulations aimed at protecting consumer health can be effective only if the risk manager is aware of the resources that companies and manufacturers can set aside for managing risks. One recommendation is thus to bring together professionals involved in drafting regulatory texts to hear their opinions.

It is recognized, in this respect, that in some cases risk evaluation cannot on its own provide all the information on which to base a risk management decision. In response to the expectations of the general public and consumers, other relevant factors should legitimately also be taken into consideration, notably social and economic factors (technical feasibility, economic impact), traditional and ethical factors (animal well-being) and environmental factors, as well as the feasibility of inspections.

WHAT IS THE ROLE OF RISK MANAGERS?

The first responsibility of professionals is the marketing of their products. They can participate in the policy to improve food safety in various ways.

1. Self-monitoring and company laboratory accreditation
2. Guides to good hygiene practice
3. The development of company certification
4. Product standardization
5. Contribution to product traceability
6. Distribution

In the risk evaluation step risk managers determine whether the current level of risk, as determined through risk assessment, is acceptable. In other words, is the risk of enough concern to warrant measures for reduction? If the risk is acceptable no further action is required. If the risk is unacceptable managers need to consider appropriate interventions. An additional complication in the management of microbiological risk is the lack of a common understanding of the process risk managers should use to evaluate risk. Zero risk is not achievable because it is generally assumed that a single pathogen has some probability of causing illness under certain circumstances. Also, illness from microorganisms in food is not a new risk that we can choose to accept or reject, but a risk we have always experienced, although perhaps not at the current level. For these reasons microbiological risk evaluation requires a more sophisticated approach than simply determining an acceptable level of risk. Some countries have set risk reduction goals with associated time frames in recognition of this difficulty.

The primary responsibility for compiling the list of available options lies with the risk manager in consultation with other stakeholders e.g. industry, consumers, regulatory food scientists, academia, etc. However, experience gained by the risk assessor may be useful in compiling the list of options and therefore the managers and assessors may interact on this basis. The risk assessors should provide good explanatory narrative in support of risk estimates generated from the risk assessment process to assist in compiling a list of options.

The primary objective of microbiological risk management option assessment is optimization of the interventions necessary to prevent and control microbiological risks. It is aimed at selecting the option or options that achieve the chosen level of public health protection for the microbiological hazard in the commodity of concern in as cost effective manner as possible within the technical feasibility of the industry. There are no hard and fast rules about how managers select options but there are a number of possibilities based on the food safety issue at hand, and the choice of a risk management approach.

The interaction between managers and assessors depends on the scope of the risk assessment. Often the risk assessment is designed to identify the stage in the food chain where interventions will most effectively reduce the public health burden attributable to the specific food and pathogen in question. A risk assessment may also be initiated to examine the cost effectiveness of current controls or to evaluate a new technology for control. In this case a list of options for consideration will be included in the scope. In an emergency situation with an emerging pathogen where the etiology of disease is not well understood the options comparison will be abbreviated.

Consideration of legitimate factors other than public health e.g. economic and social concerns will also enter the process of risk management. In most cases, the risk management process used to arrive at a decision on an appropriate level of protection and appropriate sanitary measures will be broad based, and relevant other legitimate factors will ideally have been identified before that part of risk management begins.

The degree of interaction will depend on a number of factors including:

- The scope and reason for the risk assessment.
- The judgement of an appropriate level of protection.
- Consideration of "other legitimate factors."
- Final management decision.

In most cases risk management decisions take the form of a set of sanitary measures that collectively make up the food control system. If an FSO is established, the food control system should include microbiological performance and/or process criteria that are validated as achieving, or contributing to the achievement, of the FSO.

IMPLEMENTATION OF MANAGEMENT DECISION

Interaction between risk managers and risk assessors is considered unlikely at this step.

Monitoring and Review

Assessment of effectiveness of measures taken

An essential part of risk management is gathering and analyzing data from a range of points in the food chain so as to ensure that food safety goals are being achieved on an ongoing basis. Some examples include the following:

- Prevalence of a pathogen in herds of animals or flocks of birds.
- Pathogen prevalence at the beginning and end of processing.
- Pathogen prevalence in a food commodity at retail.

Review Risk Management and/or Assessment as Necessary

Emergence of new problems, unsatisfactory epidemiological findings or high variability in compliance with required performance/process criteria might all indicate that the risk assessment process should be revisited. Risk managers may request the re-evaluation of specific inputs to the risk assessment model, or may commission a new risk assessment. Both of these requests will involve interaction between risk managers and risk assessors in formulating the statement of the problem.

Recommendations

- Food Safety Authorities in each Country should structure their food safety system(s) on a risk-based approach that includes appropriate communication and interaction between risk assessors, risk managers, and stakeholders.

- FAO and WHO should actively seek opportunities to promote collaborative international risk assessment and risk management activities among Member Countries.
- FAO and WHO should emphasize that communication has to occur frequently and iteratively while striving to ensure scientific integrity and achieve freedom from bias in risk assessments.
- National authorities should consider carefully the training needs of risk assessors and managers so that they are able to undertake the full range of their responsibilities efficiently and effectively.
- Regulation shall facilitate discussions of the nature and value of food safety objectives especially in the microbiological field.
- National governments should acknowledge the importance of functional separation between risk assessment and risk management while ensuring transparent and appropriate interaction between them.

CORRIDORS OF ANTIBACTERIALS

Modern Aquaculture Systems involving intensive farming have brought into use several antibacterial agents including pharmacologically active compounds of both natural and synthetic origin to prevent the outbreak of diseases. These drugs are mainly administered by two different routes water medication and infeed medication.

The administration of these pharmaco-dynamic agents also brings into picture the food safety aspect. As the human food safety is the foremost consideration in food animal production, international systems of legal controls have been installed to prevent residue contaminated food products from entering the human food supply chain. Safety aspect is concerned with residues, either of the drug or of its metabolites in any tissue which may be used for human food. Safety is achieved by adhering to a "withdrawal period" between the last use of the drug in the animal and the earliest time at which it may be caught and slaughtered for food. The withdrawal period is such that at the end of it the residue level of the drug or the metabolite is safe for the consumer. Thus it follows that the determination of withdrawal period requires the fixation of safe level of residue known as Maximum Residue Limit "MRL" for that drug or metabolite.

With this general background, an attempt made for a random walk through the corridor of antibacterial drugs of importance in aqua culture practices, focusing mainly on the general chemistry, uses and demerits of them. The antibacterials discussed are Tetracyclines, lactams (Penicillins), Macrolides, Sulfonamides, potentiated Sulfonamides, Quinolones and Fluoroquinolones, Nilrofurans, Chloramphenicols and Florfenicols.

Tetracyclines

Tetracyclines are broad spectrum bacteria static drugs. In this group, Oxytetracycline and Chlortetracycline are the most widely used compounds in aquaculture. The tetracyclines are yellow crystalline compounds. These are good chelating agents - forming complexes with divalent cations such as Mg++ and Ca++. However, these complexes are pharmacologically and microbiologically inert. The tetracyclines will also complex with organic matter especially proteins and with clay.

The tetracyclines are active against cold water vibriosis (in salmon), flavobacteriosis in carp, streptococcosis and strawberry disease in rainbow trout.

The widespread use of these compounds has led to the development of resistant bacterial strains. The complex formation with Ca++ and Mg++ make it an unsuitable antibiotic for fish culture in sea water. This effect is less in fresh water because of the lower concentration of cations.

β-Lactams

This group includes both penicillin and cephalosporins. All the β-lactams have on their basic unit a thiazolidine ring, and a β-lactam ring. They are active against Gram +ve bacteria. Two semi synthetic penicillins *viz*, ampicillin and amoxycillin are widely used in aquaculture. They do not form complexes with divalent ions. However, amoxycillin is liable to photodecomposition thus lowering its activity.

Amoxycillin & Ampicillin are useful against infections such as furunculosis (salmonids), Pasteurellosis, Edwardsiellosis and Streptococcosis.

Macrolides

Macrolides are antibiotics with a larger ring in their molecular structure. The mainly used macrolides in aqua culture are Erythromycin, Spiramycin and Josamycin. They are colourless crystals, poorly soluble in water. Being weak bases they are soluble both in acids and alkalies, but unstable at high and low pH. Microbiologically they are more active in slightly alkaline condition.

Macrolides are medium spectrum antibiotics active mainly against Gram +ve bacteria, *Chlamyda and Rickettsia.* Because most of the fish infections are Gram –ve, the use of macrolides are limited and specific. It is used in the prevention of transmission of bacterial kidney disease (BKD) in Salmonid species.

Sulfonamides

Sulfonamides are derivatives of Sulfanilamides. They are amphoteric but more soluble in alkali than in acids. They have a broad spectrum of activity. However they are active only at a higher does and hence this group has little use at present, except in synergistic combination with pyrimidine potentiators.

Potentiated Sulfonamides

Potentiated sulfonamides are in fact combinations of two antibacterial drugs - a Sulfonamide and a Pyrimidine potentiator. The combination is 'synergistic'. This is because they act as a competitive inhibitor of two successive steps in the synthetic pathway of Folinic Acid in bacteria. The commonly used potentiators are Trimethoprim and Ormetroprim. Romet-30 is a combination of Trimethoprim and Sulfadimethoxine. It is active against a wide range of bacterial infections in fish.

Quinolones and Fluoroquinolones

The important quinolones are Nalidixic acid, Oxolonic acid. They are used against a number of Gram +ve organisms. Quinolones act by inhibiting the bacterial, enzyme-DNA-gyrase. Floumequines & Sarafloxacins are "second generation" 4-quinolones. Their spectrum of activity covers not only Gram –ve bacteria but also fungi, protozoa and even some helminthes. These fluoroquinolones are replacing oxolonic acid in fish medicine because of its more appropriate pharmaco kinetic profile. Further their higher activity brings down the dosage level.

Nitrofurans

They all are substitution products of the 5-nitrofuran nucleus differing in substituents at position 2, with broad spectrum activity covering not only Gram +ve and Gram –ve bacteria, but also many protozoan parasites. As the Nitrofurans are carcinogenic, their use as fish medicine is banned. These drugs have less than 1 hour half life period, i.e. their concentration gets reduced by half within this period. The parent drug becomes undetectable within 5-6 hours of ingestion. But protein bound metabolites are relatively stable and can be detected by modern techniques such as LC MS-MS as their nitrophenyl derivatives at levels of 1 ppb.

Chloramphenicol

Chloramphenicol is a broad spectrum antibiotic that was first isolated from Streptomycin Venezuelae, but is now produced synthetically. It is banned both in aquaculture and veterinary practices. The main reason is that it causes aplastic anaemia. As Chloramphenicol is a molecule of zero tolerance, the detection of this molecule at very very low levels can be effected only through molecular spectrometric methods such as LC-MSMS.

Florfenicol

Thiamphenicol & florfenicol differs from Chloramphenicol in having a methyl sulfonate moiety at the place of nitro group in the molecule structure. The methyl sulfonate moiety transforms the molecule pharmacologically to a benign compound incapable of causing aplastic anaemia. These compounds are acceptable for use in food producing animals Florfenicol is considered to be a good medicine against Furunculosis infections, Pasterurellosis and Edwardsiellosis.

Analytical Problem

The extremely low levels (ppb) at which an antibacterial residues should be analyzed in a complex biological matrix is a major analytical challenge. Procedures for selective extraction of analyte, removal of co-extractants and sensitive and specific detection are especially challenging. As most of the antibacterials are either oxidized, reduced or hydrolysed in the body to produce more toxic metabolites, specific procedures have to be developed to detect the metabolites, to evaluate the residual levels.

The various analytical methods for screening identifying and quantifying antibacterial residues fall into 3 broad types - microbiological, immunochemical and physico-chemical methods. Microbiological and Immunochemical procedures are designed for screening purposes while physico-chemical procedures are used primarily for the isolation, separation, quantification and confirmation of the violative residues in samples. This requires that the sensitivity of the screening method and the confirmatory method should be compatible and matching.

Conclusion

The need to use antibacterials in aquaculture will continue to increase unabated. Therefore monitoring of edible animal products for violative residues will ever remain an area of increasing concern because of the food safety element present in it. Analysis of antibacterials residues is a challenging task because of the very low level of occurrence of the analyte of interest in a complex food matrix.

Now most of the rapid screening methods are based on immuno chemical and receptor based assays. Large scale screening test requires methods that are rapid, sensitive and specific. However these assay shall be supported by methods capable of specific isolation and quantification that can provide rapid identification and a high degree of confirmation. These challenges are increasingly met by the molecular spectrometric identification provided by GC MS-MS and LC-MS-MS instruments. Once full mass spectral data including fragmentation patterns of various drugs are made available the residue analysis programme can successfully meet the ever increasing demands of food safety standards.

ANALOGUE/TEXTURISED FISH PRODUCTS

A meat analogue is a food product that approximates the aesthetic qualities and chemical characteristics of certain types of meat. Meat analogue may also refer to a meat-based, but healthier and/or less-expensive alternative to a particular meat product. Meat analog sales are estimated to have reached $600 million annually, and are growing at 20-25% yearly. The meat analog market is one of the fast-growing areas of the food industry.

Meat from low value fish can be used for the production of analogue products having the shape and flavour of shrimp, crab etc. The important analogue products in market are crab meat, scallop and shrimp. Among these imitation crab meat is the most popular in countries like Japan, Unites States and EEC countries.

The general flow chart for moulded products is given below :

Flow diagram for manufacture of moulded and texturised products

Fish	Conveyer	Shaping/Cutting
Surimi	Gas cook	Wrapper
Silent cutter	Steam cook	Steam cooker
Fish paste	Gas cook	Cooling drier
Extruder	Cooling	Packing

Imitation Crab Meat

Imitation crab meat is a seafood product made by blending processed fish, known as surimi, with various texturizing ingredients, flavorants, and colorants. First invented in the mid-1970s, imitation crab meat has become a popular food in the United States, with annual sales of over $250 million. Surimi is the primary ingredient used to create imitation crab meat.

Table 1 : A General Composition of Imitation Crab Meat (kg)

Alaska pollack surimi	100
Potato starch	4
Starch (waxy)	2
Egg white	10-15
Sodium glutamate	1
Glycine	1-1.5
Alanine	0.5
Crab extract	1-2
Crab flavour	1
"Purura"	0.5
Water	30

Three types of imitation crab meat are popular - flake type, bite type and stick type. The flow chart for the processing of imitation crab stick is given in chart below. The starting raw material is Alaska pollack surimi. In place of Alaska Pollack Surimi, Surimi from Indian fishes like Ribbon Fish, Dhoma or Threadfin bream can be substituted. The precut block of surimi is mixed with salt, water, flavour components, egg white and other additives in a silent cutter or ball cutter. A typical recipe for the processing of imitation crab meat is given in Table 1.

The mixed meat is then fed to a moulding machine by feed pump.

Flow Chart of Imitation Crab Stick Processing

- Frozen Surimi
- Pre-cutting of Surimi blocks

- Mixing of Surimi with crab meat additives & Suwari setting
- Moulding of mixed meat to thin sheets
- Steaming
- Baking or surface drying
- Cooling
- Longitudinal cutting into noodle like lines
- Gathering & bundling process
- Colouring & inner wrapping
- Cutting
- Vacuum packing
- Sterilizing (in steam or boiling water)
- Sudden cooling
- Dewatering
- Freezing
- Cartoning and storing

In the moulding machine, the material is passed through a moulding nozzle to produce thin and wide sheets of mixed meat. After the moulding process the sheet is passed through a steaming chamber for pre-heating. The steaming is adjusted to get proper moisture content yield and the required elasticity (gel strength). It is then passed through a gas baking process where the surface of the sheet gets dried to the required level. The cooked sheet is then cooled and passed through a thin cutting machine having comb like blades to obtain noodle like lines which are still attached to each other on the skin of baked sheet.

The slitted sheet is then rolled inward to form the shape of crab leg. It is then packed in inner wraps with a coating of red dye to imitate the colour of crab stick. The wrapped material is then uniformly cut in a cutting machine. The individually wrapped crab sticks are then vacuum packed. The vacuum packed products are boiled or steamed in a chamber for sterilizing the colouring material and bundled stuff. It is then immediately cooled and dewatered in dehydrating machine. The dehydrated packets are frozen and carton packed. The standard diameter of the imitation crab leg is 18 mm. Crab flakes are also processed in a similar way but the cutting is different to make flakes. The flakes usually have a dimension of 1 × 1.7 × 30-50 mm.

Future developments in the imitation crab meat industry are likely to be found in a few key areas. One important area of research has focused on the development of surimi from different kinds of fish. These would include fish that currently have low economic value and are quite abundant. Many of these new fish have more fat and different body chemistries than the fish currently used, so the challenge will be to improve the surimi that can be made using them. In the manufacturing area, a more continuous process is being developed. These processes result in better yields of surimi. Also, environmental concerns will lead to new technologies that will minimize the amount of waste involved in manufacture. Finally, new crab meat recipes aimed at improving the nutritional value of the product will be developed.

Imitation Scallop

This is also very popular in Japan and United States. The Processing method is similar as that of imitation crab meat except the addition of 2-3% scallop extract instead of crab leg extract. The final shaping and cutting process is done to imitate the appearance of whole scallop meat. For that slits are made at 2 mm intervals by comb like blades as mentioned earlier. The sheet is then rolled inward from both the sides. The cylindrical meat thus formed is cut 500 mm long and placed in a cylindrical mould to steam al 85-90°C for 30'. It is then cooled and cut into 15 mm thick pieces. The standard diameter is 39-40 mm. These pieces are marketed either in breaded or nonbreaded form.

Imitation Shrimp Tails

Imitation Shrimp Tails are moulded out of Surimi (white pollack meats), mixed with different ingredients (shrimp extracts, shrimp flavour, wheat starch, seasoning powder etc.) in shrimp shape. It's a quality product.

Table 2 : Types of Spoilage Noticed in Packed Imitations Products.

Spoilage of Analogue Products

Types of Spoilage	*Micro organisms responsible*
Swelling	*Bacillus and clostridium*
Softening	*B. subtilis & cereus*
Brown	*B. licheniformis & B. circulans*
Spot discolouration	*B. sphericus*

The shelf life of refrigerated imitation products is less than 10 days and that of frozen products is 3 months. The types of putrefaction noticed in imitation products are given in Table 2. The main organisms are heat resistant. Some strains are also resistant to sorbic acid.

PRACTICAL ASPECTS OF HACCP IMPLEMENTATION

Over a period of time there has been considerable change in the concept of food processing. The modern food processing industry is very much sophisticated and technologically advanced. Further, several ingredients are now added to foods as additives, antioxidants, preservatives, emulsifiers, cryoprotectants and colouring materials. There are also problems of pesticide residues, toxic metals, mycotoxins, biotoxins, antibiotic residues and the like. Under these circumstances, the responsibility of the processor has become increasingly complex and hence, there is a global shift from food quality to food safety. This has resulted in the development of a safety-oriented quality system, the Hazard Analysis Critical Control Point (HACCP) System. HACCP is a globally recognized systematic and preventive approach that addresses physical, chemical and biological hazards through anticipation and prevention rather than end product inspection and testing.

Advantages of HACCP System

The HACCP food safety management system uses the approach of controlling critical points in food handling to prevent food safety problems. The system which is science based identifies specific hazards and takes measures for their control. HACCP is based on prevention and reduces reliance on end-product inspection and testing. The system can be applied in the food industry from 'farm to fork'. It enhances responsibility and degree of control at the level of food industry. Implementing HACCP does not mean undoing the quality assurance procedures or Good Manufacturing Practices already established by the company; it only requires a revision and appropriate integration with the HACCP system. Further, the HACCP system can aid inspection by the regulatory authorities and promote international trade by increasing buyers' confidence. The HACCP system should be capable of accommodating advances in equipment design, process and technology.

Application of HACCP

To apply HACCP system in any food industry sector, that sector should be operating according to the Codex General Principles of Food Hygiene and appropriate food safety legislation. Management Commitment is essential for effective implementation of HACCP system. The HACCP system works on seven principles and there are 12 steps. The logic sequence for the application of HACCP system is shown in Fig. 1.

An Important Point

HACCP system can be implemented only through the above cited 12 steps and 7 principles. There are no short-cut methods. Some processors use to enquire whether the number of steps can be reduced as they are small processors. It is made very clear in this context that all the twelve steps and seven principles are essential for proper implementation of the HACCP system irrespective of the size of the factory. Omission of any step will be a major NC (Non-Conformity.)

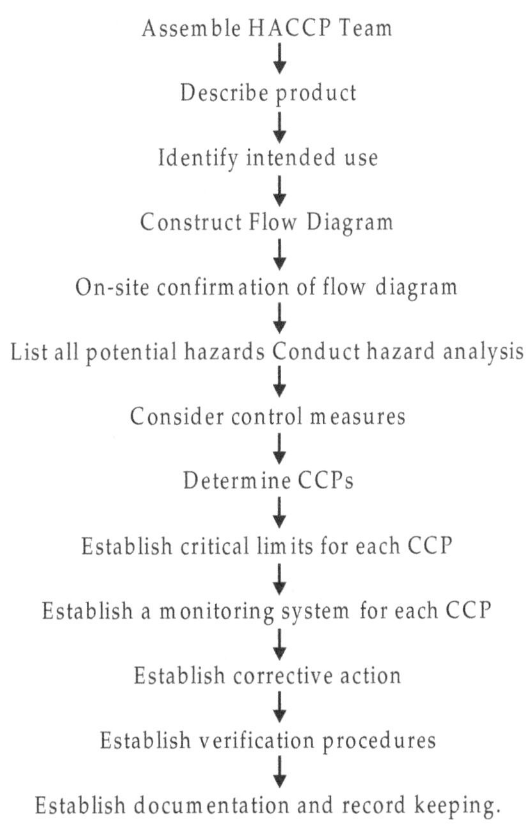

Fig. 1 : Logic sequence for application of HACCP.

Pre-requisite Programme

HACCP system is not a stand-alone programme. There are certain pre-requisite programmes that are very much essential to the successful application and effective implementation of the HACCP system. In other words, these pre-requisite programmes provide a very strong foundation for implementation of HACCP system effectively. Most of these pre-requisite programme are addressed by the Standard Sanitation Operational Procedures (SSOP) and Current Good Manufacturing

Practices (CGMPs). All food Processors are required to keep a written SSOP focusing on eight areas of sanitation such as :

1. Safety of Water/Ice.
2. Cleanliness of contact surfaces.
3. Prevention of cross contamination.
4. Employee facilities.
5. Prevention of adulteration.
6. Proper labeling storage of toxic substances.
7. Employee health.
8. Pest Management.

As there are a number of earlier publications on the various steps of HACCP implementation, they are not repeated in this paper. However, a brief account is given on various hazards that are possible in food products. Any agent which creates injury or illness in human beings is considered as a hazard. A good processor always sees that no hazard goes to the consumer from the food. If it goes, it will result in a health hazard. Hazards are classified as Physical, Chemical and Biological as given in table 3.

Table 3 : Physical, Chemical and Biological hazards in food products

	Physical Hazards	
Glass	Insulation	
Wood	Bone	
Stones	Plastic	
Metal	Personal effects	
	Chemical Hazards	
Naturally occurring Chemicals	Added chemicals	From packaging materials
Allergens	Pesticides	Vinyl chloride
Aflatoxins	Fertilizers	Printing inks
Mushroom toxins	Antibiotics	Adhesives
PSP	Growth hormones	Lead
DSP	Lead	Tin
ASP	Zinc	
Ciguatoxin	Cadmium	
	Arsenic	
	Food Additives	
	Contaminants such as:	
	Lubricants, detergents, sanitizers, paints and refrigerants.	

Biological Hazards

Bacteria (Spore forming)	*Bacteria (Non-spore forming)*	*Virus, protozoa and parasites*
Clostridium botulinum	Salmonella	Hepatitis A and E
Clostridium perfringens	Shigella	Rotavirus
Bacillus cereus	*S. aureus*	*Taenia solium*
	V. parahaemolyticus	*Taenia saginata*
	V. vulnificus	*Entamoeba histolytica*
	V. cholerae	
	L. monocytogenes	

The Food Safety Bill

The Food Safety Bill has already been placed before the Parliament for consideration. When this bill is passed, HACCP will become mandatory for all food products and the existing rules and guidelines connected with Prevention of Food Adulteration Act will be modified incorporating HACCP-based Food Safety Management System.

Food Processing Factories approved under HACCP System

As far as marine products are concerned, the system is mandatory from 1996. Those who were exporting to European Union implemented HACCP system and there are a good number of factories working under the HACCP system. As the system is going to be compulsory, the other food factories are implementing the system actively. The following is the list of food processing factories certified under the HACCP system.

Factories certified under the HACCP System*

Sea foods (Exported to EU)	152
Cold Storages	17
Curry Powder	2
Pickles	3
Ready-to-eat food items	5
Ice cream	1
Dehydrated onion	6
Air catering	4
Spices	4
Star hotels	2
Tea	1
Milk	4
Fruit Juices	
Wheat flour	1
Bakery	1
Gelatin	1

**Not a complete list*

Factories getting ready for certification	
Curry Powder	1
Tea	6
Ice cream	1
Dehydrated onion	12
Star hotels	3
Spices	2
Milk	2
Pickles	1
Coconut shell	1
Stone less rice	1
Jam and fruit juice	1
Ossein	**1**

Practical Experience in Implementing HACCP System

1. There are minimum problems in implementing the system in new factories. But, the case is different in already existing factories as the layouts in such factories are different from that envisaged in Codex General Principles of Food Hygiene. Backtracking, too much of external side opening doors, in appropriate roofing, too much of glass-paneled windows etc. are major constraints.
2. Lack of education, awareness and training seem to be the problem in certain areas of the country.
3. Lack of consumer awareness about their rights and needs seem to be blocking the rapid penetration of the safety system.
4. In certain food industries certified under the HACCP system, monitoring of the Critical Control Point is not done on a regular basis.
5. Verification of the system and validation of the HACCP system are rarely done in factories functioning under the HACCP system.
6. HACCP system insists that an internal audit is to be performed by the company once in six months and an external audit by an outside agency once in a year. This is seldom done in many food industries. However, it may be noted that some factories in the country give adequate attention to the audit aspect.
7. Scientific information regarding shelf-life, critical limits for heavy metals, pesticides bacterial load etc. are lacking for certain specific food items in the country.
8. There is no reliable recorded information in the country on food-related hazards and poisoning.
9. For many food items, HACCP is not applied from single "Farm-to-Fork".

HACCP Implementation Tips

- ❑ Use disciplined approach
- ❑ Don't make assumptions
- ❑ Challenge beliefs
- ❑ Discuss non-hierarchal
- ❑ Don't rush
- ❑ Set deadline for comments

- ❑ Keep accurate records
- ❑ Team Leader should moderate, not dominate.

Requirements for a Successful HACCP System

- ❑ Management Commitment
- ❑ Plant design as per GMP
- ❑ Insect and Pest Control
- ❑ Hygiene and Sanitation
- ❑ Trained personnel

HACCP is not a sophisticated system requiring high technology and highly educated staff. Simple tests like sensory methods, time temperature evaluation, pH determination etc. are employed in this system. The system can be run by Technical staff of the industry after they get trained adequately under an expert consultant.

REFERENCES

1. Applied Fish Pharmacology, K.M Treves-Brown Kluwer Academic Publishers (2000).
2. Determination of Veterinary Residues in Food, I.D. Morton, Ed, Ellis Horwood, England (1991).
3. Handbook of Food Analysis Vol. 2, Residues and other food component analysis, Ed: Leo ML. Nollet, Marcell Dekker Inc New York (1996).

Chapter 21

Nanotechnology in Food Design and Packaging

Nanotechnology is a field of research and innovation on building things, generally materials and devices on the scale of atoms and molecules. A nanometer is one billionth of a metre, ten times, the diameter of a hydrogen atom. The diameter of a human hair is on an average 80,000 nanometres. At such scales the ordinary rules of physics and chemistry no longer apply. For instance, materials characteristics such as their colour, strength, conductivity and reactivity can differ substantially between the nanoscale and the macro. The physical and biological structures so created have properties unique to them and are not found in the original matter or material from which the nanoparticles have been formed and the uniqueness of the particles can be attributed to their nanosize. Further nanomachines can be made using the nanostructures in order to have functions which are otherwise not possible. In other words nanotechnology is the manufacture of structures, materials, device and machines using nanoparticles with programmed precision.

What can Nanotechnology do?

Nanotechnology is hailed as having the potential to increase the efficiency of energy consumption, to help clean the environment, and solve major health problems. It is said to massively increase manufacturing production at significantly reduced costs. Products of nanotechnology will be smaller, cheaper, lighter yet more functional and require less energy and fewer raw materials to manufacture. Nanotechnology has opened a path to an unexplored science for studying individual nanoparticles and the unique way in which they interact. Nanotechnology can make products cost effective (production will be carried out by self replicating nanodevices using small amount of material, energy, low capital, less labour and land) production being more efficient decreased use of water and minimum wastages.

Applications

In addition to a handful of nanofood products that are already on the marked over 135 applications of nanotechnology in fisheries and sea food industries (primarily nutrition and cosmetics) are in various stages of development. According to Helmut Kaiser more than 200 companies including Australia's leading food corporation and Japan's largest sea food producer and processed food manufacturer are in Research & Development work on nanotechnology.

Mode of Applications

Nanotechnology has three broad areas through which it can be applied :

1. Nanotools instrumentation and equipments that could help to understand nanoword.
2. Nanomaterials that will have desired physical chemical and biological attributes.

3. Nanodevices, nanosensors and actuators that will help to manipulate and control nano particles.

Nanosensors

Nanosensors can work through a variety of methods. One variation being developed by researchers uses nanoparticles which can either be tailor made to fluoresce different colours or alternatively be manufactured out of nanoparticles. These nanoparticles can then selectively attach themselves to any of the food pathogens. The advantage of such a system is that literally hundreds and potentially thousands of nanoparticles can be placed on a single nanosensor to rapidly, accurately and affordably detect the presence of any number of different bacteria and pathogens. A second advantage of nanosensors is that given their small size, they can gain access into the tiny cervices, where to reduce the time it takes to detect the presence of microbial pathogens from two to seven days down to a few hours and ultimately minutes or even seconds. Much of the food safety problems arise due to the contamination of food processing equipment with microorganisms. Earlier it was difficult to quantify such contamination, but now-a-days it can be easily quantified with the aid of nanotools such as Atomic Force Microscope (Moraru *et. al.*, 2003).

Food Design Using Nanotechnology

In food acceptability, various factors come into play such as flavour, aroma, taste, texture, colour, appearance, nutrient content etc., Food has to be designed in a cost effective way that strikes right balance between these characteristics. Many of the molecular structures that determine these characteristics are in the nanometer range and information on the source can play an important role in the food design. Food is composed of carbohydrates, fats, proteins, minerals, vitamins and water. When food is subjected to pressure heat, temperature change and storage, the molecular structure of different moieties changes, thereby affecting the characteristics of food. The purpose is to master over the characteristics of food components in an intelligent and place them exactly where they are needed to produce the desired flavour, texture etc.

Nanotechnology in Food Packaging

The consumer demands food, to remain fresher for long time, ease in handling, safe and healthy with environmental friendly packaging.

Some of the essential characteristics desired out of a packaging material for IQF shrimp are:

1. Low water vapour transmission rate to reduce the risk of dehydration.
2. Low oxygen / gas permeability, thereby reducing the risk of oxidation and thus changes in odour and flavour, and retention of volatile flavours.
3. Flexibility to fix the contours of the food.
4. Resistance to puncture, brittleness and deterioration at low temperature and ease of filling.

Today food-packaging and monitoring are a major focus of food industry - related nanotech R&D. Packaging that incorporates nano material can be "Smart", which means that it can respond to environmental conditions or repair itself or alert a consumer to contamination and or the presence of pathogens. Through nanotechnology we can increase or decrease gas transmission rate to suit the packaging requirements of the product. Properties such as mechanical and heat resistance can be increased. Packaging materials that have improved temperature performance can be used for hot fill operations. Very thin films can be made that will offer the advantage of flexibility and functionalities like being anti-counterfett, antitamper, antimicrobial (temperature, moisture, light, decay) self heating feature can also be incorporated in the packaging material. Environment friendly, light weight packaging materials can be made for use in army rations. In future, with the aid of nanocomposites we maybe able to modify plastic into a super barrier just as glass or metal (Brody 2003 Table 1.)

Properties of packaging materials in use

Polymer	*02 transmission rate (cc/0.001 in/100 in^2/24 hrs*	*Water vapour trans-rate (g/24hr/100 In^2)*	*Use*
Nylon	2.6	24–26	Films
LDPE	250–840	1.2	Low gas barrier, High moisture barrier
EVA *(Ethylene vinyl acetate)-LDPE*	515–645	3.9	Heat Shrinkable
PVDC *(Poly Vinylidine Chloride)*	0.08–1.7	0.05–0.3	Barrier layer

Hotchkiss, I.H. (1994).

Using Clay Nanoparticles to Improve Plastic Packaging for Food Products

Chemical gaint Bayer produces a transparent plastic film (called Durethan) containing nanoparticles of clay. The nanoparticles are dispersed throughout the plastic and are able to block oxygen, carbon dioxide and moisture form reaching fresh meats or other foods. The nano clay also makes the plastic lighter, stronger and more heat resistant. By embedding nanocrystals in plastic, researchers have created a molecular barrier that helps prevent the escape of oxygen.

Using Nanotechnology Methods to Develop 'Active Packaging' and 'Anti Microbial Packaging'

Using nanotech to develop anti microbial packaging for food products that will be commercially available in 2005. 'Active Packaging' which absorbs oxygen, thereby keeping food fresh. 'Active packaging maintains, quality and extends shelf-life of product packed in it. Active packaging can interact with food to reduce O_2 levels or add flavouring/preservatives. Intelligent packaging can also change colour to help the consumer know how fresh the food is, and show if the food has been spoiled because of change in temperature or leak in the package. Some types of active packaging contain oxygen released from the food and thus inhibit rancidity development and maintain flavour. Antimicrobial packaging materials are very useful but very few such materials are at the market today. With the advancement and growth in nanoscience we can expect almost all packaging materials with antimicrobial agents incorporated in them.

Embedded Sensors in Food Packaging and 'Electronic-Tongue' Technology

Scientists are working on nano-particle films and other packaging with embedded sensors that will detect food pathogens called 'electronic tongue' technology. The sensors can detect substances in parts per trillion and would trigger a colour change in the packaging to alert the consumer if a food has become contaminated or if it has begun to spoil.

Using a Nanotech Bioswitch in 'Release on Command' Food Packaging

Further to develop intelligent packaging that will release a preservative if the food begins to spoil. This release on command preservatives packaging is operated by means of a bioswitch developed through nanotechnology. With present technologies, testing for microbial food-contamination takes two to seven days and the sensors that have been developed to date are too big to be transported easily. Several groups of researcher in the US are developing bio sensors that can detect pathogens quickly and easily, reasoning that "super sensors" would play a crucial role in the event of a terrorist attack on the food supply. Researchers are working to produce a hold-held sensor capable of detecting a specific bacteria instantaneously from any sample.

Problems in Industrial Food production that sensors and 'smart packaging' will not Address

While devices capable of detecting food-borne pathogens could be useful in monitoring the food supply, sensors and smart packaging will not address the root problems inherent in industrial food production that result in contaminated foods: faster meat assembly lines, increased mechanization, a shrinking labour force of low wage workers, fewer inspectors, the lack of corporate and government accountability and the great distance between food producers, processors and consumers.

Utilizing Solar Energy as Fuel in Sea Food Industry via Nanotechnology

At present only 13.8% of world's energy supply comes from renewable sources. However, in the coming years, nanostructured materials and nanopower isolation materials etc., will help to cut energy costs through solar cells, hydrogen and fuel cells, batteries, improved light bulbs, fossil fuels etc. Today, most sea food industries rely on carbon containing fuels for energy sources. These are generally costly, frequently non-renewable and release CO_2 as well as other waste products in to the atmosphere. If carbon containing fuels replaced with solar energy, the cost of food production will reduce and provide a cleaner environment. At present solar electricity generation depends on either photo-voltaic conversion (or) concentrating direct sunlight. With molecular manufacturing it would be possible to store energy efficient for several days in smaller energy storage systems. Nanosteller, a US-based company is developing highly efficient platinum nanocomposite catalyst and reduce the cost of automobile catalytic converter (Mettoth, 2004).

Conclusion

Nanotechnology through its multifarious applications in various fields, especially in seafood industry has proven to be a "miniscule miracle". Nanotechnology has paved a way for poverty alleviation and economic development. Nanotechnology needs very high investment, but once it gets installed, it would certainly create unforeseen, benefits which can be produced cost effectively. Though nanotechnology has unknown risks and hazards, it has the potential to shape the future, if it is judiciously used.

INDIA'S SEAFOOD EXPORT CROSS RS. 7,000/- CR. (US $ 1644 MILLION) ACHIEVES 11.21% GROWTH IN $ TERMS

India's Seafood Exports reached an all time high in 2001-06. It was for the first time that India's seafood exports crossed 1.5 billion US$ and the figure of Rs. 7,000 crore mark. The actual export made during the year 2005-2006 compared with 2004-05 is given below :

Exports During 2005-06 Compared to 2004-05

Export Details	*2005-06*	*2004-05*	*Growth %*
Quantity Tonnes	512164	461329	11.02
Value Hs.Crore	7245.30	6646.69	9.01
$ Million	1644.21	1478.48	11.21
Unit. Value $/Kg.	3.21	3.20	

After a long gap, exports achieved positive growth in two consecutive years. Frozen shrimp continues to be India's major item of export accounting for (59%), of the export value followed by Frozen fish (14%), cuttlefish (8%) Squid (8%), dried item (2%), live/chilled (2%) and other items (8%). During the year all major items have shown an increase in exports compared to the previous year. The European Union, as in the previous year, continued to be the major market for Indian seafoods with a percentage share of 29% followed by USA with 23%, Japan (16%), China (12%), South East Asia (8%), Middle East (4%) and other countries (8%) in terms of value. Among the major ports from where seafood were exported. Chennai retained its first position by exporting (19%) followed by Kochi (17%) JNP (16%), Vizag (15%), Pipanav (11%), Tuticorin (8%), Kolkata (7%) and others (6%).

While there was significant increase in India's exports to European Union, China and Middle East, there was only a marginal increase in the exports to USA. However, our seafood exports to Japan and South East Asia showed a declining trend.

However, it was gratifying to note that in spite of the Tsunami and its impact on fisheries sector in India and the adverse market situation prevailing in various overseas markets due to tariff/non tariff trade barriers and due to the unprecedented floods in east/west coasts. Indian Seafood exports could grow at a rate of 11% in value terms.

This significant achievements in exports especially with regard to the exports to the EU is due to the stringent quality up gradation steps undertaken by the industry with the active support by MPEDA. With a record number of 157 processing units and 19 Frozen Cold Storages approved for processing and exporting to European Union, India has been generally successful in crossing the stiff technical barriers to trade like stringent maximum residue limit for antibiotic, heavy metals. The MPEDA with the strong support of MoCI has strengthened the country's capabilities for testing antibiotic; residue installed HPLC MS equipment in 11 major centers in maritime states. Further the HACCP cell formed in MPEDA gives an adequate training to the seafood processors for the implementation of HACCP system in Seafood processing plants. A National Residue Control Programme is implemented by MPEDA to comply with the requirements of the European Commission and accordingly samples shrimp, fresh water fish, water and feed are collected from farmers, processing plants, feed mills all over India and analysed at MPEDA lab set up in Kochi, Nellore and Bhimavaram for Antibiotic, pesticide residue, heavy metals, stillborns, steroids, anthelmentics, dyes etc. MPEDA also extends financial assistance for setting lip mini laboratory in seafood processing plants, for construction of captive/independent pre-processing centers and to acquire quick testing kits.

Also considering the importance of hygienic pre-processing in the seafood industry, the MPEDA has given emphasis on their upgradation to bring the facilities on par with international standards. Accordingly, MPEDA extends financial assistance upto 50% of the actual expenditure.

As on date, a total area of 1,40,682 ha is under Shrimp Culture and 43,433 ha. Are under Scampi culture in coastal areas. India produced Shrimp and fresh water prawns during the year 2005-06 are as follows:

Shrimp	:	1,43,170 MT
Fresh Water Prawn	:	42,820 MT

While states like Andra Pradesh, Tamil Nadu and West Bengal are in the fore front of aquaculture activities, potential states like Gujarat, Maharashtra and Orissa are yet to be tapped. Hence MPEDA is promoting a Mission Mode programme in these states along with the State Governments to increase production of shrimps targeting both shrimp and fresh water prawns. This ensure expansion of farming in about 100,000 ha. Of land by the year 2009-2010. MPEDA also organizes field level campaign in remote farming villages against the use of prohibited antibiotics and pharmacologically active substances in coastal aquaculture. Further MPEDA -NACA (Network of Aquaculture Centre of Asian-Pacific, Bangkok) technical assistance programmes followed by field level demonstrations in the state like Andhra Pradesh have contributed significantly in shrimp health management and aquaculture production in the country. Increased quality conciousness, reduced production cost and increased higher bargaining power to the small-scale farmers is the main feature of this programme. To make shrimp farming more participatory and sustainable. MPEDA has launched a scheme to organize groups farming into the societies and to provide financial assistance for various group activities.

Exploration of Tuna resources is one of the targeted activities of MPEDA. For promoting monofilament long line technology in Tuna fishing MPEDA operates a scheme for subsidizing the cost of conversion of existing fishing vessels to Tuna long liners and for the acquisition of monofilament long line system for which a subsidy assistance up to a maximum of 7.5 lakh or 50% of the cost (whichever is less) is available. The assistance in Deep Sea fishing vessels by inviting overseas experts. Upgradation of fishing Hector is another major task being undertaken by MPEDA. It helps maritime State Governments to avail assistance from ASIDE fund and Munambam Fisheries Harbour in Kerala is one of the projects undertaken by MPEDA and the Govt. of Kerala in order to make it as a model fishing harbour in the country.

In order to develop new aquaculture technologies for diversification of aquaculture and to develop Biotechnology related techniques for mitigating crucial disease challenges, the MPEDA has setup the Rajiv Gandhi Center for Aquaculture (RGCA) with Headquarters at Thirumullaivasal in Tamilnadu. The center has taken lead in developing technology for aquaculture of modern fin fishes like Asian Sea bass and Grouper. RGCA has also taken initiative to develop culture of live groupers in open sea cages in Andaman Islands. Another major achievements of the RGCA that for the first time in the country, they have steam lined production of mud crab farming which in turn will lead to increase in the live mud crab exports. Recently RGCA has already launched a project at Andaman Islands to produce Specific Pathogen Free (SPF) Shrimp Broodstock through domestication of Tiger Shrimp MPEDA'S ANOTHER 2 Societies, *viz.* TASPARC and OSSPARC now produces high quality shrimp seeds through pro-biotic mode in a most bio-secure production technology. It is expected that this innovative approach will be a guiding force or other shrimp hatcheries in the country to produce, antibiotic free seeds for farming sector.

In order to give further boost for aquaculture activities, MPEDA holds the 3rd INDEQUA exhibition in Chennai Trade Center Complex at Chennai during 11-13 January 2007. This event highlight India's enormous aquaculture potential and its present status and future prospects. A major feature of this event an exhibition, which was highlight the latest development in the sector. Also technical sessions on current and relevant topics related to aquaculture activities been arranged and world-renowned speakers on these subjects had were participated.

MPEDA's vision is to make India one of the top seafood exporting countries in the world in terms of value and quantity by the year 2015 and to achieve an export turn over of 6 billion by the year 2015. In order to achieve this objective, MPEDA have identified 3 specific areas such as Tuna exploitation, diversification of aquaculture species and promotion of value added products export as the focus areas. A detailed road-map for achieving this goal is being prepared.

The focus areas on which MPEDA will be concentrating in next few years will be to tap the unexploited Tuna fishery resources, to expand and diversify its aquaculture species and above all to enhance the share of value added products in the seafood export basket from India.

During the year 2006-07 the MPEDA participated in 8 International Seafood Fairs and will highlight India's immense export potential, processing facilities, infrastructural developments and capability to produce ready to eat and ready to cook products etc. MPEDA also launched a logo scheme by which Indian products will get an identity of its own which in turn will pave for fetching a higher unit price realization.

EXPORT PERFORMANCE OF MARINE PRODUCTS DURING 2005–06

The Overall export of marine products has reached an ever time record of 1.6 billion US$ during the year 2005–06. The total exports aggregated to 512164 M. Tons valued at Rs. 7245.30 crore and US$ 1644.21 million against 461329 M. Tons valued at Rs. 6646.69 crore and US$ 1478.48 million during the same period of last year. The export had shown an increase of 11.02% in quantity, 9.01% in rupee value and 11.21% in US$ realization. The unit value also increased from 3.20 US$ to 3.21 US$ per Kg. The details of exports during the last five years are given below :

Export Growth of Marine Products

Q : Quantity in M. Tons V : Value Rs. Crore $: US $ Million

Year		*Export*	*Growth*	*(%)*	*Unit value*
2001–02	Q :	424470	–16003	–3.63	
	V :	5957.05	–486.84	–7.56	Rs. 140.34
	$:	1253.35	–162.97	–11.51	$: 2.95
2002–03	Q :	467297	42827	10.09	
	V :	6881.31	924.26	15.52	Rs. 147.26
	$:	1424.90	171.55	13.69	$: 3.05
2003–04	Q :	412017	–55280	–11.83	
	V :	6091.95	–789.36	–11.47	Rs. 147.86
	$:	1330.76	–94.14	–6.61	$: 3.23
2004–05	Q :	461329	49312	11.97	
	V :	6646.69	554.74	9.11	Rs. 144.08
	$:	1478.48	147.71	11.10	$: 3.20
2005–06	Q :	512164	50835	11.02	
	V :	7245.30	598.61	9.05	141.46
	$:	1644.21	165.74	11.21	$3.21

Item – Wise Export

Frozen Shrimp continued to be the largest item exported in terms of value with 59.02% and Fr. Fish continued to be the major item exported in terms of quantity with 35.60%. The export of Fr. Shrimp was increased to the tune of 5.14%, 1.20% and 3.41% in terms of quantity, value and US$ realisation. However, the unit value has shown a short fall of 1.64% in terms of US$. USA continued to be the major market for Indian frozen shrimp with 31.90% followed by European union with 27.98%, Japan with 20.25%, South East with 4.64% etc. There was a short fall in the export of frozen shrimp to Japan by 7.39% as compared to the previous year. Export of frozen shrimp to European union and USA was increased by 12.18% and 0.32% respectively. During the year 2005-06 Fr. Shrimp was exported to 62 countries.

The export of frozen fish also increased by 14.19%, 31.54%, 33.93% and 17.29% in terms of quantity, rupee value, US$ and unit value realisation. China is the main market for our frozen fish. 57.45% accounted to China, 13.65% to South East Asia, 7.31% to Middle East etc., Fr. Fish was exported to 63 countries during the year 2005–06.

Item Wise Export of Marine Products

Q : Quantity in M.T., V : Value in Rs. crore, US$ in Million, UV$: Unit value in US$/Kg

Item	*% Share to total*		*April–March 2005–06*	*April–March 2004–05*	*Variation*	*(%)*
Frozen Shrimp	28	Q :	145180	138085	7095	5.14
	58.96	V :	4272	4221	50.84	1.20
	59.02	$:	970	938	32.01	3.41
		UV$:	7	7	0	– 1.65
Frozen Fish	36	Q :	182344	159689	22654	14.19
	13.78	V :	999	759	239.44	31.54
	13.74	$:	226	169	57.24	33.93
		UV$:	1	1	0	17.29
Fr Cuttle Fish	10	Q :	49651	44239	5412	12.23
	7.58	V :	549	474	75.13	15.85
	7.57	$:	124	105	19.59	18.67
		UV$:	3	2	0	5.74
Fr Squid	10	Q :	52352	48124	4228	8.79
	7.94	V :	576	477	98.26	20.59
	7.94	$:	130	107	23.86	22.38
		UV$:	2	2	0	12.50
Dried Item	3	Q :	14167	9692	4476	46.18
	1.83	V :	133	121	11.55	9.54
	1.83	$:	30	27	2.94	10.86
		UV$:	2	3	– 1	– 24.16
Live Items	1	Q :	2568	2262	306	13.53
	0.85	V :	62	51	10.96	21.61
	0.85	$:	14	11	2.68	23.71
		UV$:	5	5	0	8.97
Chilled Items	1	Q :	5060	3988	1072	26.88
	1.13	V :	82	68	13.42	19.70
	1.12	$:	18	15	3.25	21.41
		UV$:	4	4	0	– 4.32
Others	12	Q :	60841	55250	5592	10.12
	7.93	V :	575	476	99.00	20.82
	7.93	$:	130	106	24.16	22.73
		UV$:	2	2	0	11.45
Total	100	Q :	512164	461329	50835	11.02
	100	V :	7246	6647	598.61	9.01
	100	$:	1644	1478	165.74	11.21
		UV$:	3.21	3.20	0.01	***

The export of Frozen Cuttlefish also showed a positive growth of 12.23% in quantity, 15.85% in value and 18.67% in US$ realisation. Our main market for frozen cuttlefish was European Union with 73.50%, China with 14.45% etc. Exports were made to a total 40 countries during the year 2005-06.

Frozen squid export also recorded an increase of 8.79%, 20.59%, 22.38% and 12.50% in terms of quantity, value, US$, and unit value realisation respectively. European Union continued to be the main market for Indian frozen Squid export with 69.15% followed by USA with 16.45%, Japan with 5.21% etc.

Export of Dried items recorded an increase of 46.18% in volume 9.54% in value and 10.86% in US$ realisation. Dried items were exported to 33 countries and the major players were Hong Kong (.38.98%), China (17.65%), Sri Lanka (14.35%), Singapore (8.56%). Dried items were exported to 36 countries during this year.

Live items exported to 27 countries. It recorded a marginal increase of 13.53%, 21.61%, 23.71% in quantity, value and US$ realisation respectively. Singapore was the main market for Indian live fish (52.68%) followed by Hong Kong (26.52%), Thailand (14.37%) etc.

The export of Chilled fish also increased marginal by 26.88% in volume, 19.70% in value and 21.41% in US$ realisation. Chilled fish were exported to 30 countries during the year 2005-06. Major market for chilled fish were Singapore (23.26%), UAE (22.72%), Thailand (15.22%.) etc.

Other items exported mainly to Japan (41.56%), South East Asia (17.73%), USA (16.17%) etc.

Country–Wise Export of Marine Products

Q : Quantity in M T, V : Value in Rs. Crore, $: US Dollar Million, UV$: Unit value in US$/Kg.

Country	*% Share to total*		*April–March 2005–06*	*April–March 2004–05*	*Variation*	*(%)*
Japan	12	Q :	59785	57832	1953	3.38
	15.96	V :	1156	1,202.46	- 46.49	- 3.87
	15.98	$:	263	266.96	- 4.16	- 1.56
USA	11	Q :	55817	50045	5772	11.53
	22.63	V :	1639	1,556.09	83.15	5.34
	22.66	$:	373	345.52	27.11	7.84
European Union	27	Q :	136842	117742	19100	16.22
	29.46	V :	2134	1,819.28	314.97	17.31
	29.44	$:	484	405.40	78.63	19.39
China	27	Q :	137076	124826	12250	9.81
	11.72	V :	849	693.25	156.20	22.53
	11.68	$:	192	154.10	37.89	24.59
South East Asia	12	Q :	60140	63842	- 3701	- 5.80
	8.09	V :	586	628.83	- 42.98	- 6.83
	8.07	$:	133	139.77	- 7.07	- 5.06
Middle East	4	Q :	22270	16624	5646	33.96
	4.25	V :	308	244.42	63.23	25.87
	4.24	$:	70	54.70	14.94	27.30

Contd...

Country	*% Share to total*		*April–March 2005–06*	*April–March 2004–05*	*Variation*	*(%)*
Others	8	**Q :**	40234	30418	9816	32.27
	7.91	**V :**	573	502.37	70.53	14.04
	7.93	**$:**	130	112.03	18.41	16.43
Total	**100**	**Q :**	**512164**	**461329**	**50835**	**11.02**
	100	**V :**	**7245**	**6,646.69**	**598.61**	**9.01**
	100	**$:**	**1644**	**1,478.48**	**165.74**	**11.21**

Major Markets

European union continued to be the largest market for Indian marine products during the year 2005-06 also. Its share was 26.72% in quantity, 29.46% in value, and 29.44% in US$ realisation. It has registered an export growth of 16.22% 17.31% and 19.39% in quantity, value and US$ realisation.

USA, the second largest market in terns of value had a share of 10.90%, 22.63%, 22.66% in quantity, value and US$ respectively. It recorded a growth of 11.53% in quantity 5.34% in value and 7.84% in US$ terms.

The share of Japan to our export basket was 11.67% in quantity, 15.96% in value, and 15.98% in US$. Export to Japan had shown a negative growth during the year 2005-06 by 3.87% in value and 1.56% in US$ realisation. However, exports in terms of quantity was increased by 3.38%.

Export to Middle East countries showed a tremendous growth by 33.96% in quantity, 25.87% in value and 27.30% in US$ terms. Export to China showed a marginal increase of 9.81%, 22.53% and 27.30% in quantity, value and US$.

Exports to Canada, Tunisia, Puertorico, Russia, Lithuania, Reunion, Fuji Island. Bangladesh etc. showed a positive growth, whereas export to Mexico, Cyprus, Australia, Maldives Islands etc. showed a negative trend.

Port-Wise Export of Marine Products

Q : Quantity in M T, V : Value in Rs. Crore, $: US Dollar Million UV$: Unit value in US$/Kg

Ports	*% Share to total*		*April–March 2005–06*	*April–March 2004–05*	*Variation*	*(%)*
Chennai	8.83	**Q :**	45246	42649	2597	6.09
	19	**V :**	1,382.56	1,432.87	- 50.31	- 3.51
	19	**$:**	315.13	318.56	- 3.43	- 1.08
Kochi	18.69	**Q :**	95737	86291	9446	10.95
	17	**V :**	1,218.97	1,135.70	83.27	7.33
	17	**$:**	277.06	252.44	24.62	9.75
J N P	23.53	**Q :**	120492	109430	11062	10.11
	16	**V :**	1,173.04	965.32	207.73	21.52
	16	**$:**	265.59	215.24	50.35	23.39
Vizag	7.25	**Q :**	37121	32028	5093	15.9
	15	**V :**	1,115.30	1,029.07	86.23	8.38
	15	**$:**	253.07	228.58	24.50	10.72

Contd...

Ports	% Share to total		April–March 2005–06	April–March 2004–05	Variation	(%)
Pipavav	22.47	**Q :**	115101	109597	5504	5.02
	11	**V :**	776.83	629.54	147.29	23.4
	11	**$:**	174.87	140.83	34.03	24.16
Tuticorin	5.31	**Q :**	27172	28160	– 988	– 3.51
	8	**V :**	613.17	635.19	– 22.02	– 3.47
	8	**$:**	139.27	141.01	– 1.74	– 1.24
Calcutta	3.57	**Q :**	18291	18492	– 201	– 1.09
	7	**V :**	537.95	521.13	16.82	3.23
	7	**$:**	122.18	115.66	6.52	5.64
Mundra	3.57	**Q :**	18304	2708	15596	576.02
	2	**V :**	128.67	16.16	112.51	696.23
	2	**$:**	29.27	3.56	25.71	722.61
Mangalore/ICD	3.12	**Q :**	15965	10349	5616	54.27
	1	**V :**	103.27	76.92	26.34	34.25
	1	**$:**	23.49	17.15	6.34	36.96
Mumbai	0.63	**Q :**	3224	2744	480	17.5
	1	**V :**	69.17	72.39	– 3.22	– 4.45
	1	**$:**	15.62	16.17	– 0.55	– 3.38
Goa	2.09	**Q :**	10719	10030	689	6.87
	1	**V :**	55.56	46.96	8.60	18.31
	1	**$:**	12.59	10.42	2.17	20.83
Trivandrum	0.29	**Q :**	1501	1040	461	44.32
	1	**V :**	37.58	21.70	15.87	73.14
	1	**$:**	8.47	4.84	3.63	75.15
Kandla	0.59	**Q :**	3045	6281	– 3236	– 51.52
	0	**V :**	28.32	51.44	– 23.11	– 44.94
	0	**$:**	6.49	11.29	– 4.79	– 42.46
Haldia	0.01	**Q :**	66	113	– 47	– 41.45
	0	**V :**	1.73	4.18	– 2.44	– 58.5
	0	**$:**	0.40	0.91	– 0.51	– 56.03
Calicut	0.01	**Q :**	73	47	27	56.76
	0	**V :**	1.10	0.67	0.44	65.33
	0	**$:**	0.25	0.15	0.10	65.74
Ahmedabad	0.01	**Q :**	35	0	35	***
	0	**V :**	1.06	0.00	1.06	***
	0	**$:**	0.24	0.00	0.24	***

Contd...

Ports	*% Share to total*		*April–March 2005–06*	*April–March 2004–05*	*Variation*	*(%)*
Paradeep	0.01	**Q :**	60	0	60	***
	0	**V :**	0.98	0.00	0.98	***
	0	**$:**	0.23	0.00	0.23	***
Agartala	0.00	**Q :**	11	0	11	***
	0	**V :**	0.04	0.00	0.04	***
	0	**$:**	0.01	0.00	0.01	***
NSICT	0.00	**Q :**	0	6	– 6	– 100
	0	**V :**	0.00	0.02	– 0.02	– 100
	0	**$:**	0.00	0.00	0.00	***
Porbandar	0.00	**Q :**	0	1365	– 1365	– 100
	0	**V :**	0.00	7.45	– 7.45	– 100
	0	**$:**	0.00	1.68	– 1.68	– 100
Total	**100**	**Q :**	**512164**	**461329**	**50835**	**11.02**
	100	**V :**	**7245**	**6,646.69**	**598.61**	**9.01**
	100	**$:**	**1644**	**1,478.48**	**165.74**	**11.21**

Performance of Sea Ports/Airports

Export of marine products has taken place through 18 seaports/airports during the year 2005-2006. Two new ports were emerged this period. They are Ahmedabad and Agartala. Chennai kept its position as the largest port in terms of value with a share of 19.17%. However, there was a decline in the export to the tune of 3.51% in value and 1.06% in US$. The largest port in terms of volume was JNP with a share of 23.53% and in value wise it hold third place with a share of 16.19%. Kochi continued to be the 2nd largest port in terms of value and third largest port in terms of quantity. Exports from ports like Kochi, JNP, Vizag, Pipavav, Mundra, Mangalore/ICD, Goa, Trivandrum, Calicut etc. registered a positive growth where as Chennai, Tuticorin Mumbai, Kandala, Haldia NSICT, Porbandar etc. showed a negative growth.

www.ingramcontent.com/pod-product-compliance
Lightning Source LLC
Chambersburg PA
CBHW061323190326
41458CB00011B/3869
9789351240525